ƐVOLUTION

EVOLUTION

A SCIENTIFIC AMERICAN READER

THE UNIVERSITY OF CHICAGO PRESS · CHICAGO AND LONDON

The University of Chicago Press, Chicago 60637
The University of Chicago Press, Ltd., London
© 2006 by Scientific American, Inc.
All rights reserved. Published 2006
Printed in the United States of America
15 14 13 12 11 10 09 08 07 06   1 2 3 4 5
ISBN-10: 0-226-74268-7   (cloth)
ISBN-13: 978-0-226-74268-7   (cloth)
ISBN-10: 0-226-74269-5   (paper)
ISBN-13: 978-0-226-74269-4   (paper)

Library of Congress Cataloging-in-Publication Data

Evolution : a Scientific American reader.
     p.   cm.
   Includes bibliographical references (p.   ).
     ISBN 0-226-74268-7 (cloth : alk. paper)— ISBN 0-226-74269-5 (pbk. : alk. paper)
     1. Evolution (Biology)   I. University of Chicago Press.
QH366.2.E8492 2006
576.8—dc22   2005046639

CONTENTS

THE EVOLUTION OF THE UNIVERSE

The Evolution of the Universe

P. JAMES E. PEEBLES, DAVID N. SCHRAMM, EDWIN L. TURNER
AND RICHARD G. KRON

ORIGINALLY PUBLISHED IN OCTOBER 1994

At a particular instant roughly 15 billion years ago, all the matter and energy we can observe, concentrated in a region smaller than a dime, began to expand and cool at an incredibly rapid rate. By the time the temperature had dropped to 100 million times that of the sun's core, the forces of nature assumed their present properties, and the elementary particles known as quarks roamed freely in a sea of energy. When the universe had expanded an additional 1,000 times, all the matter we can measure filled a region the size of the solar system.

At that time, the free quarks became confined in neutrons and protons. After the universe had grown by another factor of 1,000, protons and neutrons combined to form atomic nuclei, including most of the helium and deuterium present today. All of this occurred within the first minute of the expansion. Conditions were still too hot, however, for atomic nuclei to capture electrons. Neutral atoms appeared in abundance only after the expansion had continued for 300,000 years and the universe was 1,000 times smaller than it is now. The neutral atoms then began to coalesce into gas clouds, which later evolved into stars. By the time the universe had expanded to one fifth its present size, the stars had formed groups recognizable as young galaxies.

When the universe was half its present size, nuclear reactions in stars had produced most of the heavy elements from which terrestrial planets were made. Our solar system is relatively young: it formed five billion years ago, when the universe was two thirds its present size. Over time the formation of stars has consumed the supply of gas in galaxies, and hence the population of stars is waning. Fifteen billion years from now stars like our sun will be relatively rare, making the universe a far less hospitable place for observers like us.

Our understanding of the genesis and evolution of the universe is one of the great achievements of 20th-century science. This knowledge comes from decades of innovative experiments and theories. Modern telescopes

3

on the ground and in space detect the light from galaxies billions of light-years away, showing us what the universe looked like when it was young. Particle accelerators probe the basic physics of the high-energy environment of the early universe. Satellites detect the cosmic background radiation left over from the early stages of expansion, providing an image of the universe on the largest scales we can observe.

Our best efforts to explain this wealth of data are embodied in a theory known as the standard cosmological model or the big bang cosmology. The major claim of the theory is that in the large-scale average the universe is expanding in a nearly homogeneous way from a dense early state. At present, there are no fundamental challenges to the big bang theory, although there are certainly unresolved issues within the theory itself. Astronomers are not sure, for example, how the galaxies were formed, but there is no reason to think the process did not occur within the framework of the big bang. Indeed, the predictions of the theory have survived all tests to date.

Yet the big bang model goes only so far, and many fundamental mysteries remain. What was the universe like before it was expanding? (No observation we have made allows us to look back beyond the moment at which the expansion began.) What will happen in the distant future, when the last of the stars exhaust the supply of nuclear fuel? No one knows the answers yet.

Our universe may be viewed in many lights—by mystics, theologians, philosophers or scientists. In science we adopt the plodding route: we accept only what is tested by experiment or observation. Albert Einstein gave us the now well-tested and accepted Theory of General Relativity, which establishes the relations between mass, energy, space and time. Einstein showed that a homogeneous distribution of matter in space fits nicely with his theory. He assumed without discussion that the universe is static, unchanging in the large-scale average [see "How Cosmology Became a Science," by Stephen G. Brush; *Scientific American,* August 1992].

In 1922 the Russian theorist Alexander A. Friedmann realized that Einstein's universe is unstable; the slightest perturbation would cause it to expand or contract. At that time, Vesto M. Slipher of Lowell Observatory was collecting the first evidence that galaxies are actually moving apart. Then, in 1929, the eminent astronomer Edwin P. Hubble showed that the rate a galaxy is moving away from us is roughly proportional to its distance from us.

The existence of an expanding universe implies that the cosmos has evolved from a dense concentration of matter into the present broadly spread distribution of galaxies. Fred Hoyle, an English cosmologist, was the first to call this process the big bang. Hoyle intended to disparage the theory, but the name was so catchy it gained popularity. It is somewhat misleading, however, to describe the expansion as some type of explosion of matter away from some particular point in space.

That is not the picture at all: in Einstein's universe the concept of space and the distribution of matter are intimately linked; the observed expansion of the system of galaxies reveals the unfolding of space itself. An essential feature of the theory is that the average density in space declines as the universe expands; the distribution of matter forms no observable edge. In an explosion the fastest particles move out into empty space, but in the big bang cosmology, particles uniformly fill all space. The expansion of the universe has had little influence on the size of galaxies or even clusters of galaxies that are bound by gravity; space is simply opening up between them. In this sense, the expansion is similar to a rising loaf of raisin bread. The dough is analogous to space, and the raisins, to clusters of galaxies. As the dough expands, the raisins move apart. Moreover, the speed with which any two raisins move apart is directly and positively related to the amount of dough separating them.

The evidence for the expansion of the universe has been accumulating for some 60 years. The first important clue is the redshift. A galaxy emits or absorbs some wavelengths of light more strongly than others. If the galaxy is moving away from us, these emission and absorption features are shifted to longer wavelengths—that is, they become redder as the recession velocity increases. This phenomenon is known as the redshift.

Hubble's measurements indicated that the redshift of a distant galaxy is greater than that of one closer to the earth. This relation, now known as Hubble's law, is just what one would expect in a uniformly expanding universe. Hubble's law says the recession velocity of a galaxy is equal to its distance multiplied by a quantity called Hubble's constant. The redshift effect in nearby galaxies is relatively subtle, requiring good instrumentation to detect it. In contrast, the redshift of very distant objects—radio galaxies and quasars—is an awesome phenomenon; some appear to be moving away at greater than 90 percent of the speed of light.

Hubble contributed to another crucial part of the picture. He counted the number of visible galaxies in different directions in the sky and found that they appear to be rather uniformly distributed. The value of Hubble's constant seemed to be the same in all directions, a necessary consequence of uniform expansion. Modern surveys confirm the fundamental tenet that the universe is homogeneous on large scales. Although maps of the distribution of the nearby galaxies display clumpiness, deeper surveys reveal considerable uniformity.

The Milky Way, for instance, resides in a knot of two dozen galaxies; these in turn are part of a complex of galaxies that protrudes from the so-called local supercluster. The hierarchy of clustering has been traced up to dimensions of about 500 million light-years. The fluctuations in the average density of matter diminish as the scale of the structure being investigated increases. In maps that cover distances that reach close to the observable limit, the average density of matter changes by less than a tenth of a percent.

To test Hubble's law, astronomers need to measure distances to galaxies. One method for gauging distance is to observe the apparent brightness of a galaxy. If one galaxy is four times fainter in the night sky than an otherwise comparable galaxy, then it can be estimated to be twice as far away. This expectation has now been tested over the whole of the visible range of distances.

Some critics of the theory have pointed out that a galaxy that appears to be smaller and fainter might not actually be more distant. Fortunately, there is a direct indication that objects whose redshifts are larger really are more distant. The evidence comes from observations of an effect known as gravitational lensing. An object as massive and compact as a galaxy can act as a crude lens, producing a distorted, magnified image (or even many images) of any background radiation source that lies behind it. Such an object does so by bending the paths of light rays and other electromagnetic radiation. So if a galaxy sits in the line of sight between the earth and some distant object, it will bend the light rays from the object so that they are observable [see "Gravitational Lenses," by Edwin L. Turner; *Scientific American,* July 1988]. During the past decade, astronomers have discovered more than a dozen gravitational lenses. The object behind the lens is always found to have a higher redshift than the lens itself, confirming the qualitative prediction of Hubble's law.

Hubble's law has great significance not only because it describes the expansion of the universe but also because it can be used to calculate the age of the cosmos. To be precise, the time elapsed since the big bang is a function of the present value of Hubble's constant and its rate of change. Astronomers have determined the approximate rate of the expansion, but no one has yet been able to measure the second value precisely.

Still, one can estimate this quantity from knowledge of the universe's average density. One expects that because gravity exerts a force that opposes expansion, galaxies would tend to move apart more slowly now

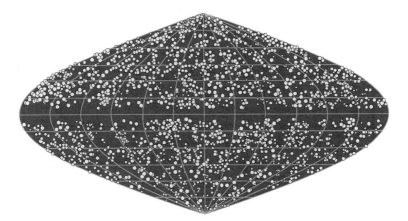

Homogeneous distribution of galaxies is apparent in a map that includes objects from 300 to 1,000 million light-years away. The only inhomogeneity, a gap near the center line, occurs because part of the sky is obscured by the Milky Way. Michael Strauss of the Institute for Advanced Study in Princeton, N.J., created the map using data from NASA's *Infrared Astronomical Satellite.*

than they did in the past. The rate of change in expansion is therefore related to the gravitational pull of the universe set by its average density. If the density is that of just the visible material in and around galaxies, the age of the universe probably lies between 12 and 20 billion years. (The range allows for the uncertainty in the rate of expansion.)

Yet many researchers believe the density is greater than this minimum value. So-called dark matter would make up the difference. A strongly defended argument holds that the universe is just dense enough that in the remote future the expansion will slow almost to zero. Under this assumption, the age of the universe decreases to the range of seven to 13 billion years.

To improve these estimates, many astronomers are involved in intensive research to measure both the distances to galaxies and the density of the universe. Estimates of the expansion time provide an important test for the big bang model of the universe. If the theory is correct, everything in the visible universe should be younger than the expansion time computed from Hubble's law.

These two timescales do appear to be in at least rough concordance. For example, the oldest stars in the disk of the Milky Way galaxy are about 9 billion years old—an estimate derived from the rate of cooling of white dwarf stars. The stars in the halo of the Milky Way are somewhat older, about 15 billion years—a value derived from the rate of nuclear fuel consumption in the cores of these stars. The ages of the oldest known chemical elements are also approximately 15 billion years—a number that comes from radioactive dating techniques. Workers in laboratories have derived these age estimates from atomic and nuclear physics. It is noteworthy that their results agree, at least approximately, with the age that astronomers have derived by measuring cosmic expansion.

Another theory, the steady state theory, also succeeds in accounting for the expansion and homogeneity of the universe. In 1946 three physicists in England—Fred Hoyle, Hermann Bondi and Thomas Gold—proposed such a cosmology. In their theory the universe is forever expanding, and matter is created spontaneously to fill the voids. As this material accumulates, they suggested, it forms new stars to replace the old. This steady-state hypothesis predicts that ensembles of galaxies close to us should look statistically the same as those far away. The big bang cosmology makes a different prediction: if galaxies were all formed long ago, distant galaxies should look younger than those nearby because light from them requires a longer time to reach us. Such galaxies should contain more short-lived stars and more gas out of which future generations of stars will form.

The test is simple conceptually, but it took decades for astronomers to develop detectors sensitive enough to study distant galaxies in detail. When astronomers examine nearby galaxies that are powerful emitters of radio wavelengths, they see, at optical wavelengths, relatively round systems of stars. Distant radio galaxies, on the other hand, appear to have elongated and sometimes irregular structures. Moreover, in most distant radio galaxies, unlike the ones nearby, the distribution of light tends to be aligned with the pattern of the radio emission.

Likewise, when astronomers study the population of massive, dense clusters of galaxies, they find differences between those that are close and those far away. Distant clusters contain bluish galaxies that show evidence of ongoing star formation. Similar clusters that are nearby contain reddish galaxies in which active star formation ceased long ago. Observations made with the Hubble Space Telescope confirm that at least some of the enhanced star formation in these younger clusters may be the result of collisions between their member galaxies, a process that is much rarer in the present epoch.

So if galaxies are all moving away from one another and are evolving from earlier forms, it seems logical that they were once crowded together in some dense sea of matter and energy. Indeed, in 1927, before much was known about distant galaxies, a Belgian cosmologist and priest, Georges Lemaître, proposed that the expansion of the universe might be traced to an exceedingly dense state he called the primeval "super-atom." It might

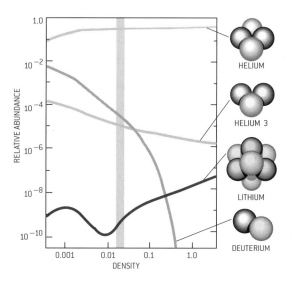

Density of neutrons and protons in the universe determined the abundances of certain elements. For a higher density universe, the computed helium abundance is little different, and the computed abundance of deuterium is considerably lower. The shaded region is consistent with the observations, ranging from an abundance of 24 percent for helium to one part in 10^{10} for the lithium isotope. This quantitative agreement is a prime success of the big bang cosmology.

even be possible, he thought, to detect remnant radiation from the primeval atom. But what would this radiation signature look like?

When the universe was very young and hot, radiation could not travel very far without being absorbed and emitted by some particle. This continuous exchange of energy maintained a state of thermal equilibrium; any particular region was unlikely to be much hotter or cooler than the average. When matter and energy settle to such a state, the result is a so-called thermal spectrum, where the intensity of radiation at each wavelength is a definite function of the temperature. Hence, radiation originating in the hot big bang is recognizable by its spectrum.

In fact, this thermal cosmic background radiation has been detected. While working on the development of radar in the 1940s, Robert H. Dicke, then at the Massachusetts Institute of Technology, invented the microwave radiometer—a device capable of detecting low levels of radiation. In the 1960s Bell Laboratories used a radiometer in a telescope that would track the early communications satellites Echo-1 and Telstar. The engineer who built this instrument found that it was detecting unexpected radiation. Arno A. Penzias and Robert W. Wilson identified the signal as the cosmic background radiation. It is interesting that Penzias and Wilson were led to this idea by the news that Dicke had suggested that one ought to use a radiometer to search for the cosmic background.

Astronomers have studied this radiation in great detail using the Cosmic Background Explorer (COBE) satellite and a number of rocket-launched, balloon-borne and ground-based experiments. The cosmic background radiation has two distinctive properties. First, it is nearly the same in all directions. (As George F. Smoot of Lawrence Berkeley Laboratory and his team discovered in 1992, the variation is just one part per 100,000.) The interpretation is that the radiation uniformly fills space, as predicted in the big bang cosmology. Second, the spectrum is very close to that of an object in thermal equilibrium at 2.726 kelvins above absolute zero. To be sure, the cosmic background radiation was produced when the universe was far hotter than 2.726 degrees, yet researchers anticipated correctly that the apparent temperature of the radiation would be low. In the 1930s Richard C. Tolman of the California Institute of Technology showed that the temperature of the cosmic background would diminish because of the universe's expansion.

The cosmic background radiation provides direct evidence that the universe did expand from a dense, hot state, for this is the condition needed to produce the radiation. In the dense, hot early universe thermonuclear reactions produced elements heavier than hydrogen, including deuterium, helium and lithium. It is striking that the computed mix of the light elements agrees with the observed abundances. That is, all evidence indicates that the light elements were produced in the hot, young universe, whereas the heavier elements appeared later, as products of the thermonuclear reactions that power stars.

The theory for the origin of the light elements emerged from the burst of research that followed the end of World War II. George Gamow and graduate student Ralph A. Alpher of George Washington University and Robert Herman of the Johns Hopkins University Applied Physics Laboratory and others used nuclear physics data from the war effort to predict what kind of nuclear processes might have occurred in the early universe and what elements might have been produced. Alpher and Herman also realized that a remnant of the original expansion would still be detectable in the existing universe.

Despite the fact that significant details of this pioneering work were in error, it forged a link between nuclear physics and cosmology. The workers demonstrated that the early universe could be viewed as a type of thermonuclear reactor. As a result, physicists have now precisely calculated the abundances of light elements produced in the big bang and how those quantities have changed because of subsequent events in the interstellar medium and nuclear processes in stars.

Our grasp of the conditions that prevailed in the early universe does not translate into a full understanding of how galaxies formed. Nevertheless, we do have quite a few pieces of the puzzle. Gravity causes the growth of density fluctuations in the distribution of matter, because it more strongly slows the expansion of denser regions, making them grow still denser. This process is observed in the growth of nearby clusters of galaxies, and the galaxies themselves were probably assembled by the same process on a smaller scale.

The growth of structure in the early universe was prevented by radiation pressure, but that changed when the universe had expanded to about 0.1 percent of its present size. At that point, the temperature was about 3,000 kelvins, cool enough to allow the ions and electrons to combine to form neutral hydrogen and helium. The neutral matter was able to slip through the radiation and to form gas clouds that could collapse into star clusters. Observations show that by the time the universe was one fifth its present size, matter had gathered into gas clouds large enough to be called young galaxies.

A pressing challenge now is to reconcile the apparent uniformity of the early universe with the lumpy distribution of galaxies in the present universe. Astronomers know that the density of the early universe did not vary by much, because they observe only slight irregularities in the cosmic background radiation. So far it has been easy to develop theories that are consistent with the available measurements, but more critical tests are in progress. In particular, different theories for galaxy formation predict quite different fluctuations in the cosmic background radiation on angular scales less than about one degree. Measurements of such tiny fluctuations have not yet been done, but they might be accomplished in the generation of experiments now under way. It will be exciting to learn whether any of the theories of galaxy formation now under consideration survive these tests.

The present-day universe has provided ample opportunity for the development of life as we know it—there are some 100 billion billion stars similar to the sun in the part of the universe we can observe. The big bang cosmology implies, however, that life is possible only for a bounded span of time: the universe was too hot in the distant past, and it has limited resources for the future. Most galaxies are still producing new stars, but many others have already exhausted their supply of gas. Thirty billion years from now, galaxies will be much darker and filled with dead or dying stars, so there will be far fewer planets capable of supporting life as it now exists.

The universe may expand forever, in which case all the galaxies and stars will eventually grow dark and cold. The alternative to this big chill is a big crunch. If the mass of the universe is large enough, gravity will eventually reverse the expansion, and all matter and energy will be re-united. During the next decade, as researchers improve techniques for measuring the mass of the universe, we may learn whether the present expansion is headed toward a big chill or a big crunch.

In the near future, we expect new experiments to provide a better understanding of the big bang. As we improve measurements of the expansion rate and the ages of stars, we may be able to confirm that the stars are indeed younger than the expanding universe. The larger telescopes recently completed or under construction may allow us to see how the mass of the universe affects the curvature of space-time, which in turn influences our observations of distant galaxies.

We will also continue to study issues that the big bang cosmology does not address. We do not know why there was a big bang or what may have existed before. We do not know whether our universe has siblings—other expanding regions well removed from what we can observe. We do not understand why the fundamental constants of nature have the values they do. Advances in particle physics suggest some interesting ways these questions might be answered; the challenge is to find experimental tests of the ideas.

In following the debate on such matters of cosmology, one should bear in mind that all physical theories are approximations of reality that can fail if pushed too far. Physical science advances by incorporating earlier theories that are experimentally supported into larger, more encompassing frameworks. The big bang theory is supported by a wealth of evidence: it explains the cosmic background radiation, the abundances of light elements and the Hubble expansion. Thus, any new cosmology surely will include the big bang picture. Whatever developments the coming decades may bring, cosmology has moved from a branch of philosophy to a physical science where hypotheses meet the test of observation and experiment.

FURTHER READING

Dennis Overbye. *Lonely Hearts of the Cosmos: The Scientific Quest for the Secret of the Universe.* Harper-Collins, 1991.

Michael Riordan and David N. Schramm. *The Shadows of Creation: Dark Matter and the Structure of the Universe.* W. H. Freeman and Company, 1991.

Michael D. Lemonick. *The Light at the Edge of the Universe: Astronomers on the Front Lines of the Cosmological Revolution.* Villard Books, 1993.

P. J. E. Peebles. *Principles of Physical Cosmology.* Princeton University Press, 1993.

The First Stars in the Universe

RICHARD B. LARSON AND VOLKER BROMM

ORIGINALLY PUBLISHED IN DECEMBER 2001

We live in a universe that is full of bright objects. On a clear night one can see thousands of stars with the naked eye. These stars occupy merely a small nearby part of the Milky Way galaxy; telescopes reveal a much vaster realm that shines with the light from billions of galaxies. According to our current understanding of cosmology, however, the universe was featureless and dark for a long stretch of its early history. The first stars did not appear until perhaps 100 million years after the big bang, and nearly a billion years passed before galaxies proliferated across the cosmos. Astronomers have long wondered: How did this dramatic transition from darkness to light come about?

After decades of study, researchers have recently made great strides toward answering this question. Using sophisticated computer simulation techniques, cosmologists have devised models that show how the density fluctuations left over from the big bang could have evolved into the first stars. In addition, observations of distant quasars have allowed scientists to probe back in time and catch a glimpse of the final days of the "cosmic dark ages."

The new models indicate that the first stars were most likely quite massive and luminous and that their formation was an epochal event that fundamentally changed the universe and its subsequent evolution. These stars altered the dynamics of the cosmos by heating and ionizing the surrounding gases. The earliest stars also produced and dispersed the first heavy elements, paving the way for the eventual formation of solar systems like our own. And the collapse of some of the first stars may have seeded the growth of supermassive black holes that formed in the hearts of galaxies and became the spectacular power sources of quasars. In short, the earliest stars made possible the emergence of the universe that we see today—everything from galaxies and quasars to planets and people.

THE DARK AGES

The study of the early universe is hampered by a lack of direct observations. Astronomers have been able to examine much of the universe's history by training their telescopes on distant galaxies and quasars that emitted their light billions of years ago. The age of each object can be determined by the redshift of its light, which shows how much the universe has expanded since the light was produced. The oldest galaxies and quasars that have been observed so far date from about a billion years after the big bang (assuming a present age for the universe of 12 billion to 14 billion years). Researchers will need better telescopes to see more distant objects dating from still earlier times.

Cosmologists, however, can make deductions about the early universe based on the cosmic microwave background radiation, which was emitted about 400,000 years after the big bang. The uniformity of this radiation indicates that matter was distributed very smoothly at that time. Because there were no large luminous objects to disturb the primordial soup, it must have remained smooth and featureless for millions of years afterward. As the cosmos expanded, the background radiation redshifted to longer wavelengths and the universe grew increasingly cold and dark. Astronomers have no observations of this dark era. But by a billion years after the big bang, some bright galaxies and quasars had already appeared, so the first stars must have formed sometime before. When did these first luminous objects arise, and how might they have formed?

Many astrophysicists, including Martin Rees of the University of Cambridge and Abraham Loeb of Harvard University, have made important contributions toward solving these problems. The recent studies begin with the standard cosmological models that describe the evolution of the universe following the big bang. Although the early universe was remarkably smooth, the background radiation shows evidence of small-scale density fluctuations—clumps in the primordial soup. The cosmological models predict that these clumps would gradually evolve into gravitationally bound structures. Smaller systems would form first and then merge into larger agglomerations. The denser regions would take the form of a network of filaments, and the first star-forming systems—small protogalaxies—would coalesce at the nodes of this network. In a similar way, the protogalaxies would then merge to form galaxies, and the galaxies would congregate into galaxy clusters. The process is ongoing: although galaxy formation is now mostly complete, galaxies are still assembling into clusters, which are in turn aggregating into a vast filamentary network that stretches across the universe.

According to the cosmological models, the first small systems capable of forming stars should have appeared between 100 million and 250 million years after the big bang. These protogalaxies would have been 100,000 to one million times more massive than the sun and would have measured 30 to 100 light-years across. These properties are similar to those of the molecular gas clouds in which stars are currently forming in the Milky Way, but the first protogalaxies would have differed in some fundamental ways. For one, they would have consisted mostly of dark matter, the putative elementary particles that are believed to make up about 90 percent of the universe's mass. In present-day large galaxies, dark matter is segregated from ordinary matter: over time, ordinary matter concentrates in the galaxy's inner region, whereas the dark matter remains scattered throughout an enormous outer halo. But in the protogalaxies, the ordinary matter would still have been mixed with the dark matter.

The second important difference is that the protogalaxies would have contained no significant amounts of any elements besides hydrogen and helium. The big bang produced hydrogen and helium, but most of the heavier elements are created only by the thermonuclear fusion reactions in stars, so they would not have been present before the first stars had formed. Astronomers use the term "metals" for all these heavier elements. The young metal-rich stars in the Milky Way are called Population I stars, and the old metal-poor stars are called Population II stars; following this terminology, the stars with no metals at all—the very first generation—are sometimes called Population III stars.

In the absence of metals, the physics of the first star-forming systems would have been much simpler than that of present-day molecular gas clouds. Furthermore, the cosmological models can provide, in principle, a complete description of the initial conditions that preceded the first generation of stars. In contrast, the stars that arise from molecular gas clouds are born in complex environments that have been altered by the effects of previous star formation. Therefore, scientists may find it easier to model the formation of the first stars than to model how stars form at present. In any case, the problem is an appealing one for theoretical study, and several research groups have used computer simulations to portray the formation of the earliest stars.

A group consisting of Tom Abel, Greg Bryan and Michael L. Norman (now at Pennsylvania State University, the Massachusetts Institute of Technology and the University of California at San Diego, respectively) has made the most realistic simulations. In collaboration with Paolo Coppi of Yale University, we have done simulations based on simpler assumptions but intended to explore a wider range of possibilities. Toru Tsuribe, now

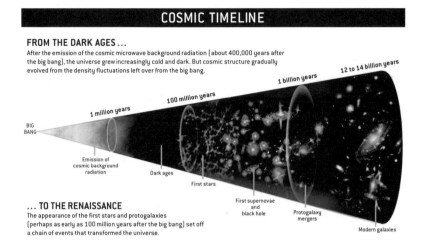

COSMIC TIMELINE

FROM THE DARK AGES . . .
After the emission of the cosmic microwave background radiation (about 400,000 years after the big bang), the universe grew increasingly cold and dark. But cosmic structure gradually evolved from the density fluctuations left over from the big bang.

12 to 14 billion years

1 billion years

100 million years

1 million years

BIG BANG

Emission of cosmic background radiation

Dark ages

First stars

First supernovae and black hole

Protogalaxy mergers

Modern galaxies

. . . TO THE RENAISSANCE
The appearance of the first stars and protogalaxies (perhaps as early as 100 million years after the big bang) set off a chain of events that transformed the universe.

at Osaka University in Japan, has made similar calculations using more powerful computers. Fumitaka Nakamura and Masayuki Umemura (now at Niigata and Tsukuba universities in Japan, respectively) have worked with a more idealized simulation, but it has still yielded instructive results. Although these studies differ in various details, they have all produced similar descriptions of how the earliest stars might have been born.

LET THERE BE LIGHT!

The simulations show that the primordial gas clouds would typically form at the nodes of a small-scale filamentary network and then begin to contract because of their gravity. Compression would heat the gas to temperatures above 1,000 kelvins. Some hydrogen atoms would pair up in the dense, hot gas, creating trace amounts of molecular hydrogen. The hydrogen molecules would then start to cool the densest parts of the gas by emitting infrared radiation after they collide with hydrogen atoms. The temperature in the densest parts would drop to about 200 to 300 kelvins, reducing the gas pressure in these regions and hence allowing them to contract into gravitationally bound clumps.

This cooling plays an essential role in allowing the ordinary matter in the primordial system to separate from the dark matter. The cooling hydrogen settles into a flattened rotating configuration that is clumpy and filamentary and possibly shaped like a disk. But because the dark-matter particles would not emit radiation or lose energy, they would remain scattered in the primordial cloud. Thus, the star-forming system would

THE BIRTH AND DEATH OF THE FIRST STARS

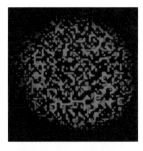

PRIMEVAL TURMOIL

The process that led to the creation of the first stars was very different from present-day star formation. But the violent deaths of some of these stars paved the way for the emergence of the universe that we see today.

1 The first star-forming systems—small protogalaxies—consisted mostly of the elementary particles known as dark matter. Ordinary matter—mainly hydrogen gas —was initially mixed with the dark matter.

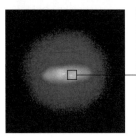

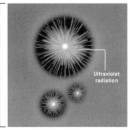

Ultraviolet radiation

2 The cooling of the hydrogen allowed the ordinary matter to contract, whereas the dark matter remained dispersed. The hydrogen settled into a disk at the center of the protogalaxy.

3 The denser regions of gas contracted into star-forming clumps, each hundreds of times as massive as the sun. Some of the clumps of gas collapsed to form very massive, luminous stars.

4 Ultraviolet radiation from the stars ionized the surrounding neutral hydrogen gas. As more and more stars formed, the bubbles of ionized gas merged and the intergalactic gas became ionized.

Supernova

Black hole

5 A few million years later, at the end of their brief lives, some of the first stars exploded as supernovae. The most massive stars collapsed into black holes.

6 Gravitational attraction pulled the protogalaxies toward one another. The collisions most likely triggered star formation, just as galactic mergers do now.

7 Black holes possibly merged to form a supermassive hole at the protogalaxy's center. Gas swirling into this hole might have generated quasarlike radiation.

come to resemble a miniature galaxy, with a disk of ordinary matter and a halo of dark matter. Inside the disk, the densest clumps of gas would continue to contract, and eventually some of them would undergo a runaway collapse and become stars.

The first star-forming clumps were much warmer than the molecular gas clouds in which most stars currently form. Dust grains and molecules containing heavy elements cool the present-day clouds much more efficiently to temperatures of only about 10 kelvins. The minimum mass that a clump of gas must have to collapse under its gravity is called the Jeans mass, which is proportional to the square of the gas temperature and inversely proportional to the square root of the gas pressure. The first star-forming systems would have had pressures similar to those of present-day molecular clouds. But because the temperatures of the first collapsing gas clumps were almost 30 times higher than those of molecular clouds, their Jeans mass would have been almost 1,000 times larger.

In molecular clouds in the nearby part of the Milky Way, the Jeans mass is roughly equal to the mass of the sun, and the masses of the prestellar clumps observed in these clouds are about the same. If we scale up by a factor of almost 1,000, we can estimate that the masses of the first star-forming clumps would have been about 500 to 1,000 solar masses. In agreement with this prediction, all the computer simulations mentioned above showed the formation of clumps with masses of several hundred solar masses or more.

Our group's calculations suggest that the predicted masses of the first star-forming clumps are not very sensitive to the assumed cosmological conditions (for example, the exact nature of the initial density fluctuations). In fact, the predicted masses depend primarily on the physics of the hydrogen molecule and only secondarily on the cosmological model or simulation technique. One reason is that molecular hydrogen cannot cool the gas below 200 kelvins, making this a lower limit to the temperature of the first star-forming clumps. Another is that the cooling from molecular hydrogen becomes inefficient at the higher densities encountered when the clumps begin to collapse. At these densities the hydrogen molecules collide with other atoms before they have time to emit an infrared photon; this raises the gas temperature and slows down the contraction until the clumps have built up to at least a few hundred solar masses.

What was the fate of the first collapsing clumps? Did they form stars with similarly large masses, or did they fragment into many smaller parts and form many smaller stars? The research groups have pushed their calculations to the point at which the clumps are well on their way to forming stars, and none of the simulations has yet revealed any tendency for the clumps to fragment. This agrees with our understanding of present-day star formation; observations and simulations show that the fragmentation of star-forming clumps is typically limited to the formation of binary systems (two stars orbiting around each other). Fragmentation seems even less likely to occur in the primordial clumps, because the inefficiency of molecular hydrogen cooling would keep the Jeans mass high. The simulations, however, have not yet determined the final outcome of collapse with certainty, and the formation of binary systems cannot be ruled out.

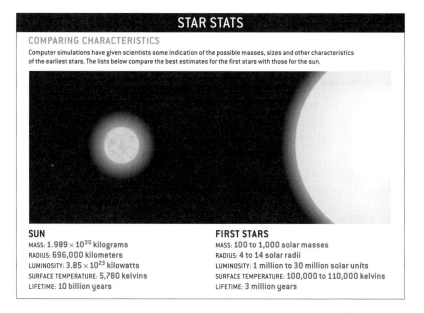

STAR STATS

COMPARING CHARACTERISTICS

Computer simulations have given scientists some indication of the possible masses, sizes and other characteristics of the earliest stars. The lists below compare the best estimates for the first stars with those for the sun.

SUN
MASS: 1.989×10^{30} kilograms
RADIUS: 696,000 kilometers
LUMINOSITY: 3.85×10^{23} kilowatts
SURFACE TEMPERATURE: 5,780 kelvins
LIFETIME: 10 billion years

FIRST STARS
MASS: 100 to 1,000 solar masses
RADIUS: 4 to 14 solar radii
LUMINOSITY: 1 million to 30 million solar units
SURFACE TEMPERATURE: 100,000 to 110,000 kelvins
LIFETIME: 3 million years

Different groups have arrived at somewhat different estimates of just how massive the first stars might have been. Abel, Bryan and Norman have argued that the stars probably had masses no greater than 300 solar masses. Our own work suggests that masses as high as 1,000 solar masses might have been possible. Both predictions might be valid in different circumstances: the very first stars to form might have had masses no larger than 300 solar masses, whereas stars that formed a little later from the collapse of larger protogalaxies might have reached the higher estimate. Quantitative predictions are difficult because of feedback effects; as a massive star forms, it produces intense radiation and matter outflows that may blow away some of the gas in the collapsing clump. But these effects depend strongly on the presence of heavy elements in the gas, and therefore they should be less important for the earliest stars. Thus, it seems safe to conclude that the first stars in the universe were typically many times more massive and luminous than the sun.

THE COSMIC RENAISSANCE

What effects did these first stars have on the rest of the universe? An important property of stars with no metals is that they have higher surface temperatures than stars with compositions like that of the sun. The production of nuclear energy at the center of a star is less efficient without

metals, and the star would have to be hotter and more compact to produce enough energy to counteract gravity. Because of the more compact
structure, the surface layers of the star would also be hotter. In collaboration with Rolf-Peter Kudritzki of the University of Hawaii and Abraham Loeb of Harvard, one of us (Bromm) devised theoretical models of
such stars with masses between 100 and 1,000 solar masses. The models
showed that the stars had surface temperatures of about 100,000 kelvins—
about 17 times higher than the sun's surface temperature. Therefore, the
first starlight in the universe would have been mainly ultraviolet radiation from very hot stars, and it would have begun to heat and ionize the
neutral hydrogen and helium gas around these stars soon after they
formed.

We call this event the cosmic renaissance. Although astronomers cannot yet estimate how much of the gas in the universe condensed into the
first stars, even a fraction as small as one part in 100,000 could have been
enough for these stars to ionize much of the remaining gas. Once the first
stars started shining, a growing bubble of ionized gas would have formed
around each one. As more and more stars formed over hundreds of millions of years, the bubbles of ionized gas would have eventually merged,
and the intergalactic gas would have become completely ionized.

Scientists from the California Institute of Technology and the Sloan
Digital Sky Survey have recently found evidence for the final stages of this
ionization process. The researchers observed strong absorption of ultraviolet light in the spectra of quasars that date from about 900 million years
after the big bang. The results suggest that the last patches of neutral hydrogen gas were being ionized at that time. Helium requires more·energy
to ionize than hydrogen does, but if the first stars were as massive as predicted, they would have ionized helium at the same time. On the other
hand, if the first stars were not quite so massive, the helium must have
been ionized later by energetic radiation from sources such as quasars. Future observations of distant objects may help determine when the universe's helium was ionized.

If the first stars were indeed very massive, they would also have had
relatively short lifetimes—only a few million years. Some of the stars
would have exploded as supernovae at the end of their lives, expelling the
metals they produced by fusion reactions. Stars that are between 100 and
250 times as massive as the sun are predicted to blow up completely in energetic explosions, and some of the first stars most likely had masses in
this range. Because metals are much more effective than hydrogen in
cooling star-forming clouds and allowing them to collapse into stars, the

production and dispersal of even a small amount could have had a major effect on star formation.

Working in collaboration with Andrea Ferrara of the University of Florence in Italy, we have found that when the abundance of metals in star-forming clouds rises above one thousandth of the metal abundance in the sun, the metals rapidly cool the gas to the temperature of the cosmic background radiation. (This temperature declines as the universe expands, falling to 19 kelvins a billion years after the big bang and to 2.7 kelvins today.) This efficient cooling allows the formation of stars with smaller masses and may also considerably boost the overall rate at which stars are born. In fact, it is possible that the pace of star formation did not accelerate until after the first metals had been produced. In this case, the second-generation stars might have been the ones primarily responsible for lighting up the universe and bringing about the cosmic renaissance.

At the start of this active period of star birth, the cosmic background temperature would have been higher than the temperature in present-day molecular clouds (10 kelvins). Until the temperature dropped to that level—which happened about two billion years after the big bang—the process of star formation may still have favored massive stars. As a result, large numbers of such stars may have formed during the early stages of galaxy building by successive mergers of protogalaxies. A similar phenomenon may occur in the modern universe when two galaxies collide and trigger a starburst—a sudden increase in the rate of star formation. Such events are now fairly rare, but some evidence suggests that they may produce relatively large numbers of massive stars.

PUZZLING EVIDENCE

This hypothesis about early star formation might help explain some puzzling features of the present universe. One unsolved problem is that galaxies contain fewer metal-poor stars than would be expected if metals were produced at a rate proportional to the star formation rate. This discrepancy might be resolved if early star formation had produced relatively more massive stars; on dying, these stars would have dispersed large amounts of metals, which would have then been incorporated into most of the low-mass stars that we now see.

Another puzzling feature is the high metal abundance of the hot x-ray-emitting intergalactic gas in clusters of galaxies. This observation could be accounted for most easily if there had been an early period of rapid formation of massive stars and a correspondingly high supernova rate that chemically enriched the intergalactic gas. The case for a high supernova rate at early times also dovetails with the recent evidence suggesting that most of the ordinary matter and metals in the universe lies in the diffuse intergalactic medium rather than in galaxies. To produce such a distribution of matter, galaxy formation must have been a spectacular process, involving intense bursts of massive star formation and barrages of supernovae that expelled most of the gas and metals out of the galaxies.

Stars that are more than 250 times more massive than the sun do not explode at the end of their lives; instead they collapse into similarly massive black holes. Several of the computer simulations mentioned above predict that some of the first stars would have had masses this great. Because the first stars formed in the densest parts of the universe, any black holes resulting from their collapse would have become incorporated, via successive mergers, into systems of larger and larger size. It is possible that some of these black holes became concentrated in the inner part of large galaxies and seeded the growth of the supermassive black holes—millions of times more massive than the sun—that are now found in galactic nuclei.

Furthermore, astronomers believe that the energy source for quasars is the gas whirling into the black holes at the centers of large galaxies. If smaller black holes had formed at the centers of some of the first protogalaxies, the accretion of matter into the holes might have generated "mini quasars." Because these objects could have appeared soon after the first stars, they might have provided an additional source of light and ionizing radiation at early times.

Thus, a coherent picture of the universe's early history is emerging, although certain parts remain speculative. The formation of the first stars and protogalaxies began a process of cosmic evolution. Much evidence suggests that the period of most intense star formation, galaxy building and quasar activity occurred a few billion years after the big bang and that all these phenomena have continued at declining rates as the universe has aged. Most of the cosmic structure building has now shifted to larger scales as galaxies assemble into clusters.

In the coming years, researchers hope to learn more about the early stages of the story, when structures started developing on the smallest scales. Because the first stars were most likely very massive and bright, instruments such as the Next Generation Space Telescope—the planned successor to the Hubble Space Telescope—might detect some of these ancient bodies. Then astronomers may be able to observe directly how a dark, featureless universe formed the brilliant panoply of objects that now give us light and life.

FURTHER READING

Martin J. Rees. *Before the Beginning: Our Universe and Others.* Perseus Books, 1998.

Richard B. Larson. "The Formation of the First Stars" in *Star Formation from the Small to the Large Scale.* Edited by F. Favata, A. A. Kaas and A. Wilson. ESA Publications, 2000. Available on the Web at www.astro.yale.edu/larson/papers/Noordwijk99.pdf

R. Barkana and A. Loeb. "In the Beginning: The First Sources of Light and the Reionization of the Universe" in *Physics Reports,* Vol. 349, No.2, pages 125–238; July 2001. Available on the Web at aps.arxiv.org/abs/astro-ph/0010468

Graphics from computer simulations of the formation of the first stars can be found at www.tomabel.com

Exploring Our Universe and Others

MARTIN REES

ORIGINALLY PUBLISHED IN DECEMBER 1999

Cosmic exploration is preeminently a 20th-century achievement. Only in the 1920s did we realize that our Milky Way, with its 100 billion stars, is just one galaxy among millions. Our empirical knowledge of the universe has been accumulating ever since. We can now set our entire solar system in a grand evolutionary context, tracing its constituent atoms back to the initial instants of the big bang. If we were ever to discover alien intelligences, one thing we might share with them—perhaps the only thing—would be a common interest in the cosmos from which we have all emerged.

Using the current generation of ground-based and orbital observatories, astronomers can look back into the past and see plain evidence of the evolution of the universe. Marvelous images from the Hubble Space Telescope reveal galaxies as they were in remote times: balls of glowing, diffuse gas dotted with massive, fast-burning blue stars. These stars transmuted the pristine hydrogen from the big bang into heavier atoms, and when the stars died they seeded their galaxies with the basic building blocks of planets and life—carbon, oxygen, iron and so on. A Creator didn't have to turn 92 different knobs to make all the naturally occurring elements in the periodic table. Instead the galaxies act as immense ecosystems, forging elements and recycling gas through successive generations of stars. The human race itself is composed of stardust—or, less romantically, the nuclear waste from the fuel that makes stars shine.

Astronomers have also learned much about the earlier, pregalactic era by studying the microwave background radiation that makes even intergalactic space slightly warm. This afterglow of creation tells us that the entire universe was once hotter than the centers of stars. Scientists can use laboratory data to calculate how much nuclear fusion would have happened during the first few minutes after the big bang. The predicted proportions of hydrogen, deuterium and helium accord well with what astronomers have observed, thereby corroborating the big bang theory.

At first sight, attempts to fathom the cosmos might seem presumptuous and premature, even in the closing days of the 20th century. Cosmologists have, nonetheless, made real progress in recent years. This is because what makes things baffling is their degree of complexity, not their sheer size—and a star is simpler than an insect. The fierce heat within stars, and in the early universe, guarantees that everything breaks down into its simplest constituents. It is the biologists, whose role it is to study the intricate multilayered structure of trees, butterflies and brains, who face the tougher challenge.

The progress in cosmology has brought new mysteries into sharper focus and raised questions that will challenge astronomers well into the next century. For example, why does our universe contain its observed mix of ingredients? And how, from its dense beginnings, did it heave itself up to such a vast size? The answers will take us beyond the physics with which we are familiar and will require new insights into the nature of space and time. To truly understand the history of the universe, scientists must discover the profound links between the cosmic realm of the very large and the quantum world of the very small.

It is embarrassing to admit, but astronomers still don't know what our universe is made of. The objects that emit radiation that we can observe—such as stars, quasars and galaxies—constitute only a small fraction of the universe's matter. The vast bulk of matter is dark and unaccounted for. Most cosmologists believe dark matter is composed of weakly interacting particles left over from the big bang, but it could be something even more exotic. Whatever the case, it is clear that galaxies, stars and planets are a mere afterthought in a cosmos dominated by quite different stuff. Searches for dark matter, mainly via sensitive underground experiments designed to detect elusive subatomic particles, will continue apace in the coming decade. The stakes are high: success would not only tell us what most of the universe is made of but would also probably reveal some fundamentally new kinds of particles.

Astronomers are also unsure how much dark matter there is. The ultimate fate of our universe—whether it continues expanding indefinitely or eventually changes course and collapses to the so-called big crunch—depends on the total amount of dark matter and the gravity it exerts. Current data indicate that the universe contains only about 30 percent of the matter that would be needed to halt the expansion. (In cosmologists' jargon, omega—the ratio of observed density to critical density—is 0.3.) The odds favoring perpetual growth have recently strengthened further: tantalizing observations of distant supernovae suggest that the expansion of the

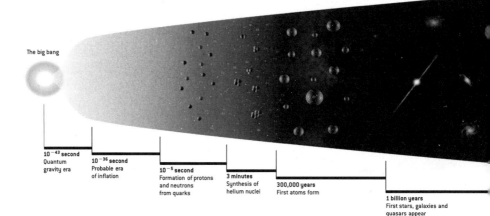

The big bang

10^{-43} second
Quantum
gravity era

10^{-36} second
Probable era
of inflation

10^{-5} second
Formation of protons
and neutrons
from quarks

3 minutes
Synthesis of
helium nuclei

300,000 years
First atoms form

1 billion years
First stars, galaxies and
quasars appear

universe may be speeding up rather than slowing down. Some astronomers say the observations are evidence of an extra repulsive force that overwhelms gravity on cosmic scales—what Albert Einstein called the cosmological constant. The jury is still out on this issue, but if the existence of the repulsive force is confirmed, physicists will learn something radically new about the energy latent in empty space.

Research is also likely to focus on the evolution of the universe's large-scale structure. If one had to answer the question "What's been happening since the big bang?" in just one sentence, the best response might be to take a deep breath and say, "Ever since the beginning, gravity has been amplifying inhomogeneities, building up structures and enhancing temperature contrasts—a prerequisite for the emergence of the complexity that lies around us now and of which we're a part." Astronomers are now learning more about this 10-billion-year process by creating "virtual universes" on their computers. In the coming years, they will be able to simulate the history of the universe with ever improving realism and then compare the results with what telescopes reveal.

Questions of structure have preoccupied astronomers since the time of Isaac Newton, who wondered why all the planets circled the sun in the same direction and in almost the same plane. In his 1704 work *Opticks* he wrote: "Blind fate could never make all the planets move one and the same way in orbits concentrick." Such a wonderful uniformity in the planetary system, Newton believed, must be the effect of divine providence.

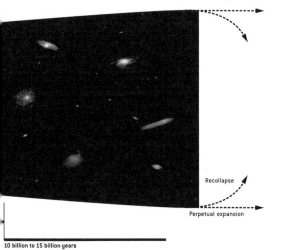

Recollapse

Perpetual expansion

10 billion to 15 billion years
Modern galaxies appear

Cosmic timeline shows the evolution of our universe from the big bang to the present day. In the first instant of creation—the epoch of inflation—the universe expanded at a staggering rate. After about three minutes, the plasma of particles and radiation cooled enough to allow the formation of simple atomic nuclei; after another 300,000 years, atoms of hydrogen and helium began to form. The first stars and galaxies appeared about a billion years later. The ultimate fate of the universe—whether it will expand forever or recollapse—is still unknown, although current evidence favors perpetual expansion.

Now astronomers know that the coplanarity of the planets is a natural outcome of the solar system's origin as a spinning disk of gas and dust. Indeed, we have extended the frontiers of our knowledge to far earlier times; cosmologists can roughly outline the history of the universe back to the very first second after the big bang. Conceptually, however, we're in little better shape than Newton was. Our understanding of the causal chain of events now stretches further back in time, but we still run into a barrier, just as surely as Newton did. The great mystery for cosmologists is the series of events that occurred less than one millisecond after the big bang, when the universe was extraordinarily small, hot and dense. The laws of physics with which we are familiar offer little firm guidance for explaining what happened during this critical period.

To unravel this mystery, cosmologists must first pin down—by improving and refining current observations—some of the characteristics of the universe when it was only one second old: its expansion rate, the size of its density fluctuations, and its proportions of ordinary atoms, dark matter and radiation. But to comprehend why our universe was set up this way, we must probe further back, to the very first tiny fraction of a microsecond. Such an effort will require theoretical advances. Physicists must discover a way to relate Einstein's theory of general relativity, which governs large-scale interactions in the cosmos, with the quantum principles that apply at very short distances [see "A Unified Physics by 2050?," by Steven Weinberg; _Scientific American,_ December 1999]. A unified theory would be needed to explain what happened in the first crucial moments after the big bang, when the entire universe was squeezed into a space smaller than a single atom.

Astronomy is a subject in which observation is king. Now the same is true for cosmology—in contrast with the pre-1965 era, when speculation was largely unconstrained. The answers to many of cosmology's long-standing questions are most likely to come from the new telescopes now going into use. The two Keck Telescopes on Mauna Kea in Hawaii are far more sensitive than earlier observatories and thus can glimpse fainter objects. Still more impressive is the Very Large Telescope being built in northern Chile, which will be the world's premier optical facility when it is completed. Astronomers can take advantage of the Chandra X-ray Observatory, launched into orbit this past summer, and several new radio arrays on the ground. And a decade from now next-generation space telescopes will carry the enterprise far beyond what the Hubble can achieve.

Well before 2050 we are likely to see the construction of giant observatories in space or perhaps on the far side of the moon. The sensitivity and imaging power of these arrays will vastly surpass that of any instruments now in use. The new telescopes will target black holes and planets in other solar systems. They will also provide snapshots of every cosmological era going back to the very first light, when the earliest stars (or maybe quasars) condensed out of the expanding debris from the big bang. Some of these observatories may even be able to measure gravitational waves, allowing scientists to probe vibrations in the fabric of space-time itself.

The amount of data provided by all these instruments will be so colossal that the entire process of analysis and discovery will most likely be automated. Astronomers will focus their attention on heavily processed statistics for each population of objects they are studying and in this way find the best examples—for instance, the planets in other solar systems that are most like Earth. Researchers will also concentrate on extreme objects that may hold clues to physical processes that are not yet fully understood. One such object is the gamma-ray burster, which emits, for a few seconds, as much power as a billion galaxies. Increasingly, astronomers will use the heavens as a cosmic laboratory to probe phenomena that cannot be simulated on Earth.

Another benefit of automation will be open access to astronomical data that in the past were available only to a privileged few. Detailed maps of the sky will be available to anyone who can access or download them. Enthusiasts anywhere in the world will be able to check their own hunches, seek new patterns and discover unusual objects.

INTIMATIONS OF A MULTIVERSE?

Cosmologists view the universe as an intricate tapestry that has evolved from initial conditions that were imprinted in the first microsecond after the big bang. Complex structures and phenomena have unfolded from simple physical laws—we wouldn't be here if they hadn't. Simple laws, however, do not necessarily lead to complex consequences. Consider an analogue from the field of fractal mathematics: the Mandelbrot set, a pattern with an infinite depth of structure, is encoded by a short algorithm, but other simple algorithms that are superficially similar yield very boring patterns.

Our universe could not have become structured if it were not expanding at a special rate. If the big bang had produced fewer density fluctuations, the universe would have remained dark and featureless, with no galaxies or stars. And there are other prerequisites for complexity. If our universe had more than three spatial dimensions, planets could not stay in orbits around stars. If gravity were much stronger, it would crush living organisms of human size, and stars would be small and short-lived. If nuclear forces were a few percent weaker, only hydrogen would be stable: there would be no periodic table, no chemistry and no life. On the other hand, if nuclear forces were slightly stronger, hydrogen itself could not exist.

Some would argue that this fine-tuning of the universe, which seems so providential, is nothing to be surprised about, because we could not exist otherwise. There is, however, another interpretation: many universes may exist, but only some would allow creatures like us to emerge, and we obviously find ourselves in one of that subset. The seemingly designed features of our universe need then occasion no surprise.

Perhaps, then, our big bang wasn't the only one. This speculation dramatically enlarges our concept of reality. The entire history of our universe becomes just an episode, a single facet, of the infinite multiverse. Some universes might resemble ours, but most would be "stillborn." They would recollapse after a brief existence, or the laws governing them would not permit complex consequences.

Some cosmologists, especially Andrei Linde of Stanford University and Alex Vilenkin of Tufts University, have already shown how certain mathematical assumptions lead, at least in theory, to the creation of a multiverse. But such ideas will remain on the speculative fringe of cosmology until we really understand—rather than just guess at—the extreme physics that prevailed immediately after the big bang. Will the long-awaited unified theory uniquely determine the masses of particles and the strengths of the basic forces? Or are these properties in some sense accidental outcomes of how our universe cooled—secondary manifestations of still deeper laws governing an entire ensemble of universes?

This topic might seem arcane, but the status of multiverse ideas affects how we should place our bets in some ongoing cosmological controversies. Some theorists have a strong preference for the simplest picture of the cosmos, which would require an omega of 1—the universe would be just dense enough to halt its own expansion. They are unhappy with observations suggesting that the universe is not nearly so dense and with extra complications such as the cosmological constant. Perhaps we should draw a lesson from 17th-century astronomers Johannes Kepler and Galileo Galilei, who were upset to find that planetary orbits were elliptical. Circles, they thought, were simpler and more beautiful. But Newton later explained all orbits in terms of a simple, universal law of gravity. Had Galileo still been alive, he would have surely been joyfully reconciled to ellipses.

The parallel is obvious. If a low-density universe with a cosmological constant seems ugly, maybe this shows our limited vision. Just as Earth follows one of the few Keplerian orbits around the sun that allow it to be habitable, our universe may be one of the few habitable members of a grander ensemble.

A CHALLENGE FOR THE NEW MILLENNIUM

As the 21st century dawns, scientists are expanding humanity's store of knowledge on three great frontiers: the very big, the very small and the very complex. Cosmology involves them all. In the coming years, researchers will focus their efforts on pinning down the basic universal constants, such as omega, and on discovering what dark matter is. I think there is a good chance of achieving both goals within 10 years. Maybe everything will fit the standard theoretical framework, and we will successfully determine not only the relative abundance of ordinary atoms and dark matter in the universe but also the cosmological constant and the primordial density fluctuations. If that happens, we will have taken the measure of our universe just as, over the past few centuries, we have learned the size and shape of Earth and our sun. On the other hand, our universe may turn out to be too complicated to fit into the standard framework. Some may describe the first outcome as optimistic; others may prefer to inhabit a more complicated and challenging universe!

In addition, theorists must elucidate the exotic physics of the very earliest moments of the universe. If they succeed, we will learn whether there are many universes and which features of our universe are mere contingencies rather than the necessary outcomes of the deepest laws. Our understanding will still have limits, however. Physicists may someday discover a unified theory that governs all of physical reality, but they will never be able to tell us what breathes fire into their equations and what actualizes them in a real cosmos.

Cosmology is not only a fundamental science; it is also the grandest of the environmental sciences. How did a hot amorphous fireball evolve, over 10 to 15 billion years, into our complex cosmos of galaxies, stars and planets? How did atoms assemble—here on Earth and perhaps on other worlds—into living beings intricate enough to ponder their own origins? These questions are a challenge for the new millennium. Answering them may well be an unending quest.

FURTHER READING

Stephen P. Maran, ed. *The Astronomy and Astrophysics Encyclopedia.* Van Nostrand Reinhold, 1992.
Ken Croswell. *Planet Quest: The Epic Discovery of Alien Solar Systems.* Free Press, 1997.
Craig J. Hogan. *The Little Book of the Big Bang: A Cosmic Primer.* Copernicus, 1998.

Searching for Life in Our Solar System

BRUCE M. JAKOSKY

ORIGINALLY PUBLISHED IN MAY 1998

Since antiquity, human beings have imagined life spread far and wide in the universe. Only recently has science caught up, as we have come to understand the nature of life on Earth and the possibility that life exists elsewhere. Recent discoveries of planets orbiting other stars and of possible fossil evidence in Martian meteorites have gained considerable public acclaim. And the scientific case for life elsewhere has grown stronger during the past decade. There is now a sense that we are verging on the discovery of life on other planets.

To search for life in our solar system, we need to start at home. Because Earth is our only example of a planet endowed with life, we can use it to understand the conditions needed to spawn life elsewhere. As we define these conditions, though, we need to consider whether they are specific to life on Earth or general enough to apply anywhere.

Our geologic record tells us that life on Earth started shortly after life's existence became possible—only after protoplanets (small, planetlike objects) stopped bombarding our planet near the end of its formation. The last "Earth-sterilizing" giant impact probably occurred between 4.4 and 4.0 billion years ago. Fossil microscopic cells and carbon isotopic evidence suggest that life had grown widespread some 3.5 billion years ago and may have existed before 3.85 billion years ago.

Once it became safe for life to exist, no more than half a billion years— and perhaps as little as 100 million to 200 million years—passed before life rooted itself firmly on Earth. This short time span indicates that life's origin followed a relatively straightforward process, the natural consequence of chemical reactions in a geologically active environment. Equally important, this observation tells us that life may originate along similar lines in any place with chemical and environmental conditions akin to those of Earth.

The standard wisdom of the past 40 years holds that prebiological organic molecules formed in a so-called reducing atmosphere, with energy sources such as lightning triggering chemical reactions to combine gaseous molecules. A more recent theory offers a tantalizing alternative. As water circulates through ocean-floor volcanic systems, it heats to temperatures above 400 degrees Celsius (720 degrees Fahrenheit). When that superhot water returns to the ocean, it can chemically reduce agents, facilitating the formation of organic molecules. This reducing environment also provides an energy source to help organic molecules combine into larger structures and to foster primitive metabolic reactions.

WHERE DID LIFE ORIGINATE?

The significance of hydrothermal systems in life's history appears in the "tree of life," constructed recently from genetic sequences in RNA molecules, which carry forward genetic information. This tree arises from differences in RNA sequences common to all of Earth's living organisms. Organisms evolving little since their separation from their last common ancestor have similar RNA base sequences. Those organisms closest to the "root"—or last common ancestor of all living organisms—are hyperthermophiles, which live in hot water, possibly as high as 115 degrees C. This relationship indicates either that terrestrial life "passed through" hydrothermal systems at some early time or that life's origin took place within such systems. Either way, the earliest history of life reveals an intimate connection to hydrothermal systems.

As we consider possible occurrences of life elsewhere in the solar system, we can generalize environmental conditions required for life to emerge and flourish. We assume that liquid water is necessary—a medium through which primitive organisms can gain nutrients and disperse waste. Although other liquids, such as methane or ammonia, could serve the same function, water is likely to have been much more abundant, as well as chemically better for precipitating reactions necessary to spark biological activity.

To create the building blocks from which life can assemble itself, one needs access to biogenic elements. On Earth, these elements include carbon, hydrogen, oxygen, nitrogen, sulfur and phosphorus, among the two dozen or so others playing a pivotal role in life. Although life elsewhere might not use exactly the same elements, we would expect it to use many of them. Life on Earth utilizes carbon (over silicon, for example) because of its versatility in forming chemical bonds, rather than strictly its abundance. Carbon also exists readily as carbon dioxide, available as a gas or dissolved in water. Silicon dioxide, on the other hand, exists plentifully in neither form and would be much less accessible. Given the ubiquity of carbon-containing organic molecules throughout the universe, we would expect carbon to play a role in life anywhere.

Of course, an energy source must drive chemical disequilibrium, which fosters the reactions necessary to spawn living systems. On Earth today, nearly all of life's energy comes from the sun, through photosynthesis. Yet chemical energy sources suffice—and would be more readily available for early life. These sources would include geochemical energy from hydrothermal systems near volcanoes or chemical energy from the weathering of minerals at or near a planet's surface.

POSSIBILITIES FOR LIFE ON MARS

Looking beyond Earth, two planets show strong evidence for having had environmental conditions suitable to originate life at some time in their history—Mars and Europa. (For this purpose, we will consider Europa, a moon of Jupiter, to be a planetary body.)

Mars today is not very hospitable. Daily average temperatures rarely rise much above 220 kelvins, some 53 kelvins below water's freezing point. Despite this drawback, abundant evidence suggests that liquid water has existed on Mars's surface in the past and probably is present within its crust today.

Networks of dendritic valleys on the oldest Martian surfaces look like those on Earth formed by running water. The water may have come from atmospheric precipitation or "sapping," released from a crustal aquifer. Regardless of where it came from, liquid water undoubtedly played a role. The valleys' dendritic structure indicates that they formed gradually, meaning that water once may have flowed on Mars's surface, although we do not observe such signs today.

In addition, ancient impact craters larger than about 15 kilometers (nine miles) in diameter have degraded heavily, showing no signs of ejecta blankets, the raised rims or central peaks typically present on fresh craters. Some partly eroded craters display gullies on their walls, which look water-carved. Craters smaller than about 15 kilometers have eroded away entirely. The simplest explanation holds that surface water eroded the craters.

Although the history of Mars's atmosphere is obscure, the atmosphere may have been denser during the earliest epochs, 3.5 to 4.0 billion years ago. Correspondingly, a denser atmosphere could have yielded a strong greenhouse effect, which would have warmed the planet enough to permit liquid water to remain stable. Subsequent to 3.5 billion years ago, evidence tells us that the planet's crust did contain much water. Evidently, catastrophic floods, bursting from below the planet's surface, carved out great flood channels. These floods occurred periodically over geologic time. Based on this evidence, liquid water should exist several kilometers underground, where geothermal heating would raise temperatures to the melting point of ice.

Mars also has had rich energy sources throughout time. Volcanism has supplied heat from the earliest epochs to the recent past, as have impact events. Additional energy to sustain life can come from the weathering of volcanic rocks. Oxidation of iron within basalt, for example, releases energy that organisms can use.

The plentiful availability of biogenic elements on Mars's surface completes life's requirements. Given the presence of water and energy, Mars may well have independently originated life. Moreover, even if life did not originate on Mars, life still could be present there. Just as high-velocity impacts have jettisoned Martian surface rocks into space—only to fall on Earth as Martian meteorites—rocks from Earth could similarly have landed on the red planet. Should they contain organisms that survive the journey and should they land in suitable Martian habitats, the bacteria could survive. Or, for all we know, life could have originated on Mars and been transplanted subsequently to Earth.

An inventory of energy available on Mars suggests that enough is present to support life. Whether photosynthesis evolved, and thereby allowed life to move into other ecological niches, remains uncertain. Certainly, data returned from the Viking spacecraft during the 1970s presented no evidence that life is widespread on Mars. Yet it is possible that some Martian life currently exists, cloistered in isolated, energy-rich and water-laden niches—perhaps in volcanically heated, subsurface hydrothermal systems or merely underground, drawing energy from chemical interactions of liquid water and rock.

Recent analysis of Martian meteorites found on Earth has led many scientists to conclude that life may have once thrived on Mars—based on fossil remnants seen within the rock. Yet this evidence does not definitively indicate biological activity; indeed, it may result from natural geochemical processes. Even if scientists determine that these rocks contain no evidence of Martian life, life on the red planet might still be possible—but in locations not yet searched. To draw a definitive conclusion, we must study those places where life (or evidence of past life) will most likely appear.

EUROPA

Europa, on the other hand, presents a different possible scenario for life's origin. At first glance, Europa seems an unlikely place for life. The largest of Jupiter's satellites, Europa is a little bit smaller than our moon, and its surface is covered with nearly pure ice. Yet Europa's interior may be less frigid, warmed by a combination of radioactive decay and tidal heating, which could raise the temperature above the melting point of ice at relatively shallow depths. Because the layer of surface ice stands 150 to 300 kilometers thick, a global, ice-covered ocean of liquid water may exist underneath.

Recent images of Europa's surface from the Galileo spacecraft reveal the possible presence of at least transient pockets of liquid water. Globally, the surface appears covered with long grooves or cracks. On a smaller scale, these quasilinear features show detailed structures indicating local ice-related tectonic activity and infilling from below. On the smallest scale, blocks of ice are present. By tracing the crisscrossing grooves, the blocks clearly have moved with respect to the larger mass. They appear similar to sea ice on Earth—as if large ice blocks had broken off the main mass, floated a small distance away and then frozen in place. Unfortunately, we cannot yet determine if the ice blocks floated through liquid water or slid on relatively warm, soft ice. The dearth of impact craters on the ice indicates that fresh ice continually resurfaces Europa. It is also likely that liquid water is present at least on an intermittent basis.

If Europa has liquid water at all, then that water probably exists at the interface between the ice and underlying rocky interior. Europa's rocky center probably has had volcanic activity—perhaps at a level similar to that of Earth's moon, which rumbled with volcanism until about 3.0 billion years ago. The volcanism within its core would create an energy source for possible life, as would the weathering of minerals reacting with water. Thus, Europa has all the ingredients from which to spark life. Of course, less chemical energy is likely to exist on Europa than Mars, so we should not expect to see an abundance of life, if any. Although the Galileo space probe has detected organic molecules and frozen water on Callisto and Ganymede, two of Jupiter's four Galilean satellites, these moons lack the energy sources that life would require to take hold. Only Io, also a Galilean satellite, has volcanic heat—yet it has no liquid water, necessary to sustain life as we know it.

Mars and Europa stand today as the only places in our solar system that we can identify as having (or having had) all ingredients necessary to spawn life. Yet they are not the only planetary bodies in our solar system relevant to exobiology. In particular, we can look at Venus and at Titan, Saturn's largest moon. Venus currently remains too hot to sustain life, with scorching surface temperatures around 750 kelvins, sustained by greenhouse warming from carbon dioxide and sulfur dioxide gases. Any liquid water has long since disappeared into space.

VENUS AND TITAN

Why are Venus and Earth so different? If Earth orbited the sun at the same distance that Venus does, then Earth, too, would blister with heat—causing more water vapor to fill the atmosphere and augmenting the greenhouse effect. Positive feedback would spur this cycle, with more water, greater greenhouse warming and so on saturating the atmosphere and sending temperatures soaring. Because temperature plays such a strong role in determining the atmosphere's water content, both Earth and Venus have a temperature threshold, above which the positive feedback of an increasing greenhouse effect takes off. This feedback loop would load Venus's atmosphere with water, which in turn would catapult its temperatures to very high values. Below this threshold, its climate would have been more like that of Earth.

Venus, though, may not always have been so inhospitable. Four billion years ago the sun emitted about 30 percent less energy than it does today. With less sunlight, the boundary between clement and runaway climates may have been inside Venus's orbit, and Venus may have had surface temperatures only 100 degrees C above Earth's current temperature. Life could survive quite readily at those temperatures—as we observe with certain bacteria and bioorganisms living near hot springs and undersea vents. As the sun became hotter, Venus would have warmed gradually until it would have undergone a catastrophic transition to a thick, hot atmosphere. It is possible that Venus originated life several billion years ago but that high temperatures and geologic activity have since obliterated all evidence of a biosphere. As the sun continues to heat up, Earth may undergo a similar catastrophic transition only a couple of billion years from now.

Titan intrigues us because of abundant evidence of organic chemical activity in its atmosphere, similar to what might have occurred on the early Earth if its atmosphere had potent abilities to reduce chemical agents. Titan is about as big as Mercury, with an atmosphere thicker than Earth's, consisting predominantly of nitrogen, methane and ethane. Methane must be continually resupplied from the surface or subsurface, because photochemical reactions in the atmosphere drive off hydrogen (which is lost to space) and convert the methane to longer chains of organic molecules. These longer-chain hydrocarbons are thought to provide the dense haze that obscures Titan's surface at visible wavelengths.

Surface temperatures on Titan stand around 94 kelvins, too cold to sustain either liquid water or nonphotochemical reactions that could produce biological activity—although Titan apparently had some liquid water during its early history. Impacts during its formation would have deposited enough heat (from the kinetic energy of the object) to melt frozen water locally. Deposits of liquid water might have persisted for thousands of years before freezing. Every part of Titan's surface probably has melted at least once. The degree to which biochemical reactions may have proceeded during such a short time interval is uncertain, however.

EXPLORATORY MISSIONS

Clearly, the key ingredients needed for life have been present in our solar system for a long time and may be present today outside of Earth. At one time or another, four planetary bodies may have contained the necessary conditions to generate life.

We can determine life's actual existence elsewhere only empirically, and the search for life has taken center stage in the National Aeronautics and Space Administration's ongoing science missions. The Mars Surveyor series of missions, scheduled to take place during the coming decade, aims to determine if Mars ever had life. This series will culminate in a mission currently scheduled for launch in 2005, to collect Martian rocks from regions of possible biological relevance and return them to Earth for detailed analysis. The Cassini spacecraft currently is en route to Saturn. There the Huygens probe will enter Titan's atmosphere, its goal to decipher Titan's composition and chemistry. A radar instrument, too, will map Titan's surface, looking both for geologic clues to its history and evidence of exposed lakes or oceans of methane and ethane.

Moreover, the Galileo orbiter of Jupiter is focusing its extended mission on studying the surface and interior of Europa. Plans are under way to launch a spacecraft mission dedicated to Europa, to discern its geologic and geochemical history and to determine if a global ocean lies underneath its icy shell.

Of course, it is possible that, as we plumb the depths of our own solar system, no evidence of life will turn up. If life assembles itself from basic building blocks as easily as we believe it does, then life should turn up elsewhere. Indeed, life's absence would lead us to question our understanding of life's origin here on Earth. Whether or not we find life, we will gain a tremendous insight into our own history and whether life is rare or widespread in our galaxy.

The Fate of Life in the Universe

LAWRENCE M. KRAUSS AND GLENN D. STARKMAN

ORIGINALLY PUBLISHED IN NOVEMBER 1999

Eternal life is a core belief of many of the world's religions. Usually it is extolled as a spiritual Valhalla, an existence without pain, death, worry or evil, a world removed from our physical reality. But there is another sort of eternal life that we hope for, one in the temporal realm. In the conclusion to *Origin of Species,* Charles Darwin wrote: "As all the living forms of life are the lineal descendants of those which lived before the Cambrian epoch, we may feel certain that the ordinary succession by generation has never once been broken. . . . Hence we may look with some confidence to a secure future of great length." The sun will eventually exhaust its hydrogen fuel, and life as we know it on our home planet will eventually end, but the human race is resilient. Our progeny will seek new homes, spreading into every corner of the universe just as organisms have colonized every possible niche of the earth. Death and evil will take their toll, pain and worry may never go away, but somewhere we expect that some of our children will carry on.

Or maybe not. Remarkably, even though scientists fully understand neither the physical basis of life nor the unfolding of the universe, they can make educated guesses about the destiny of living things. Cosmological observations now suggest the universe will continue to expand forever—rather than, as scientists once thought, expand to a maximum size and then shrink. Therefore, we are not doomed to perish in a fiery "big crunch" in which any vestige of our current or future civilization would be erased. At first glance, eternal expansion is cause for optimism. What could stop a sufficiently intelligent civilization from exploiting the endless resources to survive indefinitely?

Yet life thrives on energy and information, and very general scientific arguments hint that only a finite amount of energy and a finite amount of information can be amassed in even an infinite period. For life to persist, it would have to make do with dwindling resources and limited knowledge. We have concluded that no meaningful form of consciousness could exist forever under these conditions.

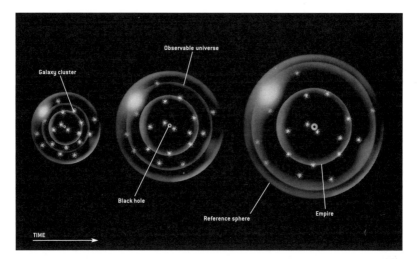

Energy collection strategy devised by physicist Steven Frautschi illustrates how difficult it will be to survive in the far future, 10^{100} or so years from now. In many cosmological scenarios, resources multiply as the universe—and any arbitrary reference sphere within it (*outer sphere*)—expands and an increasing fraction of it becomes observable (*middle sphere*). A civilization could use a black hole to convert matter—plundered from its empire (*inner sphere*)—into energy. But as the empire grows, the cost of capturing new territory increases; the conquest can barely keep pace with the dilution of matter. In fact, matter will become so diluted that the civilization will not be able to safely build a black hole large enough to collect it.

THE DESERTS OF VAST ETERNITY

Over the past century, scientific eschatology has swung between optimism and pessimism. Not long after Darwin's confident prediction, Victorian-era scientists began to fret about the "heat death," in which the whole cosmos would come to a common temperature and thereafter be incapable of change. The discovery of the expansion of the universe in the 1920s allayed this concern, because expansion prevents the universe from reaching such an equilibrium. But few cosmologists thought through the other implications for life in an ever expanding universe, until a classic paper in 1979 by physicist Freeman Dyson of the Institute for Advanced Study in Princeton, N.J., itself motivated by earlier work by Jamal Islam, now at the University of Chittagong in Bangladesh. Since Dyson's paper, physicists and astronomers have periodically reexamined the topic [see "The Future of the Universe," by Duane A. Dicus, John R. Letaw, Doris C. Teplitz and Vigdor L. Teplitz; *Scientific American*, March 1983]. A year ago, spurred on by new observations that suggest a drastically different long-term future for the universe than that previously envisaged, we decided to take another look.

Over the past 12 billion years or so, the universe has passed through many stages. At the earliest times for which scientists now have empirical information, it was incredibly hot and dense. Gradually, it expanded and cooled. For hundreds of thousands of years, radiation ruled; the famous cosmic microwave background radiation is thought to be a vestige of this era. Then matter started to dominate, and progressively larger astronomical structures condensed out. Now, if recent cosmological observations are correct, the expansion of the universe is beginning to accelerate—a sign that a strange new type of energy, perhaps springing from space itself, may be taking over.

Life as we know it depends on stars. But stars inevitably die, and their birth rate has declined dramatically since an initial burst about 10 billion years ago. About 100 trillion years from now, the last conventionally formed star will wink out, and a new era will commence. Processes currently too slow to be noticed will become important: the dispersal of planetary systems by stellar close encounters, the possible decay of ordinary and exotic matter, the slow evaporation of black holes.

Assuming that intelligent life can adapt to the changing circumstances, what fundamental limits does it face? In an eternal universe, potentially of infinite volume, one might hope that a sufficiently advanced civilization could collect an infinite amount of matter, energy and information. Surprisingly, this is not true. Even after an eternity of hard and well-planned labor, living beings could accumulate only a finite number of particles, a finite quantity of energy and a finite number of bits of information. What makes this failure all the more frustrating is that the number of available particles, ergs and bits may grow without bound. The problem is not necessarily the lack of resources but rather the difficulty in collecting them.

The culprit is the very thing that allows us to contemplate an eternal tenure: the expansion of the universe. As the cosmos grows in size, the average density of ordinary sources of energy declines. Doubling the radius of the universe decreases the density of atoms eightfold. For light waves, the decline is even more precipitous. Their energy density drops by a factor of 16 because the expansion stretches them and thereby saps their energy.

As a result of this dilution, resources become ever more time-consuming to collect. Intelligent beings have two distinct strategies: let the material come to them or try to chase it down. For the former, the best approach in the long run is to let gravity do the work. Of all the forces of nature, only gravity and electromagnetism can draw things in from arbitrarily far away. But the latter gets screened out: oppositely charged particles balance one another, so that the typical object is neutral and hence immune

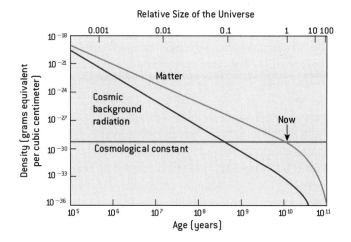

Dilution of the cosmos by the expansion of space affects different forms of energy in different ways. Ordinary matter thins out in direct proportion to volume, whereas the cosmic background radiation weakens even faster as it is stretched from light into microwaves and beyond. The energy density represented by a cosmological constant does not change, at least according to present theories.

to long-range electrical and magnetic forces. Gravity, on the other hand, cannot be screened out, because particles of matter and radiation only attract gravitationally; they do not repel.

SURRENDER TO THE VOID

Even gravity, however, must contend with the expansion of the universe, which pulls objects apart and thereby weakens their mutual attraction. In all but one scenario, gravity eventually becomes unable to pull together larger quantities of material. Indeed, our universe may have already reached this point; clusters of galaxies may be the largest bodies that gravity will ever be able to bind together [see "The Evolution of Galaxy Clusters," by J. Patrick Henry, Ulrich G. Briel and Hans Böhringer; *Scientific American,* December 1998]. The lone exception occurs if the universe is poised between expansion and contraction, in which case gravity continues indefinitely to assemble ever greater amounts of matter. But that scenario is now thought to contradict observations, and in any event it poses its own difficulty: after 10^{33} years or so, the accessible matter will become so concentrated that most of it will collapse into black holes, sweeping up any life-forms. Being inside a black hole is not a happy condition. On the earth, all roads may lead to Rome, but inside a black hole, all roads lead in a finite amount of time to the center of the hole, where death and dismemberment are certain.

Sadly, the strategy of actively seeking resources fares no better than the passive approach does. The expansion of the universe drains away kinetic energy, so prospectors would have to squander their booty to maintain their speed. Even in the most optimistic scenario—in which the energy is traveling toward the scavenger at the speed of light and is collected without loss—a civilization could garner limitless energy only in or near a black hole. The latter possibility was explored by Steven Frautschi of the California Institute of Technology in 1982. He concluded that the energy available from the holes would dwindle more quickly than the costs of scavenging. We recently reexamined this possibility and found that the predicament is even worse than Frautschi thought. The size of a black hole required to sweep up energy forever exceeds the extent of the visible universe.

The cosmic dilution of energy is truly dire if the universe is expanding at an accelerating rate. All distant objects that are currently in view will eventually move away from us faster than the speed of light and, in doing so, disappear from view. The total resources at our disposal are therefore limited by what we can see today, at most.

Not all forms of energy are equally subject to the dilution. The universe might, for example, be filled with a network of cosmic strings infinitely long, thin concentrations of energy that could have developed as the early universe cooled unevenly. The energy per unit length of a cosmic string remains unchanged despite cosmic expansion [see "Cosmic Strings," by Alexander Vilenkin; *Scientific American,* December 1987]. Intelligent beings might try to cut one, congregate around the loose ends and begin consuming it. If the string network is infinite, they might hope to satisfy their appetite forever. The problem with this strategy is that whatever lifeforms can do, natural processes can also do. If a civilization can figure out a way to cut cosmic strings, then the string network will fall apart of its own accord. For example, black holes may spontaneously appear on the strings and devour them. Therefore, the beings could swallow only a finite amount of string before running into another loose end. The entire string network would eventually disappear, leaving the civilization destitute.

What about mining the quantum vacuum? After all, the cosmic acceleration may be driven by the so-called cosmological constant, a form of energy that does not dilute as the universe expands [see "Cosmological Antigravity," by Lawrence M. Krauss; *Scientific American*, January 1999]. If so, empty space is filled with a bizarre type of radiation, called Gibbons-Hawking or de Sitter radiation. Alas, it is impossible to extract energy from

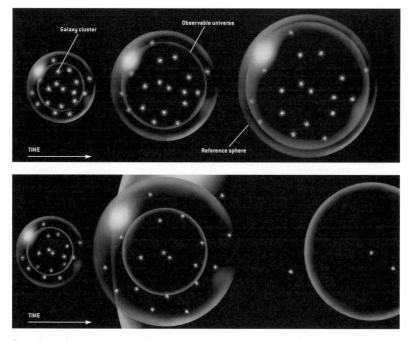

Expanding universe looks dramatically different depending on whether the growth is decelerating (*upper sequence*) or accelerating (*lower sequence*). In both these cases, the universe is infinite, but any patch of space—demarcated by a reference sphere that represents the distance to particular galaxies—enlarges (*outer sphere*). We can see only a limited volume, which grows steadily as light signals have time to propagate (*inner sphere*). If expansion is decelerating, we can see an increasing fraction of the cosmos. More and more galaxies fill the sky. But if expansion is accelerating, we can see a decreasing fraction of the cosmos. Space seems to empty out.

this radiation for useful work. If the vacuum yielded up energy, it would drop into a lower energy state, yet the vacuum is already the lowest energy state there is.

No matter how clever we try to be and how cooperative the universe is, we will someday have to confront the finiteness of the resources at our disposal. Even so, are there ways to cope forever?

The obvious strategy is to learn to make do with less, a scheme first discussed quantitatively by Dyson. In order to reduce energy consumption and keep it low despite exertion, we would eventually have to reduce our body temperature. One might speculate about genetically engineered humans who function at somewhat lower temperatures than 310 kelvins (98.6 degrees Fahrenheit). Yet the human body temperature cannot be reduced arbitrarily; the freezing point of blood is a firm lower limit. Ultimately, we will need to abandon our bodies entirely.

While futuristic, the idea of shedding our bodies presents no fundamental difficulties. It presumes only that consciousness is not tied to a particular set of organic molecules but rather can be embodied in a multitude of different forms, from cyborgs to sentient interstellar clouds [see "Will Robots Inherit the Earth?" by Marvin Minsky; *Scientific American*, October 1994]. Most modern philosophers and cognitive scientists regard conscious thought as a process that a computer could perform. The details need not concern us here (which is convenient, as we are not competent to discuss them). We still have many billions of years to design new physical incarnations to which we will someday transfer our conscious selves. These new "bodies" will need to operate at cooler temperatures and at lower metabolic rates—that is, lower rates of energy consumption.

Dyson showed that if organisms could slow their metabolism as the universe cooled, they could arrange to consume a finite total amount of energy over all of eternity. Although the lower temperatures would also slow consciousness—the number of thoughts per second—the rate would remain large enough for the total number of thoughts, in principle, to be unlimited. In short, intelligent beings could survive forever, not just in absolute time but in subjective time. As long as organisms were guaranteed to have an infinite number of thoughts, they would not mind a languid pace of life. When billions of years stretch out before you, what's the rush?

At first glance, this might look like a case of something for nothing. But the mathematics of infinity can defy intuition. For an organism to maintain the same degree of complexity, Dyson argued, its rate of information processing must be directly proportional to body temperature, whereas the rate of energy consumption is proportional to the square of the temperature (the additional factor of temperature comes from basic thermodynamics). Therefore, the power requirements slacken faster than cognitive alacrity does. At 310 kelvins, the human body expends approximately 100 watts. At 155 kelvins, an equivalently complex organism could think at half the speed but consume a quarter of the power. The trade-off is acceptable because physical processes in the environment slow down at a similar rate.

TO SLEEP, TO DIE

Unfortunately, there is a catch. Most of the power is dissipated as heat, which must escape—usually by radiating away—if the object is not to heat up. Human skin, for example, glows in infrared light. At very low temperatures, the most efficient radiator would be a dilute gas of electrons.

Yet the efficiency even of this optimal radiator declines as the cube of the temperature, faster than the decrease in the metabolic rate. A point would come when organisms could not lower their temperature further. They would be forced instead to reduce their complexity—to dumb down. Before long, they could no longer be regarded as intelligent.

To the timid, this might seem like the end. But to compensate for the inefficiency of radiators, Dyson boldly devised a strategy of hibernation. Organisms would spend only a small fraction of their time awake. While sleeping, their metabolic rates would drop, but—crucially—they would continue to dissipate heat. In this way, they could achieve an ever lower average body temperature. In fact, by spending an increasing fraction of their time asleep, they could consume a finite amount of energy yet exist forever and have an infinite number of thoughts. Dyson concluded that eternal life is indeed possible.

Since his original paper, several difficulties with his plan have emerged. For one, Dyson assumed that the average temperature of deep space—currently 2.7 kelvins, as set by the cosmic microwave background radiation—would always decrease as the cosmos expands, so that organisms could continue to decrease their temperature forever. But if the universe has a cosmological constant, the temperature has an absolute floor fixed by the Gibbons-Hawking radiation. For current estimates of the value of the cosmological constant, this radiation has an effective temperature of about 10^{-29} kelvin. As was pointed out independently by cosmologists J. Richard Gott II, John Barrow, Frank Tipler and us, once organisms had cooled to this level, they could not continue to lower their temperature in order to conserve energy.

The second difficulty is the need for alarm clocks to wake the organisms periodically. These clocks would have to operate reliably for longer and longer times on less and less energy. Quantum mechanics suggests that this is impossible. Consider, for example, an alarm clock that consists of two small balls that are taken far apart and then aimed at each other and released. When they collide, they ring a bell. To lengthen the time between alarms, organisms would release the balls at a slower speed. But eventually the clock will run up against constraints from Heisenberg's uncertainty principle, which prevents the speed and position of the balls from both being specified to arbitrary precision. If one or the other is sufficiently inaccurate, the alarm clock will fail, and hibernation will turn into eternal rest.

One might imagine other alarm clocks that could forever remain above the quantum limit and might even be integrated into the organism itself.

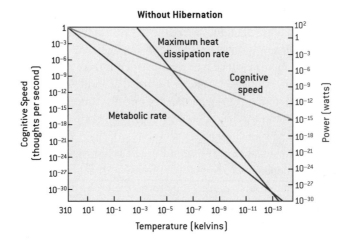

Without Hibernation

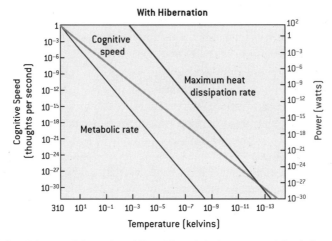

With Hibernation

Eternal life on finite energy? If a new form of life could lower its body temperature below the human value of 310 kelvins (98.6 degrees Fahrenheit), it would consume less power, albeit at the cost of thinking more sluggishly (*upper graph*). Because metabolism would decline faster than cognition, the life-form could arrange to have an infinite number of thoughts on limited resources. One caveat is that its ability to dissipate waste heat would also decline, preventing it from cooling below about 10^{-13} kelvin. Hibernation (*lower graph*) might eliminate the problem of heat disposal. As the life-form cooled, it would spend an increasing fraction of its time dormant, further reducing its average metabolic rate and cognitive speed. In this way, the power consumption could always remain lower than the maximum rate of heat dissipation, while still allowing for an infinite number of thoughts. But such a scheme might run afoul of other problems, such as quantum limits.

Nevertheless, no one has yet come up with a specific mechanism that could reliably wake an organism while consuming finite energy.

THE ETERNAL RECURRENCE OF THE SAME

The third and most general doubt about the long-term viability of intelligent life involves fundamental limitations on computation. Computer scientists once thought it was impossible to compute without expending a certain minimum amount of energy per operation, an amount that is directly proportional to the temperature of the computer. Then, in the early 1980s, researchers realized that certain physical processes, such as quantum effects or the random Brownian motion of a particle in a fiuid, could serve as the basis for a lossless computer [see "The Fundamental Physical Limits of Computation," by Charles H. Bennett and Rolf Landauer; *Scientific American,* July 1985]. Such computers could operate with an arbitrarily small amount of energy. To use less, they simply slow down—a trade-off that eternal organisms may be able to make. There are only two conditions. First, they must remain in thermal equilibrium with their environment. Second, they must never discard information. If they did, the computation would become irreversible, and thermodynamically an irreversible process must dissipate energy.

Unhappily, those conditions become insurmountable in an expanding universe. As cosmic expansion dilutes and stretches the wavelength of light, organisms become unable to emit or absorb the radiation they would need to establish thermal equilibrium with their surroundings. And with a finite amount of material at their disposal, and hence a finite memory, they would eventually have to forget an old thought in order to have a new one. What kind of perpetual existence could such organisms have, even in principle? They could collect only a finite number of particles and a finite amount of information. Those particles and bits could be configured in only a finite number of ways. Because thoughts are the reorganization of information, finite information implies a finite number of thoughts. All organisms would ever do is relive the past, having the same thoughts over and over again. Eternity would become a prison, rather than an endlessly receding horizon of creativity and exploration. It might be nirvana, but would it be living?

It is only fair to point out that Dyson has not given up. In his correspondence with us, he has suggested that life can avoid the quantum constraints on energy and information by, for example, growing in size or using different types of memory. As he puts it, the question is whether life is "analog" or "digital"—that is, whether continuum physics or quantum physics sets its limits. We believe that over the long haul life is digital.

Is there any other hope for eternal life? Quantum mechanics, which we argue puts such unbending limits on life, might come to its rescue in another guise. For example, if the quantum mechanics of gravity allows the existence of stable wormholes, life-forms might circumvent the barriers erected by the speed of light, visit parts of the universe that are otherwise inaccessible, and collect infinite amounts of energy and information. Or perhaps they could construct "baby" universes [see "The Self-Reproducing Inflationary Universe," by Andrei Linde; *Scientific American,* November 1994] and send themselves, or at least a set of instructions to reconstitute themselves, through to the baby universe. In that way, life could carry on.

The ultimate limits on life will in any case become significant only on timescales that are truly cosmic. Still, for some it may seem disturbing that life, certainly in its physical incarnation, must come to an end. But to us, it is remarkable that even with our limited knowledge, we can draw conclusions about such grand issues. Perhaps being cognizant of our fascinating universe and our destiny within it is a greater gift than being able to inhabit it forever.

FURTHER READING

Freeman J. Dyson. "Time without End: Physics and Biology in an Open Universe" in *Reviews of Modern Physics,* Vol. 51, No. 3, pages 447–460; July 1979.

John D. Barrow and Frank J. Tipler. *The Anthropic Cosmological Principle.* Oxford University Press, 1988.

Paul C. W. Davies. *The Last Three Minutes: Conjectures about the Ultimate Fate of the Universe.* Harper Collins, 1997.

Fred Adams and Greg Laughlin. *The Five Ages of the Universe: Inside the Physics of Eternity.* Free Press, 1999.

Lawrence M. Krauss. *Quintessence: The Mystery of the Missing Mass.* Basic Books, 1999.

Lawrence M. Krauss and Glenn D. Starkman. "Life, the Universe, and Nothing: Life and Death in an Ever-Expanding Universe" in *Astrophysical Journal,* Vol. 531, pages 22–30; 2000.

Life's Rocky Start

ROBERT M. HAZEN

ORIGINALLY PUBLISHED IN APRIL 2001

No one knows how life arose on the desolate young earth, but one thing is certain: life's origin was a chemical event. Once the earth formed 4.5 billion years ago, asteroid impacts periodically shattered and sterilized the planet's surface for another half a billion years. And yet, within a few hundred million years of that hellish age, microscopic life appeared in abundance. Sometime in the interim, the first living entity must have been crafted from air, water and rock.

Of those three raw materials, the atmosphere and oceans have long enjoyed the starring roles in origins-of-life scenarios. But rocks, and the minerals of which they are made, have been called on only as bit players or simply as props. Scientists are now realizing that such limited casting is a mistake. Indeed, a recent flurry of fascinating experiments is revealing that minerals play a crucial part in the basic chemical reactions from which life must have arisen.

The first act of life's origin story must have introduced collections of carbon-based molecules that could make copies of themselves. Achieving even this nascent step in evolution entailed a sequence of chemical transformations, each of which added a level of structure and complexity to a group of organic molecules. The most abundant carbon-based compounds available on the ancient earth were gases with only one atom of carbon per molecule, namely, carbon dioxide, carbon monoxide and methane. But the essential building blocks of living organisms—energy-rich sugars, membrane-forming lipids and complex amino acids—may include more than a dozen carbon atoms per molecule. Many of these molecules, in turn, must bond together to form chainlike polymers and other molecular arrays in order to accomplish life's chemical tasks. Linking small molecules into these complex, extended structures must have been especially difficult in the harsh conditions of the early earth, where intense ultraviolet radiation tended to break down clusters of molecules as quickly as they could form.

Carbon-based molecules needed protection and assistance to enact this drama. It turns out that minerals could have served at least five significant functions, from passive props to active players, in life-inducing chemical reactions. Tiny compartments in mineral structures can shelter simple molecules, while mineral surfaces can provide the scaffolding on which those molecules assemble and grow. Beyond these sheltering and supportive functions, crystal faces of certain minerals can actively select particular molecules resembling those that were destined to become biologically important. The metallic ions in other minerals can jump-start meaningful reactions like those that must have converted simple molecules into self-replicating entities. Most surprising, perhaps, are the recent indications that elements of dissolved minerals can be incorporated into biological molecules. In other words, minerals may not have merely helped biological molecules come together, they might have become part of life itself.

PROTECTION FROM THE ELEMENTS

For the better part of a century, following the 1859 publication of Charles Darwin's *On the Origin of Species,* a parade of scientists speculated on life's chemical origins. Some even had the foresight to mention rocks and minerals in their inventive scenarios. But experimental evidence only sporadically buttressed these speculations.

One of the most famous experiments took place at the University of Chicago in 1953. That year chemist Harold C. Urey's precocious graduate student Stanley L. Miller attempted to mimic the earth's primitive oceans and atmosphere in a bottle. Miller enclosed methane, ammonia and other gases thought to be components of the early atmosphere in a glass flask partially filled with water. When he subjected the gas to electric sparks to imitate a prehistoric lightning storm, the clear water turned pink and then brown as it became enriched with amino acids and other essential organic molecules. With this simple yet elegant procedure, Miller transformed origins-of-life research from a speculative philosophical game to an exacting experimental science. The popular press sensationalized the findings by suggesting that synthetic bugs might soon be crawling out of test tubes. The scientific community was more restrained, but many workers sensed that the major obstacle to creating life in the laboratory had been solved.

It did not take long to disabuse researchers of that notion. Miller may have discovered a way to make many of life's building blocks out of the earth's early supply of water and gas, but he had not discovered how or where these simple units would have linked into the complex molecular structures—such as proteins and DNA—that are intrinsic to life.

To answer that riddle, Miller and other origins scientists began proposing rocks as props. They speculated that organic molecules, floating in seawater, might have splashed into tidal pools along rocky coastlines. These molecules would have become increasingly concentrated through repeated cycles of evaporation, like soup thickening in a heated pot.

In recent years, however, researchers have envisioned that life's ingredients might have accumulated in much smaller containers. Some rocks, like gray volcanic pumice, are laced with air pockets created when gases expanded inside the rock while it was still molten. Many common minerals, such as feldspar, develop microscopic pits during weathering. Each tiny chamber in each rock on the early earth could have housed a separate experiment in molecular self-organization. Given enough time and enough chambers, serendipity might have produced a combination of molecules that would eventually deserve to be called "living."

Underlying much of this speculation was the sense that life was so fragile that it depended on rocks for survival. But in 1977 a startling discovery challenged conventional wisdom about life's fragility and, perhaps, its origins. Until then, most scientists had assumed that life spawned at or near the benign ocean surface as a result of chemistry powered by sunlight. That view began to change when deep-ocean explorers first encountered diverse ecosystems thriving at the superheated mouths of volcanic vents on the seafloor. These extreme environments manage to support elaborate communities of living creatures in isolation from the sun. In these dark realms, much of the energy that organisms need comes not from light but from the earth's internal heat. With this knowledge in mind, a few investigators began to wonder whether organic reactions relevant to the origins of life might occur in the intense heat and pressure of these so-called hydrothermal vents.

Miller and his colleagues have objected to the hydrothermal origins hypothesis in part because amino acids decompose rapidly when they are heated. This objection, it turns out, may be applicable only when key minerals are left out of the equation. The idea that minerals might have sheltered the ingredients of life received a boost from recent experiments conducted at my home base, the Carnegie Institution of Washington's Geophysical Laboratory. As a postdoctoral researcher at Carnegie, my colleague Jay A. Brandes (now at the University of Texas Marine Sciences Institute in Port Aransas) proposed that minerals help delicate amino acids remain intact. In 1998 we conducted an experiment in which the amino acid leucine broke down within a matter of minutes in pressurized water at 200 degrees Celsius—just as Miller and his colleagues predicted. But when Brandes added to the mix an iron sulfide mineral of the type commonly found in and around hydrothermal vents, the amino acid stayed intact for days—plenty of time to react with other critical molecules.

A ROCK TO STAND ON

Even if the right raw materials were contained in a protected place—whether it was a tidal pool, a microscopic pit in a mineral surface or somewhere inside the plumbing of a seafloor vent—the individual molecules would still be suspended in water. These stray molecules needed a support structure—some kind of scaffolding—where they could cling and react with one another.

One easy way to assemble molecules from a dilute solution is to concentrate them on a flat surface. Errant molecules might have been drawn to the calm surface of a tidal pool or perhaps to a primitive "oil slick" of compounds trapped at the water's surface. But such environments would have posed a potentially fatal hazard to delicate molecules. Harsh lightning storms and ultraviolet radiation accosted the young earth in doses many times greater than they do today. Such conditions would have quickly broken the bonds of complex chains of molecules.

Origins scientists with a penchant for geology have long recognized that minerals might provide attractive alternative surfaces where important molecules could assemble. Like the container idea, this notion was born half a century ago. At that time, a few scientists had begun to suspect that clays have special abilities to attract organic molecules. These ubiquitous minerals feel slick when wet because their atoms form flat, smooth layers. The surfaces of these layers frequently carry an electric charge, which might be able to attract organic molecules and hold them in place. Experiments later confirmed these speculations. In the late 1970s an Israeli research group demonstrated that amino acids can concentrate on clay surfaces and then link up into short chains that resemble biological proteins. These chemical reactions occurred when the investigators evaporated a water-based solution containing amino acids from a vessel containing clays—a situation not unlike the evaporation of a shallow pond or tidal pool with a muddy bottom.

More recently, separate research teams led by James P. Ferris of the Rensselaer Polytechnic Institute and by Gustaf Arrhenius of the Scripps Institution of Oceanography demonstrated that clays and other layered minerals can attract and assemble a variety of organic molecules. In a tour de force series of experiments during the past decade, the team at Rensselaer found that clays can act as scaffolds for the building blocks of RNA, the molecule in living organisms that translates genetic instructions into proteins.

Once organic molecules had attached themselves to a mineral scaffold, various types of complex molecules could have been forged. But only a chosen few were eventually incorporated into living cells. That means that some kind of template must have selected the primitive molecules that would become biologically important. Recent experiments show, once again, that minerals may have played a central role in this task.

PREFERENTIAL TREATMENT

Perhaps the most mysterious episode of selection left all living organisms with a strange predominance of one type of amino acid. Like many organic molecules, amino acids come in two forms. Each version comprises the same types of atoms, but the two molecules are constructed as mirror images of each other. The phenomenon is called chirality, but for simplicity's sake scientists refer to the two versions as "left-handed" (or "L") and "right-handed" (or "D"). Organic synthesis experiments like Miller's invariably produce 50–50 mixtures of L and D molecules, but the excess of left-handed amino acids in living organisms is nearly 100 percent.

Researchers have proposed a dozen theories—from the mundane to the exotic—to account for this bizarre occurrence. Some astrophysicists have argued that the earth might have formed with an excess of L amino acids—a consequence of processes that took place in the cloud of dust and gas that became the solar system. The main problem with this theory is that in most situations such processes yield only the slightest excess—less than 1 percent—of L or D molecules.

Alternatively, the world might have started with a 50–50 mixture of L and D amino acids, and then some important feature of the physical environment selected one version over the other. To me, the most obvious candidates for this specialized physical environment are crystal faces whose surface structures are mirror images of each other. Last spring I narrowed in on calcite, the common mineral that forms limestone and marble, in part because it often displays magnificent pairs of mirror-image faces. The chemical structure of calcite in many mollusk shells bonds strongly to amino acids. Knowing this, I began to suspect that calcite surfaces may feature chemical bonding sites that are ideally suited to only one type of amino acid or the other. With the help of my Carnegie colleague Timothy Filley (now at Purdue University) and Glenn Goodfriend of George Washington University, I ran more than 100 tests of this hypothesis.

Our experiments were simple in concept, although they required meticulous clean-room procedures to avoid contamination by the amino acids that exist everywhere in the environment. We immersed a well-formed, fist-size crystal of calcite into a 50–50 solution of aspartic acid, a common amino acid. After 24 hours we removed the crystal from this solution, washed it in water and carefully collected all the molecules that had adhered to specific crystal faces. In one experiment after another we observed that calcite's "left-handed" faces selected L-amino acids, and vice versa, with excesses approaching 40 percent in some cases.

Curiously, calcite faces with finely terraced surfaces displayed the greatest selectivity. This outcome led us to speculate that these terraced edges might force the L and D amino acids to line up in neat rows on their respective faces. Under the right environmental conditions, these organized rows of amino acids might chemically join to form proteinlike molecules— some made entirely of L amino acids, others entirely of D. If protein formation can indeed occur, this result becomes even more exciting, because recent experiments by other investigators indicate that some proteins can self-replicate. In the earth's early history, perhaps a self-replicating protein formed on the face of a calcite crystal.

Left- and right-handed crystal faces occur in roughly equal numbers, so chiral selection of L amino acids probably did not happen everywhere in the world at once. Our results and predictions instead suggest that the first successful set of self-replicating molecules—the precursor to all the varied life-forms on the earth today—arose at a specific time and place. It was purely chance that the successful molecule developed on a crystal face that preferentially selected left-handed amino acids over their right-handed counterparts.

Minerals undoubtedly could have acted as containers, scaffolds and templates that helped to select and organize the molecular menagerie of the primitive earth. But many of us in origins research suspect that minerals played much more active roles, catalyzing key synthesis steps that boosted the earth's early inventory of complex biological molecules.

GETTING A JUMP ON THE ACTION

Experiments led by Carnegie researcher Brandes in 1997 illustrate this idea. Biological reactions require nitrogen in the form of ammonia, but the only common nitrogen compound thought to have been available on the primitive earth is nitrogen gas. Perhaps, Brandes thought, the environment at hydrothermal vents mimics an industrial process in which ammonia is synthesized by passing nitrogen and hydrogen over a hot metallic surface. Sure enough, when we subjected hydrogen, nitrogen and the iron oxide mineral magnetite to the pressures and temperatures characteristic of a seafloor vent, the mineral catalyzed the synthesis of ammonia.

The idea that minerals may have triggered life's first crucial steps has emerged most forcefully from the landmark theory of chemist Günter Wächtershäuser, a German patent lawyer with a deep interest in life's origins. In 1988 Wächtershäuser advanced a sweeping theory of organic evolution in which minerals—mostly iron and nickel sulfides that abound at deep-sea hydrothermal vents—could have served as the template, the catalyst and the energy source that drove the formation of biological molecules. Indeed, he has argued that primitive living entities were molecular coatings that adhered to the positively charged surfaces of pyrite, a mineral composed of iron and sulfur. These entities, he further suggests, obtained energy from the chemical reactions that produce pyrite. This hypothesis makes sense in part because some metabolic enzymes—the molecules that help living cells process energy—have at their core a cluster of metal and sulfur atoms.

For much of the past three years, Wächtershäuser's provocative theory has influenced our experiments at Carnegie. Our team, including geochemist George Cody and petrologist Hatten S. Yoder, has focused on the possibility that metabolism can proceed without enzymes in the presence of minerals—especially oxides and sulfides. Our simple strategy, much in the spirit of Miller's famous experiment, has been to subject ingredients known to be available on the young earth—water, carbon dioxide and minerals—to a controlled environment. In our case, we try to replicate the bone-crushing pressures and scalding temperatures typical of a deep-sea hydrothermal vent. Most of our experiments test the interactions among ingredients enclosed in welded gold capsules, which are roughly the size of a daily vitamin pill. We place as many as six capsules into Yoder's "bomb"—a massive steel pressure chamber that squeezes the tiny capsules to pressures approaching 2,000 atmospheres and heats them to about 250 degrees C.

One of our primary goals in these organic-synthesis experiments—and one of life's fundamental chemical reactions—is carbon fixation, the process of producing molecules with an increasing number of carbon atoms in their chemical structure. Such reactions follow two different paths depending on the mineral we use. We find that many common minerals, including most oxides and sulfides of iron, copper and zinc, promote carbon addition by a routine industrial process known as Fischer-Tropsch (F-T) synthesis.

This process can build chainlike organic molecules from carbon monoxide and hydrogen. First, carbon monoxide and hydrogen react to form methane, which has one carbon atom. Adding more carbon monoxide and hydrogen to the methane produces ethane, a two-carbon molecule, and then the reaction repeats itself, adding a carbon atom each time. In the chemical industry, researchers have harnessed this reaction to manufacture molecules with virtually any desired number of carbon atoms. Our first organic-synthesis experiments in 1996, and much more extensive research by Thomas McCollom of the Woods Hole Oceanographic Institution, demonstrate that F-T reactions can build molecules with 30 or more carbon atoms under some hydrothermal-vent conditions in less than a day. If this process manufactures large organic molecules from simple inorganic chemicals throughout the earth's hydrothermal zones today, then it very likely did so in the planet's prebiological past.

When we conduct experiments using nickel or cobalt sulfides, we see that carbon addition occurs primarily by carbonylation—the insertion of a carbon and oxygen molecule, or carbonyl group. Carbonyl groups readily attach themselves to nickel or cobalt atoms, but not so strongly that they cannot link to other molecules and jump ship to form larger molecules. In one series of experiments, we observed the lengthening of the nine-carbon molecule nonyl thiol to form 10-carbon decanoic acid, a compound similar to the acids that drive metabolic reactions in living cells. What is more, all the reactants in this experiment—a thiol, carbon monoxide and water—are readily available near sulfide-rich hydrothermal vents. By repeating these simple kinds of reactions—adding a carbonyl group here or a hydroxide group there—we can synthesize a rich variety of complex organic molecules.

Our 1,500 hydrothermal organic synthesis experiments at Carnegie have done more than supplement the catalogue of interesting molecules that must have been produced on the early earth. These efforts reveal another, more complex behavior of minerals that may have significant consequences for the chemistry of life. Most previous origins-of-life studies have treated minerals as solid and unchanging—stable platforms where organic molecules could assemble. But we are finding that in the presence of hot water at high pressure, minerals start to dissolve. In the process, the liberated atoms and molecules from the minerals can become crucial reactants in the primordial soup.

THE HEART OF THE MATTER

Our first discovery of minerals as reactants was an unexpected result of our recent catalysis experiments led by Cody. As expected, carbonylation reactions produced 10-carbon decanoic acid from a mixture of simple molecules inside our gold capsules. But significant quantities of elemental sulfur, organic sulfides, methyl thiol and other sulfur compounds appeared as well. The sulfur in all these products must have been liberated from the iron sulfide mineral.

Even more striking was the liberation of iron, which brilliantly colored the water-based solutions inside the capsules. As the mineral dissolved, the iron formed bright red and orange organometallic complexes in which iron atoms are surrounded by various organic molecules. We are now investigating the extent to which these potentially reactive complexes might act as enzymes that promote the synthesis of molecular structures.

The role of minerals as essential chemical ingredients of life is not entirely unexpected. Hydrothermal fluids are well known to dissolve and concentrate mineral matter. At deep-sea vents, spectacular pillars of sulfide grow dozens of feet tall as plumes of hot, mineral-laden water rise from below the seafloor, contact the frigid water of the deep ocean and deposit new layers of minerals on the growing pillar. But the role of these dissolved minerals has not yet figured significantly in origins scenarios. Whatever their behavior, dissolved minerals seem to make the story of life's emergence much more interesting.

When we look beyond the specifics of prebiological chemistry, it is clear that the origin of life was far too complex to imagine as a single event. Rather we must work from the assumption that it was a gradual sequence of more modest events, each of which added a degree of order and complexity to the world of prebiological molecules. The first step must have been the synthesis of the basic building blocks. Half a century of research reveals that the molecules of life were manufactured in abundance—in the nebula that formed our solar system, at the ocean's surface, and near hydrothermal vents. The ancient earth suffered an embarrassment of riches—a far greater diversity of molecules than life could possibly employ.

Minerals helped to impose order on this chaos. First by confining and concentrating molecules, then by selecting and arranging those molecules, minerals may have jump-started the first self-replicating molecular systems. Such a system would not have constituted life as we know it, but it could have, for the first time, displayed a key property of life. In this scenario, a self-replicating molecular system began to use up the resources of its environment. As mutations led to slightly different variants, competition for limited resources initiated and drove the process of molecular natural selection. Self-replicating molecular systems began to evolve, inevitably becoming more efficient and more complex.

A long-term objective for our work at the Carnegie Institution is to demonstrate simple chemical steps that could lead to a self-replicating system—perhaps one related to the metabolic cycles common to all living cells. Scientists are far from creating life in the laboratory, and it may never be possible to prove exactly what chemical transformations gave rise to life on earth. What we can say for sure is that minerals played a much more complex and integral part in the origin of life than most scientists ever suspected. By being willing to cast minerals in starring roles in experiments that address life's beginnings, researchers may come closer to answering one of science's oldest questions.

FURTHER READING

Harold J. Morowitz. *Beginnings of Cellular Life.* Yale University Press, 1992.
David W. Deamer and Gail R. Fleischaker. *Origins of Life: The Central Concepts.* Jones and Bartlett, 1994.
John H. Holland, *Emergence: From Chaos to Order.* Helix Books, 1998.
Noam Lahav. *Biogenesis: Theories of Life's Origin.* Oxford University Press. 1999.

Misconceptions about the Big Bang

CHARLES H. LINEWEAVER AND TAMARA M. DAVIS

ORIGINALLY PUBLISHED IN MARCH 2005

The expansion of the universe may be the most important fact we have ever discovered about our origins. You would not be reading this article if the universe had not expanded. Human beings would not exist. Cold molecular things such as life-forms and terrestrial planets could not have come into existence unless the universe, starting from a hot big bang, had expanded and cooled. The formation of all the structures in the universe, from galaxies and stars to planets and *Scientific American* articles, has depended on the expansion.

Forty years ago this July, scientists announced the discovery of definitive evidence for the expansion of the universe from a hotter, denser, primordial state. They had found the cool afterglow of the big bang: the cosmic microwave background radiation. Since this discovery, the expansion and cooling of the universe has been the unifying theme of cosmology, much as Darwinian evolution is the unifying theme of biology. Like Darwinian evolution, cosmic expansion provides the context within which simple structures form and develop over time into complex structures. Without evolution and expansion, modern biology and cosmology make little sense.

The expansion of the universe is like Darwinian evolution in another curious way: most scientists think they understand it, but few agree on what it really means. A century and a half after *On the Origin of Species,* biologists still debate the mechanisms and implications (though not the reality) of Darwinism, while much of the public still flounders in pre-Darwinian cluelessness. Similarly, 75 years after its initial discovery, the expansion of the universe is still widely misunderstood. A prominent cosmologist involved in the interpretation of the cosmic microwave background, James Peebles of Princeton University, wrote in 1993: "The full extent and richness of this picture [the hot big bang model] is not as well understood as I think it ought to be . . . even among those making some of the most stimulating contributions to the flow of ideas."

Renowned physicists, authors of astronomy textbooks and prominent popularizers of science have made incorrect, misleading or easily misinterpreted statements about the expansion of the universe. Because expansion is the basis of the big bang model, these misunderstandings are fundamental. Expansion is a beguilingly simple idea, but what exactly does it mean to say the universe is expanding? What does it expand into? Is Earth expanding, too? To add to the befuddlement, the expansion of the universe now seems to be accelerating, a process with truly mind-stretching consequences.

WHAT IS EXPANSION, ANYWAY?

When some familiar object expands, such as a sprained ankle or the Roman Empire or a bomb, it gets bigger by expanding into the space around it. Ankles, empires and bombs have centers and edges. Outside the edges, there is room to expand into. The universe does not seem to have an edge or a center or an outside, so how can it expand?

A good analogy is to imagine that you are an ant living on the surface of an inflating balloon. Your world is two-dimensional; the only directions you know are left, right, forward and backward. You have no idea what "up" and "down" mean. One day you realize that your walk to milk your aphids is taking longer than it used to: five minutes one day, six minutes the next day, seven minutes the next. The time it takes to walk to other familiar places is also increasing. You are sure that you are not walking more slowly and that the aphids are milling around randomly in groups, not systematically crawling away from you.

This is the important point: the distances to the aphids are increasing even though the aphids are not walking away. They are just standing there, at rest with respect to the rubber of the balloon, yet the distances to them and between them are increasing. Noticing these facts, you conclude that the ground beneath your feet is expanding. That is very strange because you have walked around your world and found no edge or "outside" for it to expand into.

The expansion of our universe is much like the inflation of a balloon. The distances to remote galaxies are increasing. Astronomers casually say that distant galaxies are "receding" or "moving away" from us, but the galaxies are not traveling through space away from us. They are not fragments of a big bang bomb. Instead the space between the galaxies and us is expanding. Individual galaxies move around at random within clusters, but the clusters of galaxies are essentially at rest. The term "at rest" can be defined rigorously. The microwave background radiation fills the universe and defines a universal reference frame, analogous to the rubber of the balloon, with respect to which motion can be measured.

This balloon analogy should not be stretched too far. From our point of view outside the balloon, the expansion of the curved two-dimensional rubber is possible only because it is embedded in three-dimensional space. Within the third dimension, the balloon has a center, and its surface expands into the surrounding air as it inflates. One might conclude that the expansion of our three-dimensional space requires the presence of a fourth dimension. But in Einstein's general theory of relativity, the foundation of modern cosmology, space is dynamic. It can expand, shrink and curve without being embedded in a higher-dimensional space.

In this sense, the universe is self-contained. It needs neither a center to expand away from nor empty space on the outside (wherever that is) to expand into. When it expands, it does not claim previously unoccupied space from its surroundings. Some newer theories such as string theory do postulate extra dimensions, but as our three-dimensional universe expands, it does not need these extra dimensions to spread into.

UBIQUITOUS COSMIC TRAFFIC JAM

In our universe, as on the surface of the balloon, everything recedes from everything else. Thus, the big bang was not an explosion *in* space; it was more like an explosion *of* space. It did not go off at a particular location and spread out from there into some imagined preexisting void. It occurred everywhere at once.

If one imagines running the clock backward in time, any given region of the universe shrinks and all galaxies in it get closer and closer until they smash together in a cosmic traffic jam—the big bang. This traffic-jam analogy might imply local congestion that you could avoid if you listened to the traffic report on the radio. But the big bang was an unavoidable traffic jam. It was like having the surface of Earth and all its highways shrink while cars remained the same size. Eventually the cars will be bumper to bumper on every road. No radio broadcast is going to help you around that kind of traffic jam. The congestion is everywhere.

Similarly, the big bang happened everywhere—in the room in which you are reading this article, in a spot just to the left of Alpha Centauri, everywhere. It was not a bomb going off at a particular spot that we can identify as the center of the explosion. Likewise, in the balloon analogy, there is no special place on the surface of the balloon that is the center of the expansion.

This ubiquity of the big bang holds no matter how big the universe is

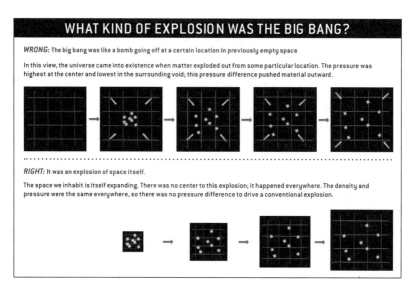

WHAT KIND OF EXPLOSION WAS THE BIG BANG?

WRONG: The big bang was like a bomb going off at a certain location in previously empty space

In this view, the universe came into existence when matter exploded out from some particular location. The pressure was highest at the center and lowest in the surrounding void; this pressure difference pushed material outward.

RIGHT: It was an explosion of space itself.

The space we inhabit is itself expanding. There was no center to this explosion; it happened everywhere. The density and pressure were the same everywhere, so there was no pressure difference to drive a conventional explosion.

or even whether it is finite or infinite in size. Cosmologists sometimes state that the universe used to be the size of a grapefruit, but what they mean is that the part of the universe we can now see—our observable universe—used to be the size of a grapefruit.

Observers living in the Andromeda galaxy and beyond have their own observable universes that are different from but overlap with ours. Andromedans can see galaxies we cannot, simply by virtue of being slightly closer to them, and vice versa. Their observable universe also used to be the size of a grapefruit. Thus, we can conceive of the early universe as a pile of overlapping grapefruits that stretches infinitely in all directions. Correspondingly, the idea that the big bang was "small" is misleading. The totality of space could be infinite. Shrink an infinite space by an arbitrary amount, and it is still infinite.

RECEDING FASTER THAN LIGHT

Another set of misconceptions involves the quantitative description of expansion. The rate at which the distance between galaxies increases follows a distinctive pattern discovered by American astronomer Edwin Hubble in 1929: the recession velocity of a galaxy away from us (v) is directly proportional to its distance from us (d), or $v = Hd$. The proportionality constant, H, is known as the Hubble constant and quantifies how fast space is stretching—not just around us but around any observer in the universe.

Some people get confused by the fact that some galaxies do not obey Hubble's law. Andromeda, our nearest large galactic neighbor, is actually moving toward us, not away. Such exceptions arise because Hubble's law describes only the average behavior of galaxies. Galaxies can also have modest local motions as they mill around and gravitationally pull on one another—as the Milky Way and Andromeda are doing. Distant galaxies also have small local velocities, but from our perspective (at large values of d) these random velocities are swamped by large recession velocities (v). Thus, for those galaxies, Hubble's law holds with good precision.

Notice that, according to Hubble's law, the universe does not expand at a single speed. Some galaxies recede from us at 1,000 kilometers per second, others (those twice as distant) at 2,000 km/s, and so on. In fact, Hubble's law predicts that galaxies beyond a certain distance, known as the Hubble distance, recede faster than the speed of light. For the measured value of the Hubble constant, this distance is about 14 billion light-years.

Does this prediction of faster-than-light galaxies mean that Hubble's law is wrong? Doesn't Einstein's special theory of relativity say that nothing can have a velocity exceeding that of light? This question has confused generations of students. The solution is that special relativity applies only to "normal" velocities—motion through space. The velocity in Hubble's law is a recession velocity caused by the expansion of space, not a motion through space. It is a general relativistic effect and is not bound by the special relativistic limit. Having a recession velocity greater than the speed of light does not violate special relativity. It is still true that nothing ever overtakes a light beam.

STRETCHING AND COOLING

The primary observation that the universe is expanding emerged between 1910 and 1930. Atoms emit and absorb light of specific wavelengths, as measured in laboratory experiments. The same patterns show up in the light from distant galaxies, except that the patterns have been shifted to longer wavelengths. Astronomers say that the galactic light has been redshifted. The explanation is straightforward: As space expands, light waves get stretched. If the universe doubles in size during the waves' journey, their wavelengths double and their energy is halved.

This process can be described in terms of temperature. The photons emitted by a body collectively have a temperature—a certain distribution of energy that reflects how hot the body is. As the photons travel through expanding space, they lose energy and their temperature decreases. In this way, the universe cools as it expands, much as compressed air in a scuba

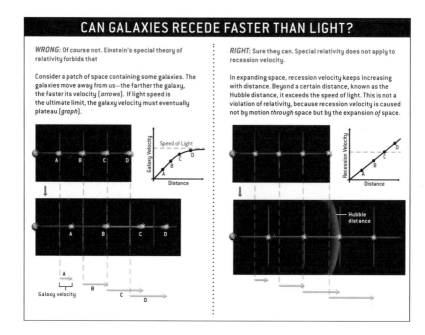

CAN GALAXIES RECEDE FASTER THAN LIGHT?

WRONG: Of course not. Einstein's special theory of relativity forbids that

Consider a patch of space containing some galaxies. The galaxies move away from us—the farther the galaxy, the faster its velocity (*arrows*). If light speed is the ultimate limit, the galaxy velocity must eventually plateau (*graph*).

RIGHT: Sure they can. Special relativity does not apply to recession velocity.

In expanding space, recession velocity keeps increasing with distance. Beyond a certain distance, known as the Hubble distance, it exceeds the speed of light. This is not a violation of relativity, because recession velocity is caused not by motion *through* space but by the expansion *of* space.

tank cools when it is released and allowed to expand. For example, the microwave background radiation currently has a temperature of about three kelvins, whereas the process that released the radiation occurred at a temperature of about 3,000 kelvins. Since the time of the emission of this radiation, the universe has increased in size by a factor of 1,000, so the temperature of the photons has decreased by the same factor. By observing the gas in distant galaxies, astronomers have directly measured the temperature of the radiation in the distant past. These measurements confirm that the universe has been cooling with time.

Misunderstandings about the relation between redshift and velocity abound. The redshift caused by the expansion is often confused with the more familiar redshift generated by the Doppler effect. The normal Doppler effect causes sound waves to get longer if the source of the sound is moving away—for example, a receding ambulance siren. The same principle also applies to light waves, which get longer if the source of the light is moving through space away from us.

This is similar, but not identical, to what happens to the light from distant galaxies. The cosmological redshift is not a normal Doppler shift. Astronomers frequently refer to it as such, and in doing so they have done their students a serious disservice. The Doppler redshift and the cosmological redshift are governed by two distinct formulas. The first comes from special relativity, which does not take into account the expansion of space, and the second comes from general relativity, which does. The two formulas are nearly the same for nearby galaxies but diverge for distant galaxies.

According to the usual Doppler formula, objects whose velocity through space approaches light speed have redshifts that approach infinity. Their wavelengths become too long to observe. If that were true for galaxies, the most distant visible objects in the sky would be receding at velocities just shy of the speed of light. But the cosmological redshift formula leads to a different conclusion. In the current standard model of cosmology, galaxies with a redshift of about 1.5—that is, whose light has a wavelength 150 percent longer than the laboratory reference value—are receding at the speed of light. Astronomers have observed about 1,000 galaxies with redshifts larger than 1.5. That is, they have observed about 1,000 objects receding from us faster than the speed of light. Equivalently, we are receding from those galaxies faster than the speed of light. The radiation of the cosmic microwave background has traveled even farther and has a redshift of about 1,000. When the hot plasma of the early universe emitted the radiation we now see, it was receding from our location at about 50 times the speed of light.

RUNNING TO STAY STILL

The idea of seeing faster-than-light galaxies may sound mystical, but it is made possible by changes in the expansion rate. Imagine a light beam that is farther than the Hubble distance of 14 billion light-years and trying to travel in our direction. It is moving toward us at the speed of light with respect to its local space, but its local space is receding from us faster than the speed of light. Although the light beam is traveling toward us at the maximum speed possible, it cannot keep up with the stretching of space. It is a bit like a child trying to run the wrong way on a moving sidewalk. Photons at the Hubble distance are like the Red Queen and Alice, running as fast as they can just to stay in the same place.

One might conclude that the light beyond the Hubble distance would never reach us and that its source would be forever undetectable. But the Hubble distance is not fixed, because the Hubble constant, on which it depends, changes with time. In particular, the constant is proportional to the rate of increase in the distance between two galaxies, divided by that distance. (Any two galaxies can be used for this calculation.) In models of the universe that fit the observational data, the denominator increases faster than the numerator, so the Hubble constant decreases. In this way, the Hubble distance gets larger. As it does, light that was initially

HOW LARGE IS THE OBSERVABLE UNIVERSE?

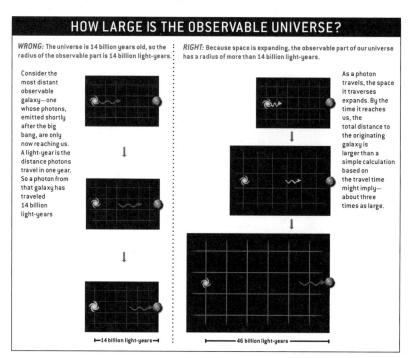

WRONG: The universe is 14 billion years old, so the radius of the observable part is 14 billion light-years.

Consider the most distant observable galaxy—one whose photons, emitted shortly after the big bang, are only now reaching us. A light-year is the distance photons travel in one year. So a photon from that galaxy has traveled 14 billion light-years

├─14 billion light-years─┤

RIGHT: Because space is expanding, the observable part of our universe has a radius of more than 14 billion light-years.

As a photon travels, the space it traverses expands. By the time it reaches us, the total distance to the originating galaxy is larger than a simple calculation based on the travel time might imply—about three times as large.

├────── 46 billion light-years ──────┤

just outside the Hubble distance and receding from us can come within the Hubble distance. The photons then find themselves in a region of space that is receding slower than the speed of light. Thereafter they can approach us.

The galaxy they came from, though, may continue to recede superluminally. Thus, we can observe light from galaxies that have always been and will always be receding faster than the speed of light. Another way to put it is that the Hubble distance is not fixed and does not mark the edge of the observable universe.

What does mark the edge of observable space? Here again there has been confusion. If space were not expanding, the most distant object we could see would now be about 14 billion light-years away from us, the distance light could have traveled in the 14 billion years since the big bang. But because the universe is expanding, the space traversed by a photon expands behind it during the voyage. Consequently, the current distance to the most distant object we can see is about three times farther, or 46 billion light-years.

The recent discovery that the rate of cosmic expansion is accelerating makes things even more interesting. Previously, cosmologists thought that we lived in a decelerating universe and that ever more galaxies would

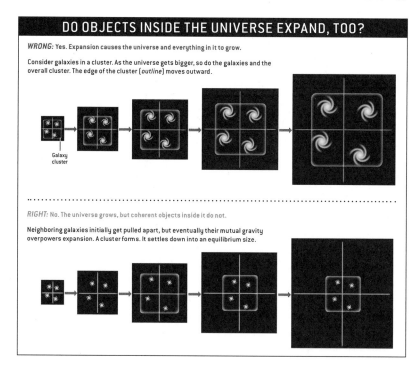

DO OBJECTS INSIDE THE UNIVERSE EXPAND, TOO?

WRONG: Yes. Expansion causes the universe and everything in it to grow.

Consider galaxies in a cluster. As the universe gets bigger, so do the galaxies and the overall cluster. The edge of the cluster (*outline*) moves outward.

Galaxy cluster

RIGHT: No. The universe grows, but coherent objects inside it do not.

Neighboring galaxies initially get pulled apart, but eventually their mutual gravity overpowers expansion. A cluster forms. It settles down into an equilibrium size.

come into view. In an accelerating universe, however, we are surrounded by a boundary beyond which occur events we will never see—a cosmic event horizon. If light from galaxies receding faster than light is to reach us, the Hubble distance has to increase, but in an accelerating universe, it stops increasing. Distant events may send out light beams aimed in our direction, but this light is trapped beyond the Hubble distance by the acceleration of the expansion.

An accelerating universe, then, resembles a black hole in that it has an event horizon, an edge beyond which we cannot see. The current distance to our cosmic event horizon is 16 billion light-years, well within our observable range. Light emitted from galaxies that are now beyond the event horizon will never be able to reach us; the distance that currently corresponds to 16 billion light-years will expand too quickly. We will still be able to see events that took place in those galaxies before they crossed the horizon, but subsequent events will be forever beyond our view.

IS BROOKLYN EXPANDING?

In *Annie Hall,* the movie character played by the young Woody Allen explains to his doctor and mother why he can't do his homework. "The universe is expanding. . . . The universe is everything, and if it's expanding, someday it will break apart and that would be the end of everything!" But his mother knows better: "You're here in Brooklyn. Brooklyn is not expanding!"

His mother is right. Brooklyn is not expanding. People often assume that as space expands, everything in it expands as well. But this is not true. Expansion by itself—that is, a coasting expansion neither accelerating nor decelerating—produces no force. Photon wavelengths expand with the universe because, unlike atoms and cities, photons are not coherent objects whose size has been set by a compromise among forces. A changing rate of expansion does add a new force to the mix, but even this new force does not make objects expand or contract.

For example, if gravity got stronger, your spinal cord would compress until the electrons in your vertebrae reached a new equilibrium slightly closer together. You would be a shorter person, but you would not continue to shrink. In the same way, if we lived in a universe dominated by the attractive force of gravity, as most cosmologists thought until a few years ago, the expansion would decelerate, putting a gentle squeeze on bodies in the universe, making them reach a smaller equilibrium size. Having done so, they would not keep shrinking.

In fact, in our universe the expansion is accelerating, and that exerts a gentle outward force on bodies. Consequently, bound objects are slightly larger than they would be in a non-accelerating universe, because the equilibrium among forces is reached at a slightly larger size. At Earth's surface, the outward acceleration away from the planet's center equals a tiny fraction (10^{-30}) of the normal inward gravitational acceleration. If this acceleration is constant, it does not make Earth expand; rather the planet simply settles into a static equilibrium size slightly larger than the size it would have attained.

This reasoning changes if acceleration is not constant, as some cosmologists have speculated. If the acceleration itself increased, it could eventually grow strong enough to tear apart all structures, leading to a "big rip." But this rip would occur not because of expansion or acceleration per se but because of an accelerating acceleration.

The big bang model is based on observations of expansion, the cosmic microwave background, the chemical composition of the universe and the clumping of matter. Like all scientific ideas, the model may one day be superseded. But it fits the current data better than any other model we have. As new precise measurements enable cosmologists to understand expansion and acceleration better, they can ask even more fundamental questions about the earliest times and largest scales of the universe. What caused the expansion? Many cosmologists attribute it to a process known as inflation, a type of accelerating expansion. But that can only be a partial answer, because it seems that to start inflating, the universe already had to be expanding. And what about the largest scales, beyond what we can see? Do different parts of the universe expand by different amounts, such that our universe is a single inflationary bubble of a much larger multiverse? Nobody knows. Although many questions remain, increasingly precise observations suggest that the universe will expand forever. We hope, though, the confusion about the expansion will shrink.

FURTHER READING

Edward R. Harrison. *Cosmology: The Science of the Universe.* Cambridge University Press, 2000.

R. Srianand, P. Petitjean and C. Ledoux. "The Cosmic Microwave Background Radiation Temperature at a Redshift of 2.34" in *Nature,* Vol. 408, No. 6815, pages 931–935; December 21, 2000. Available online at arxiv.org/abs/astro-ph/0012222

Tamara M. Davis, Charles H. Lineweaver and John K. Webb. "Solutions to the Tethered Galaxy Problem in an Expanding Universe and the Observation of Receding Blueshifted Objects" in *American Journal of Physics,* Vol. 71, No. 4, pages 358–364; April 2003. astro-ph/0104349

Tamara M. Davis and Charles H. Lineweaver. "Expanding Confusion: Common Misconceptions of Cosmological Horizons and the Superluminal Expansion of the Universe" in *Publications of the Astronomical Society of Australia,* Vol. 21, No. 1, pages 97–109; February 2004. astro-ph/0310808

An excellent resource for dispelling cosmological misconceptions is Ned Wright's Cosmology Tutorial at www.astro.ucla.edu/~wright/cosmolog.htm

The Evolution of the Earth

CLAUDE J. ALLÈGRE AND STEPHEN H. SCHNEIDER

ORIGINALLY PUBLISHED IN OCTOBER 1994

Like the lapis lazuli gem it resembles, the blue, cloud-enveloped planet that we recognize immediately from satellite pictures seems remarkably stable. Continents and oceans, encircled by an oxygen-rich atmosphere, support familiar life-forms. Yet this constancy is an illusion produced by the human experience of time. The earth and its atmosphere are continuously altered. Plate tectonics shift the continents, raise mountains and move the ocean floor while processes that no one fully comprehends alter the climate.

Such constant change has characterized the earth since its beginning some 4.5 billion years ago. From the outset, heat and gravity shaped the evolution of the planet. These forces were gradually joined by the global effects of the emergence of life. Exploring this past offers us the only possibility of understanding the origin of life and, perhaps, its future.

Scientists used to believe the rocky planets, including the earth, Mercury, Venus and Mars, were created by the rapid gravitational collapse of a dust cloud, a deflation giving rise to a dense orb. In the 1960s the Apollo space program changed this view. Studies of moon craters revealed that these gouges were caused by the impact of objects that were in great abundance about 4.5 billion years ago. Thereafter, the number of impacts appeared to have quickly decreased. This observation rejuvenated the theory of accretion postulated by Otto Schmidt. The Russian geophysicist had suggested in 1944 that planets grew in size gradually, step by step.

According to Schmidt, cosmic dust lumped together to form particulates, particulates became gravel, gravel became small balls, then big balls, then tiny planets, or planetesimals, and, finally, dust became the size of the moon. As the planetesimals became larger, their numbers decreased. Consequently, the number of collisions between planetesimals, or meteorites, decreased. Fewer items available for accretion meant that it took a long time to build up a large planet. A calculation made by George W. Wetherill of the Carnegie Institution of Washington suggests that about 100 million years could pass between the formation of an object measuring 10 kilometers in diameter and an object the size of the earth.

The process of accretion had significant thermal consequences for the earth, consequences that have forcefully directed its evolution. Large bodies slamming into the planet produced immense heat in the interior, melting the cosmic dust found there. The resulting furnace—situated some 200 to 400 kilometers underground and called a magma ocean—was active for millions of years, giving rise to volcanic eruptions. When the earth was young, heat at the surface caused by volcanism and lava flows from the interior was supplemented by the constant bombardment of huge planetesimals, some of them perhaps the size of the moon or even Mars. No life was possible during this period.

Beyond clarifying that the earth had formed through accretion, the Apollo program compelled scientists to try to reconstruct the subsequent temporal and physical development of the early earth. This undertaking had been considered impossible by founders of geology, including Charles Lyell, to whom the following phrase is attributed: No vestige of a beginning, no prospect for an end. This statement conveys the idea that the young earth could not be recreated because its remnants were destroyed by its very activity. But the development of isotope geology in the 1960s had rendered this view obsolete. Their imaginations fired by Apollo and the moon findings, geochemists began to apply this technique to understand the evolution of the earth.

Dating rocks using so-called radioactive clocks allows geologists to work on old terrains that do not contain fossils. The hands of a radioactive clock are isotopes—atoms of the same element that have different atomic weights—and geologic time is measured by the rate of decay of one isotope into another [see "The Earliest History of the Earth," by Derek York; *Scientific American,* January 1993]. Among the many clocks, those based on the decay of uranium 238 into lead 206 and of uranium 235 into lead 207 are special. Geochronologists can determine the age of samples by analyzing only the daughter product—in this case, lead—of the radioactive parent, uranium.

Isotope geology has permitted geologists to determine that the accretion of the earth culminated in the differentiation of the planet: the creation of the core—the source of the earth's magnetic field—and the beginning of the atmosphere. In 1953 the classic work of Claire C. Patterson of the California Institute of Technology used the uranium-lead clock to establish an age of 4.55 billion years for the earth and many of the meteorites that formed it. Recent work by one of us (Allègre) on lead isotopes, however, led to a somewhat new interpretation. As Patterson argued, some meteorites were indeed formed about 4.56 billion years ago, and their debris constituted the earth. But the earth continued to grow through the bombardment of planetesimals until some 120 to 150 million years later. At that time—4.44 to 4.41 billion years ago—the earth began to retain its atmosphere and create its core. This possibility had already been suggested by Bruce R. Doe and Robert E. Zartman of the U.S. Geological Survey in Denver a decade ago and is in agreement with Wetherill's estimates.

The emergence of the continents came somewhat later. According to the theory of plate tectonics, these land-masses are the only part of the earth's crust that is not recycled and, consequently, destroyed during the geothermal cycle driven by the convection in the mantle. Continents thus provide a form of memory because the record of early life can be read in their rocks. The testimony, however, is not extensive. Geologic activity, including plate tectonics, erosion and metamorphism, has destroyed almost all the ancient rocks. Very few fragments have survived this geologic machine.

Nevertheless, in recent years, several important finds have been made, again using isotope geochemistry. One group, led by Stephen Moorbath of the University of Oxford, discovered terrain in West Greenland that is between 3.7 and 3.8 billion years old. In addition, Samuel A. Bowring of the Massachusetts Institute of Technology explored a small area in North America—the Acasta gneiss—that is 3.96 billion years old.

Ultimately, a quest for the mineral zircon led other researchers to even more ancient terrain. Typically found in continental rocks, zircon is not dissolved during the process of erosion but is deposited as particles in sediment. A few pieces of zircon can therefore survive for billions of years and can serve as a witness to the earth's more ancient crust. The search for old zircons started in Paris with the work of Annie Vitrac and Joël R. Lancelot, now at the University of Marseilles and the University of Montpellier, respectively, as well as with the efforts of Moorbath and Allègre. It was a group at the Australian National University in Canberra, directed by William Compston, that was finally successful. The team discovered zircons in western Australia that were between 4.1 and 4.3 billion years old.

Zircons have been crucial not only for understanding the age of the continents but for determining when life first appeared. The earliest fossils of undisputed age were found in Australia and South Africa. These relics of blue-green algae are about 3.5 billion years old. Manfred Schidlowski of the Max Planck Institute for Chemistry in Mainz has studied the Isua formation in West Greenland and argues that organic matter existed as many as

3.8 billion years ago. Because most of the record of early life has been destroyed by geologic activity, we cannot say exactly when it first appeared—
perhaps it arose very quickly, maybe even 4.2 billion years ago.

One of the most important aspects of the earth's evolution is the formation of the atmosphere, because it is this assemblage of gases that allowed
life to crawl out of the oceans and to be sustained. Researchers have hypothesized since the 1950s that the terrestrial atmosphere was created by
gases emerging from the interior of the planet. When a volcano spews
gases, it is an example of the continuous outgassing, as it is called, of the
earth. But scientists have questioned whether this process occurred suddenly about 4.4 billion years ago when the core differentiated or whether
it took place gradually over time.

   To answer this question, Allègre and his colleagues studied the isotopes of rare gases. These gases—including helium, argon and xenon—
have the peculiarity of being chemically inert, that is, they do not react in
nature with other elements. Two of them are particularly important for
atmospheric studies: argon and xenon. Argon has three isotopes, of which
argon 40 is created by the decay of potassium 40. Xenon has nine, of which
129 has two different origins. Xenon 129 arose as the result of nucleosyn-
thesis before the earth and solar system were formed. It was also created
from the decay of radioactive iodine 129, which does not exist on the
earth anymore. This form of iodine was present very early on but has died
since, and xenon 129 has grown at its expense.

   Like most couples, both argon 40 and potassium 40 and xenon 129 and
iodine 129 have stories to tell. They are excellent chronometers. Although
the atmosphere was formed by the outgassing of the mantle, it does not
contain any potassium 40 or iodine 129. All argon 40 and xenon 129,
formed in the earth and released, are found in the atmosphere today.
Xenon was expelled from the mantle and retained in the atmosphere;
therefore, the atmosphere-mantle ratio of this element allows us to eval-
uate the age of differentiation. Argon and xenon trapped in the mantle
evolved by the radioactive decay of potassium 40 and iodine 129. Thus, if
the total outgassing of the mantle occurred at the beginning of the earth's
formation, the atmosphere would not contain any argon 40 but would
contain xenon 129.

   The major challenge facing an investigator who wants to measure such
ratios of decay is to obtain high concentrations of rare gases in mantle rocks
because they are extremely limited. Fortunately, a natural phenomenon
occurs at mid-oceanic ridges during which volcanic lava transfers some

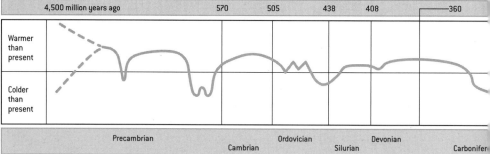

	4,500 million years ago	570	505	438	408	360

Warmer than present						
Colder than present						

	Precambrian		Ordovician		Devonian	
		Cambrian		Silurian		Carboniferous

Climate fluctuations are apparent over time. Although the earth's early temperature record is quite uncertain, good estimates can be made starting 400 million years ago, when fossils were more abundantly preserved. As climate shifted, so did life—suggesting feedback between the two. The dates of these evolutions remain unclear as well, but their order is more apparent. First a primordial soup formed, then primitive organisms, such as algae, stromatolites and jellyfish, arose; spiny fish were followed by the ichthyostega, perhaps the first creature to crawl from ocean onto land. The rest of the story is well known: dinosaurs appeared and died out, their place taken by mammals.

silicates from the mantle to the surface. The small amounts of gases trapped in mantle minerals rise with the melt to the surface and are concentrated in small vesicles in the outer glassy margin of lava flows. This process serves to concentrate the amounts of mantle gases by a factor of 10^4 or 10^5. Collecting these rocks by dredging the seafloor and then crushing them under vacuum in a sensitive mass spectrometer allow geochemists to determine the ratios of the isotopes in the mantle. The results are quite surprising. Calculations of the ratios indicate that between 80 and 85 percent of the atmosphere was outgassed in the first one million years; the rest was released slowly but constantly during the next 4.4 billion years.

The composition of this primitive atmosphere was most certainly dominated by carbon dioxide, with nitrogen as the second most abundant gas. Trace amounts of methane, ammonia, sulfur dioxide and hydrochloric

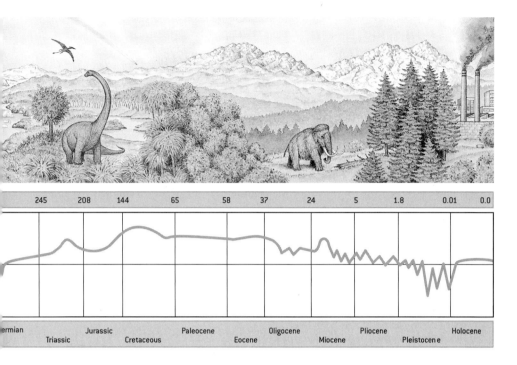

| 245 | 208 | 144 | 65 | 58 | 37 | 24 | 5 | 1.8 | 0.01 | 0.0 |

| ermian | Jurassic | Paleocene | Oligocene | Pliocene | Holocene |
| Triassic | Cretaceous | Eocene | Miocene | Pleistocene |

acid were also present, but there was no oxygen. Except for the presence of abundant water, the atmosphere was similar to that of Venus or Mars. The details of the evolution of the original atmosphere are debated, particularly because we do not know how strong the sun was at that time. Some facts, however, are not disputed. It is evident that carbon dioxide played a crucial role. In addition, many scientists believe the evolving atmosphere contained sufficient quantities of gases like ammonia and methane to give rise to organic matter.

Still, the problem of the sun remains unresolved. One hypothesis holds that during the Archean era, which lasted from about 4.5 to 2.5 billion years ago, the sun's power was only 75 percent of what it is today. This possibility raises a dilemma: How could life have survived in the relatively cold climate that should accompany a weaker sun? A solution to the faint early sun paradox, as it is called, was offered by Carl Sagan and George Mullen of Cornell University in 1970. The two scientists suggested that methane and ammonia, which are very effective at trapping infrared radiation, were quite abundant. These gases could have created a super-greenhouse effect. The idea was criticized on the basis that such gases were highly reactive and have short lifetimes in the atmosphere.

700 million
years ago

600 million
years ago

500 million
years ago

400 million
years ago

300 million
years ago

200 million
years ago

100 million
years ago

Present

Continental shift has altered the face of the planet for nearly a billion years, as can be seen in the differences between the positions of the continents that we know today and those of 700 million years ago. Pangaea, the superaggregate of early continents, came together about 200 million years ago and then promptly, in geologic terms, broke apart. This series was compiled with the advice of Christopher R. Scotese of the University of Texas at Arlington.

In the late 1970s Veerabhadran Ramanathan, now at the Scripps Institution of Oceanography, and Robert D. Cess and Tobias Owen of the State University of New York at Stony Brook proposed another solution. They postulated that there was no need for methane in the early atmosphere because carbon dioxide was abundant enough to bring about the super-greenhouse effect. Again this argument raised a different question: How much carbon dioxide was there in the early atmosphere? Terrestrial carbon dioxide is now buried in carbonate rocks, such as limestone, although it is not clear when it became trapped there. Today calcium carbonate is created primarily during biological activity; in the Archean period, carbon may have been primarily removed during inorganic reactions.

The rapid outgassing of the planet liberated voluminous quantities of water from the mantle, creating the oceans and the hydrologic cycle. The acids that were probably present in the atmosphere eroded rocks, forming carbonate-rich rocks. The relative importance of such a mechanism is, however, debated. Heinrich D. Holland of Harvard University believes the amount of carbon dioxide in the atmosphere rapidly decreased during the Archean and stayed at a low level.

Understanding the carbon dioxide content of the early atmosphere is pivotal to understanding the mechanisms of climatic control. Two sometimes conflicting camps have put forth ideas on how this process works. The first group holds that global temperatures and carbon dioxide were controlled by inorganic geochemical feedbacks; the second asserts that they were controlled by biological removal.

James C. G. Walker, James F. Kasting and Paul B. Hays, then at the University of Michigan, proposed the inorganic model in 1981. They postulated that levels of the gas were high at the outset of the Archean and did not fall precipitously. The trio suggested that as the climate warmed, more water evaporated, and the hydrologic cycle became more vigorous, increasing precipitation and runoff. The carbon dioxide in the atmosphere mixed with rainwater to create carbonic acid runoff, exposing minerals at the surface to weathering. Silicate minerals combined with carbon that had been in the atmosphere, sequestering it in sedimentary rocks. Less carbon dioxide in the atmosphere meant, in turn, less of a greenhouse effect. The inorganic negative feedback process offset the increase in solar energy.

This solution contrasts with a second paradigm: biological removal. One theory advanced by James E. Lovelock, an originator of the Gaia hypothesis, assumed that photosynthesizing microorganisms, such as phytoplankton, would be very productive in a high–carbon dioxide environment. These creatures slowly removed carbon dioxide from the air and oceans, converting it into calcium carbonate sediments. Critics retorted that phytoplankton had not even evolved for most of the time that the earth has had life. (The Gaia hypothesis holds that life on the earth has the capacity to regulate temperature and the composition of the earth's surface and to keep it comfortable for living organisms.)

More recently, Tyler Volk of New York University and David W. Schwartzman of Howard University proposed another Gaian solution. They noted that bacteria increase carbon dioxide content in soils by breaking down organic matter and by generating humic acids. Both activities accelerate weathering, removing carbon dioxide from the atmosphere. On this point, however, the controversy becomes acute. Some geochemists, including Kasting, now at Pennsylvania State University, and Holland, postulate that while life may account for some carbon dioxide removal after the Archean, inorganic geochemical processes can explain most of the sequestering. These researchers view life as a rather weak climatic stabilizing mechanism for the bulk of geologic time.

The issue of carbon remains critical to the story of how life influenced the atmosphere. Carbon burial is a key to the vital process of building up atmospheric oxygen concentrations—a pre-requisite for the development of certain life-forms. In addition, global warming may be taking place now as a result of humans releasing this carbon. For one or two billion years, algae in the oceans produced oxygen. But because this gas is highly reactive and because there were many reduced minerals in the ancient oceans—iron, for example, is easily oxidized—much of the oxygen produced by living creatures simply got used up before it could reach the atmosphere, where it would have encountered gases that would react with it.

Even if evolutionary processes had given rise to more complicated life-forms during this anaerobic era, they would have had no oxygen. Furthermore, unfiltered ultraviolet sunlight would have likely killed them if they left the ocean. Researchers such as Walker and Preston Cloud, then at the University of California at Santa Barbara, have suggested that only about two billion years ago, after most of the reduced minerals in the sea were oxidized, did atmospheric oxygen accumulate. Between one and two billion years ago oxygen reached current levels, creating a niche for evolving life.

By examining the stability of certain minerals, such as iron oxide or uranium oxide, Holland has shown that the oxygen content of the Archean atmosphere was low, before two billion years ago. It is largely agreed that the present-day oxygen content of 20 percent is the result of photosynthetic activity. Still, the question is whether the oxygen content in the atmosphere increased gradually over time or suddenly. Recent studies indicate that the increase of oxygen started abruptly between 2.1 and 2.03 billion years ago, and the present situation was reached 1.5 billion years ago.

The presence of oxygen in the atmosphere had another major benefit for an organism trying to live at or above the surface: it filtered ultraviolet radiation. Ultraviolet radiation breaks down many molecules—from DNA and oxygen to the chlorofluorocarbons that are implicated in stratospheric ozone depletion. Such energy splits oxygen into the highly unstable atomic form O, which can combine back into O_2 and into the very special molecule O_3, or ozone. Ozone, in turn, absorbs ultraviolet radiation. It was not until oxygen was abundant enough in the atmosphere to allow the formation of ozone that life even had a chance to get a root-hold or a foothold on land. It is not a coincidence that the rapid evolution of life from prokaryotes (single-celled organisms with no nucleus) to eukaryotes (single-celled organisms with a nucleus) to metazoa (multicelled organisms) took place in the billion-year-long era of oxygen and ozone.

Although the atmosphere was reaching a fairly stable level of oxygen during this period, the climate was hardly uniform. There were long

OXYGEN RISING, CO$_2$ FALLING

ATMOSPHERIC COMPOSITION, shown by the relative concentration of various gases, has been greatly influenced by life on Earth. The early atmosphere had fairly high concentrations of water and carbon dioxide and, some experts believe, methane, ammonia and nitrogen. But levels of those gases have plummeted since then. After the emergence of living organisms, the oxygen that is so vital to our survival became more plentiful. Today carbon dioxide, methane and water exist only in trace amounts in the atmosphere.

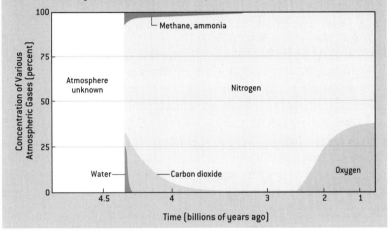

stages of relative warmth or coolness during the transition to modern geologic time. The composition of fossil plankton shells that lived near the ocean floor provides a measure of bottom water temperatures. The record suggests that over the past 100 million years bottom waters cooled by nearly 15 degrees Celsius. Sea levels dropped by hundreds of meters, and continents drifted apart. Inland seas mostly disappeared, and the climate cooled an average of 10 to 15 degrees C. Roughly 20 million years ago permanent ice appears to have built up on Antarctica.

About two to three million years ago the paleoclimatic record starts to show significant expansions and contractions of warm and cold periods on 40,000-year or so cycles. This periodicity is interesting because it corresponds to the time it takes the earth to complete an oscillation of the tilt of its axis of rotation. It has long been speculated, and recently calculated, that known changes in orbital geometry could alter the amount of sunlight coming in between winter and summer by about 10 percent or so and could be responsible for initiating or ending ice ages.

Most interesting and perplexing is the discovery that between 600,000 and 800,000 years ago the dominant cycle switched from 40,000-year periods to

100,000-year intervals with very large fluctuations. The last major phase of glaciation ended about 10,000 years ago. At its height 20,000 years ago, ice sheets a mile thick covered much of northern Europe and North America. Glaciers expanded in high plateaus and mountains throughout the world. Enough ice was locked up on land to cause sea levels to drop more than 100 meters below where they are today. Massive ice sheets scoured the land and revamped the ecological face of the earth, which was five degrees C cooler on average than it is currently.

The precise causes of these changes are not yet sorted out. Volcanic eruptions may have played a significant role as shown by the effect of El Chichón in Mexico and Mount Pinatubo in the Philippines. Tectonic events, such as the development of the Himalayas, may influence world climate. Even the impact of comets can influence short-term climatic trends with catastrophic consequences for life [see "What Caused the Mass Extinction? An Extraterrestrial Impact," by Walter Alvarez and Frank Asaro; and "What Caused the Mass Extinction? A Volcanic Eruption," by Vincent E. Courtillot; *Scientific American,* October 1990]. It is remarkable that despite violent, episodic perturbations, the climate has been buffered enough to sustain life for 3.5 billion years.

One of the most pivotal climatic discoveries of the past 20 years has come from ice cores in Greenland and Antarctica. When snow falls on these frozen continents, the air between the snow grains is trapped as bubbles. The snow is gradually compressed into ice, along with its captured gases. Some of these records can go back as far as 200,000 years; scientists can analyze the chemical content of ice and bubbles from sections of ice that lie as deep as 2,000 meters below the surface.

The ice-core borers have determined that the air breathed by ancient Egyptians and Anasazi Indians was very similar to that which we inhale today—except for a host of air pollutants introduced over the past 100 or 200 years. Principal among these added gases, or pollutants, are extra carbon dioxide and methane. The former has increased 25 percent as a result of industrialization and deforestation; the latter has doubled because of agriculture, land use and energy production. The concern that increased amounts of these gases might trap enough heat to cause global warming is at the heart of the climate debate [see "The Changing Climate," by Stephen H. Schneider; *Scientific American,* September 1989].

The ice cores have shown that sustained natural rates of worldwide temperature change are typically about one degree C per millennium. These shifts are still significant enough to have radically altered where species live and to have potentially contributed to the extinction of such charismatic megafauna as mammoths and saber-toothed tigers. But a most extraordinary story from the ice cores is not the relative stability of the climate during the past 10,000 years. It appears that during the height of the last ice age 20,000 years ago there was between 30 and 40 percent less carbon dioxide and 50 percent less methane in the air than there has been during our period, the Holocene. This finding suggests a positive feedback between carbon dioxide, methane and climatic change.

The reasoning that supports the idea of this destabilizing feedback system goes as follows. When the world was colder, there was less concentration of greenhouse gases, and so less heat was trapped. As the earth warmed up, carbon dioxide and methane levels increased, accelerating the warming. If life had a hand in this story, it would have been to drive, rather than to oppose, climatic change. Once again, though, this picture is incomplete.

Nevertheless, most scientists would agree that life could well be the principal factor in the positive feedback between climatic change and greenhouse gases. One hypothesis suggests that increased nutrient runoff from the continental shelves that were exposed as sea levels fell fertilized phytoplankton. This nutrient input could have created a larger biomass of such marine species. Because the calcium carbonate shell makes up most of the mass of many phytoplankton species, increased productivity would remove carbon dioxide from the oceans and eventually the atmosphere. At the same time, boreal forests that account for about 10 to 20 percent of the carbon in the atmosphere were decimated during the ice ages. Carbon from these high-latitude forests could have been released to the atmosphere, yet the atmosphere had less of this gas then. Thus, the positive feedback system of the ocean's biological pump may have offset the negative feedback caused by the destruction of the forests. Great amounts of carbon can be stored in soils, however, so the demise of forests may have led to the sequestering of carbon in the ground.

What is significant is the idea that the feedback was positive. By studying the transition from the high–carbon dioxide, low-oxygen atmosphere of the Archean to the era of great evolutionary progress about a half a billion years ago, it becomes clear that life may have been a factor in the stabilization of climate. In another example—during the ice ages and interglacial cycles—life seems to have the opposite function: accelerating the change rather than diminishing it. This observation has led one of us (Schneider) to contend that climate and life coevolved rather than life serving solely as a negative feedback on climate.

If we humans consider ourselves part of life—that is, part of the natural system—then it could be argued that our collective impact on the earth means we may have a significant coevolutionary role in the future of the planet. The current trends of population growth, the demands for increased standards of living and the use of technology and organizations to attain these growth-oriented goals all contribute to pollution. When the price of polluting is low and the atmosphere is used as a free sewer, carbon dioxide, methane, chlorofluorocarbons, nitrous oxides, sulfur oxides and other toxics can build up.

The theory of heat trapping—codified in mathematical models of the climate—suggests that if carbon dioxide levels double sometime in the middle of the next century, the world will warm between one and five degrees C. The mild end of that range entails warming at the rate of one degree per 100 years—a factor of 10 faster than the one degree per 1,000 years that has historically been the average rate of natural change on a global scale. Should the higher end of the range occur, then we could see rates of climatic change 50 times faster than natural average conditions. Change at this rate would almost certainly force many species to attempt to move their ranges, just as they did from the ice age–interglacial transition between 10,000 and 15,000 years ago. Not only would species have to respond to climatic change at rates 10 to 50 times faster, but few would have undisturbed, open migration routes as they did at the end of the ice age and the onset of the interglacial era. It is for these reasons that it is essential to understand whether doubling carbon dioxide will warm the earth by one degree or five.

To make the critical projections of future climatic change needed to understand the fate of ecosystems on this earth, we must dig through land, sea and ice to uncover as much of the geologic, paleoclimatic and paleoecological records as we can. These records provide the backdrop against which to calibrate the crude instruments we must use to peer into a shadowy environmental future, a future increasingly influenced by us.

FURTHER READING

Wallace Broecker. *How to Build a Habitable Planet.* Lamont-Doherty Geological Observatory Press, 1990.

Stephen H. Schneider and Penelope J. Boston, eds. *Scientists on Gaia.* MIT Press, 1991.

Claude J. Allègre. *From Stone to Star: A View of Modern Geology.* Harvard University Press, 1992.

James F. Kasting. "Earth's Early Atmosphere" in *Science*, Vol. 259, pages 920–926; February 12, 1993.

CELLULAR EVOLUTION

Uprooting the Tree of Life

W. FORD DOOLITTLE

ORIGINALLY PUBLISHED IN FEBRUARY 2000

Charles Darwin contended more than a century ago that all modern species diverged from a more limited set of ancestral groups, which themselves evolved from still fewer progenitors and so on back to the beginning of life. In principle, then, the relationships among all living and extinct organisms could be represented as a single genealogical tree.

Most contemporary researchers agree. Many would even argue that the general features of this tree are already known, all the way down to the root—a solitary cell, termed life's last universal common ancestor, that lived roughly 3.5 to 3.8 billion years ago. The consensus view did not come easily but has been widely accepted for more than a decade.

Yet ill winds are blowing. To everyone's surprise, discoveries made in the past few years have begun to cast serious doubt on some aspects of the tree, especially on the depiction of the relationships near the root.

THE FIRST SKETCHES

Scientists could not even begin to contemplate constructing a universal tree until about 35 years ago. From the time of Aristotle to the 1960s, researchers deduced the relatedness of organisms by comparing their anatomy or physiology, or both. For complex organisms, they were frequently able to draw reasonable genealogical inferences in this way. Detailed analyses of innumerable traits suggested, for instance, that hominids shared a common ancestor with apes, that this common ancestor shared an earlier one with monkeys, and that *that* precursor shared an even earlier forebear with prosimians, and so forth.

Microscopic single-celled organisms, however, often provided too little information for defining relationships. That paucity was disturbing because microbes were the only inhabitants of the earth for the first half to two thirds of the planet's history; the absence of a clear phylogeny (family tree) for microorganisms left scientists unsure about the sequence in

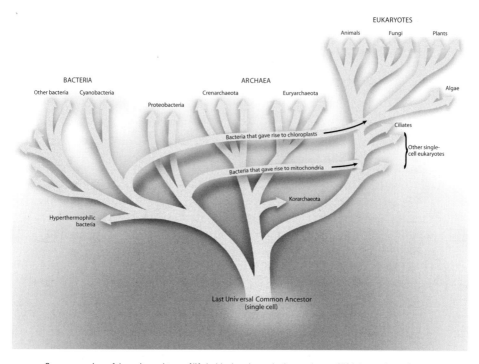

EUKARYOTES

Animals   Fungi   Plants

BACTERIA   ARCHAEA   Algae

Other bacteria   Cyanobacteria   Crenarchaeota   Euryarchaeota

Proteobacteria

Ciliates

Bacteria that gave rise to chloroplasts

Other single-
cell eukaryotes

Bacteria that gave rise to mitochondria

Korarchaeota

Hyperthermophilic
bacteria

Last Universal Common Ancestor
(single cell)

Consensus view of the universal tree of life holds that the early descendants of life's last universal common ancestor—a small cell with no nucleus—divided into two prokaryotic (nonnucleated) groups: the
bacteria and the archaea. Later, the archaea gave rise to organisms having complex cells containing a nucleus: the eukaryotes. Eukaryotes gained valuable energy-generating organelles—mitochondria and, in
the case of plants, chloroplasts—by taking up, and retaining, certain bacteria.

which some of the most radical innovations in cellular structure and function occurred. For example, between the birth of the first cell and the appearance of multicellular fungi, plants and animals, cells grew bigger and
more complex, gained a nucleus and a cytoskeleton (internal scaffolding),
and found a way to eat other cells.

In the mid-1960s Emile Zuckerkandl and Linus Pauling of the California Institute of Technology conceived of a revolutionary strategy that
could supply the missing information. Instead of looking just at anatomy
or physiology, they asked, why not base family trees on differences in the
order of the building blocks in selected genes or proteins?

Their approach, known as molecular phylogeny, is eminently logical.
Individual genes, composed of unique sequences of nucleotides, typically
serve as the blueprints for making specific proteins, which consist of
particular strings of amino acids. All genes, however, mutate (change in

sequence), sometimes altering the encoded protein. Genetic mutations that have no effect on protein function or that improve it will inevitably accumulate over time. Thus, as two species diverge from an ancestor, the sequences of the genes they share will also diverge. And as time passes, the genetic divergence will increase. Investigators can therefore reconstruct the evolutionary past of living species—can construct their phylogenetic trees—by assessing the sequence divergence of genes or proteins isolated from those organisms.

Thirty-five years ago scientists were just becoming proficient at identifying the order of amino acids in proteins and could not yet sequence genes. Protein studies completed in the 1960s and 1970s demonstrated the general utility of molecular phylogeny by confirming and then extending the family trees of well-studied groups such as the vertebrates. They also lent support to some hypotheses about the links among certain bacteria—showing, for instance, that bacteria capable of producing oxygen during photosynthesis form a group of their own (cyanobacteria).

As this protein work was progressing, Carl R. Woese of the University of Illinois was turning his attention to a powerful new yardstick of evolutionary distances: small subunit ribosomal RNA (SSU rRNA). This genetically specified molecule is a key constituent of ribosomes, the "factories" that construct proteins in cells, and cells throughout time have needed it to survive. These features suggested to Woese in the late 1960s that variations in SSU rRNA (or more precisely in the genes encoding it) would reliably indicate the relatedness among any life-forms, from the plainest bacteria to the most complex animals. Small subunit ribosomal RNA could thus serve, in Woese's words, as a "universal molecular chronometer."

Initially the methods available for the project were indirect and laborious. By the late 1970s, though, Woese had enough data to draw some important inferences. Since then, phylogeneticists studying microbial evolution, as well as investigators concerned with higher sections of the universal tree, have based many of their branching patterns on sequence analyses of SSU rRNA genes. This accumulation of rRNA data helped greatly to foster consensus about the universal tree in the late 1980s. Today investigators have rRNA sequences for several thousands of species.

From the start, the rRNA results corroborated some already accepted ideas, but they also produced an astonishing surprise. By the 1960s microscopists had determined that the world of living things could be divided into two separate groups, eukaryotes and prokaryotes, depending on the structure of the cells that composed them. Eukaryotic organisms (animals, plants, fungi and many unicellular life-forms) were defined as those composed of cells that contained a true nucleus—a membrane-bound organelle housing the chromosomes. Eukaryotic cells also displayed other prominent features, among them a cytoskeleton, an intricate system of internal membranes and, usually, mitochondria (organelles that perform respiration, using oxygen to extract energy from nutrients). In the case of algae and higher plants, the cells also contained chloroplasts (photosynthetic organelles).

Prokaryotes, thought at the time to be synonymous with bacteria, were noted to consist of smaller and simpler nonnucleated cells. They are usually enclosed by both a membrane and a rigid outer wall.

Woese's early data supported the distinction between prokaryotes and eukaryotes, by establishing that the SSU rRNAs in typical bacteria were more similar in sequence to one another than to the rRNA of eukaryotes. The initial rRNA findings also lent credence to one of the most interesting notions in evolutionary cell biology: the endosymbiont hypothesis. This conception aims to explain how eukaryotic cells first came to possess mitochondria and chloroplasts [see "The Birth of Complex Cells," by Christian de Duve, this volume].

On the way to becoming a eukaryote, the hypothesis proposes, some ancient anaerobic prokaryote (unable to use oxygen for energy) lost its cell wall. The more flexible membrane underneath then began to grow and fold in on itself. This change, in turn, led to formation of a nucleus and other internal membranes and also enabled the cell to engulf and digest neighboring prokaryotes, instead of gaining nourishment entirely by absorbing small molecules from its environment.

At some point, one of the descendants of this primitive eukaryote took up bacterial cells of the type known as alpha-proteobacteria, which are proficient at respiration. But instead of digesting this "food," the eukaryote settled into a mutually beneficial (symbiotic) relationship with it. The eukaryote sheltered the internalized cells, and the "endosymbionts" provided extra energy to the host through respiration. Finally, the endosymbionts lost the genes they formerly used for independent growth and transferred others to the host's nucleus—becoming mitochondria in the process. Likewise, chloroplasts derive from cyanobacteria that an early, mitochondria-bearing eukaryote took up and kept.

Mitochondria and chloroplasts in modern eukaryotes still retain a small number of genes, including those that encode SSU rRNA. Hence, once the right tools became available in the mid-1970s, investigators decided to see if those RNA genes were inherited from alpha-proteobacteria and

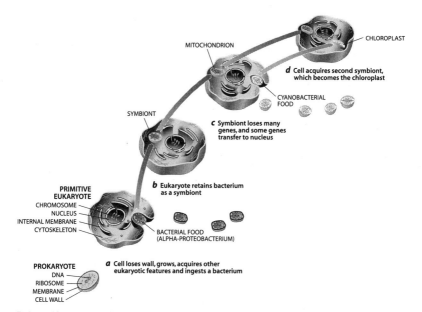

CHLOROPLAST

MITOCHONDRION

d **Cell acquires second symbiont, which becomes the chloroplast**

CYANOBACTERIAL FOOD

SYMBIONT

c **Symbiont loses many genes, and some genes transfer to nucleus**

PRIMITIVE EUKARYOTE
CHROMOSOME
NUCLEUS
INTERNAL MEMBRANE
CYTOSKELETON

b **Eukaryote retains bacterium as a symbiont**

BACTERIAL FOOD (ALPHA-PROTEOBACTERIUM)

PROKARYOTE
DNA
RIBOSOME
MEMBRANE
CELL WALL

a **Cell loses wall, grows, acquires other eukaryotic features and ingests a bacterium**

Endosymbiont hypothesis proposes that mitochondria formed after a prokaryote that had evolved into an early eukaryote engulfed (*a*) and then kept (*b*) one or more alpha-proteobacteria cells. Eventually, the bacterium gave up its ability to live on its own and transferred some of its genes to the nucleus of the host (*c*), becoming a mitochondrion. Later, some mitochondrion-bearing eukaryote ingested a cyanobacterium that became the chloroplast (*d*).

cyanobacteria, respectively—as the endosymbiont hypothesis would predict. They were.

One deduction, however, introduced a discordant note into all this harmony. In the late 1970s Woese asserted that the two-domain view of life, dividing the world into bacteria and eukaryotes, was no longer tenable; a three-domain construct had to take its place.

Certain prokaryotes classified as bacteria might look like bacteria but, he insisted, were genetically much different. In fact, their rRNA supported an early separation. Many of these species had already been noted for displaying unusual behavior, such as favoring extreme environments, but no one had disputed their status as bacteria. Now Woese claimed that they formed a third primary group—the archaea—as different from bacteria as bacteria are from eukaryotes.

ACRIMONY, THEN CONSENSUS

At first, the claim met enormous resistance. Yet eventually most scientists became convinced, in part because the overall structures of certain molecules in archaeal species corroborated the three-group arrangement. For instance, the cell membranes of all archaea are made up of unique lipids (fatty substances) that are quite distinct—in their physical properties, chemical constituents and linkages—from the lipids of bacteria.

Similarly, the archaeal proteins responsible for several crucial cellular processes have a distinct structure from the proteins that perform the same tasks in bacteria. Gene transcription and translation are two of those processes. To make a protein, a cell first copies, or transcribes, the corresponding gene into a strand of messenger RNA. Then ribosomes translate the messenger RNA codes into a specific string of amino acids. Biochemists found that archaeal RNA polymerase, the enzyme that carries out gene transcription, more resembles its eukaryotic than its bacterial counterparts in complexity and in the nature of its interactions with DNA. The protein components of the ribosomes that translate archaeal messenger RNAs are also more like the ones in eukaryotes than those in bacteria.

Once scientists accepted the idea of three domains of life instead of two, they naturally wanted to know which of the two structurally primitive groups—bacteria or archaea—gave rise to the first eukaryotic cell. The studies that showed a kinship between the transcription and translation machinery in archaea and eukaryotes implied that eukaryotes diverged from the archaeans.

This deduction gained added credibility in 1989, when groups led by J. Peter Gogarten of the University of Connecticut and Takashi Miyata, then at Kyushu University in Japan, used sequence information from genes for other cellular components to "root" the universal tree. Comparisons of SSU rRNA can indicate which organisms are closely related to one another but, for technical reasons, cannot by themselves indicate which groups are oldest and therefore closest to the root of the tree. The DNA sequences encoding two essential cellular proteins agreed that the last common ancestor spawned both the bacteria and the archaea; then the eukaryotes branched from the archaea.

Since 1989 a host of discoveries have supported that depiction. In the past five years, sequences of the full genome (the total complement of genes) in half a dozen archaea and more than 15 bacteria have become available. Comparisons of such genomes confirm earlier suggestions that many genes involved in transcription and translation are much the same in eukaryotes and archaea and that these processes are performed very similarly in the two domains. Further, although archaea do not have nuclei, under certain experimental conditions their chromosomes resemble those of eukaryotes: the DNA appears to be associated with eukaryote-type

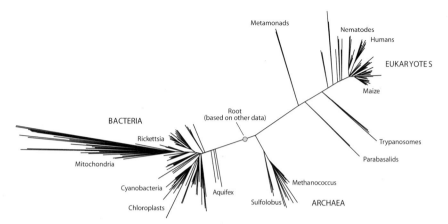

Relationships among ribosomal RNAs (rRNAs) from almost 600 species are depicted. A single line represents the rRNA sequence in one species or a group; many of the lines reflect rRNAs encoded by nuclear genes, but others reflect rRNAs encoded by chloroplast or mitochondrial genes. The mitochondrial lines are relatively long because mitochondrial genes evolve rapidly. Trees derived from rRNA data are rootless; other data put the root at the dot, corresponding to the lowest part of the tree shown on page 88.

proteins called histones, and the chromosomes can adopt a eukaryotic "beads-on-a-string" structure. These chromosomes are replicated by a suite of proteins, most of which are found in some form in eukaryotes but not in bacteria.

NEVERTHELESS, DOUBTS

The accumulation of all these wonderfully consistent data was gratifying and gave rise to the now accepted arrangement of the universal genealogical tree. This phylogeny indicates that life diverged first into bacteria and archaea. Eukaryotes then evolved from an archaealike precursor. Subsequently, eukaryotes took up genes from bacteria twice, obtaining mitochondria from alpha-proteobacteria and chloroplasts from cyanobacteria.

Still, as DNA sequences of complete genomes have become increasingly available, my group and others have noted patterns that are disturbingly at odds with the prevailing beliefs. If the consensus tree were correct, researchers would expect the only bacterial genes in eukaryotes to be those in mitochondrial or chloroplast DNA or to be those that were transferred to the nucleus from the alpha-proteobacterial or cyanobacterial precursors of these organelles. The transferred genes, moreover, would be ones involved in respiration or photosynthesis, not in cellular processes that would already be handled by genes inherited from the ancestral archaean.

Those expectations have been violated. Nuclear genes in eukaryotes often derive from bacteria, not solely from archaea. A good number of those bacterial genes serve nonrespiratory and nonphotosynthetic processes that are arguably as critical to cell survival as are transcription and translation.

The classic tree also indicates that bacterial genes migrated only to a eukaryote, not to any archaea. Yet we are seeing signs that many archaea possess a substantial store of bacterial genes. One example among many is *Archaeoglobus fulgidus*. This organism meets all the criteria for an archaean (it has all the proper lipids in its cell membrane and the right transcriptional and translational machinery), but it uses a bacterial form of the enzyme HMGCoA reductase for synthesizing membrane lipids. It also has numerous bacterial genes that help it to gain energy and nutrients in one of its favorite habitats: undersea oil wells.

The most reasonable explanation for these various contrarian results is that the pattern of evolution is not as linear and treelike as Darwin imagined it. Although genes are passed vertically from generation to generation, this vertical inheritance is not the only important process that has affected the evolution of cells. Rampant operation of a different process— lateral, or horizontal, gene transfer—has also affected the course of that evolution profoundly. Such transfer involves the delivery of single genes, or whole suites of them, not from a parent cell to its offspring but across species barriers.

Lateral gene transfer would explain how eukaryotes that supposedly evolved from an archaeal cell obtained so many bacterial genes important to metabolism: the eukaryotes picked up the genes from bacteria and kept those that proved useful. It would likewise explain how various archaea came to possess genes usually found in bacteria.

Some molecular phylogenetic theorists—among them, Mitchell L. Sogin of the Marine Biological Laboratory in Woods Hole, Mass., and Russell F. Doolittle (my very distant relative) of the University of California at San Diego—have also invoked lateral gene transfer to explain a longstanding mystery. Many eukaryotic genes turn out to be unlike those of any known archaea or bacteria; they seem to have come from nowhere. Notable in this regard are the genes for the components of two defining eukaryotic features, the cytoskeleton and the system of internal membranes. Sogin and Doolittle suppose that some fourth domain of organisms, now extinct, slipped those surprising genes into the eukaryotic nuclear genome horizontally.

In truth, microbiologists have long known that bacteria exchange genes horizontally. Gene swapping is clearly how some disease-causing bacteria

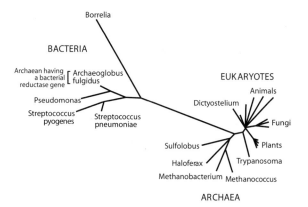

Mini phylogenetic tree groups species according to differences in a gene coding for the enzyme HMGCoA reductase. It shows that the reductase gene in *Archaeoglobus fulgidus,* a definite archaean, came from a bacterium, not from an archaean ancestor. This finding is part of growing evidence indicating that the evolution of unicellular life has long been influenced profoundly by lateral gene transfer [occurring between contemporaries]. The consensus universal tree does not take that influence into account.

give the gift of antibiotic resistance to other species of infectious bacteria. But few researchers suspected that genes essential to the very survival of cells traded hands frequently or that lateral transfer exerted great influence on the early history of microbial life. Apparently, we were mistaken.

CAN THE TREE SURVIVE?

What do the new findings say about the structure of the universal tree of life? One lesson is that the neat progression from archaea to eukaryote in the consensus tree is oversimplified or wrong. Plausibly, eukaryotes emerged not from an archaean but from some precursor cell that was the product of any number of horizontal gene transfers—events that left it part bacterial and part archaean and maybe part other things.

The weight of evidence still supports the likelihood that mitochondria in eukaryotes derived from alpha-proteobacterial cells and that chloroplasts came from ingested cyanobacteria, but it is no longer safe to assume that those were the only lateral gene transfers that occurred after the first eukaryotes arose. Only in later, multicellular eukaryotes do we know of definite restrictions on horizontal gene exchange, such as the advent of separated (and protected) germ cells.

The standard depiction of the relationships within the prokaryotes seems too pat as well. A host of genes and biochemical features do unite the prokaryotes that biologists now call archaea and distinguish those organisms from the prokaryotes we call bacteria, but bacteria and archaea (as well as species within each group) have clearly engaged in extensive gene swapping.

Researchers might choose to define evolutionary relationships within the prokaryotes on the basis of genes that seem least likely to be transferred. Indeed, many investigators still assume that genes for SSU rRNA and the proteins involved in transcription and translation are unlikely to be moveable and that the phylogenetic tree based on them thus remains valid. But this nontransferability is largely an untested assumption, and in any case, we must now admit that any tree is at best a description of the evolutionary history of only part of an organism's genome. The consensus tree is an overly simplified depiction.

What would a truer model look like? At the top, treelike branching would continue to be apt for multicellular animals, plants and fungi. And gene transfers involved in the formation of bacteria-derived mitochondria and chloroplasts in eukaryotes would still appear as fusions of major branches. Below these transfer points (and continuing up into the modern bacterial and archaeal domains), we would, however, see a great many additional branch fusions. Deep in the realm of the prokaryotes and perhaps at the base of the eukaryotic domain, designation of any trunk as the main one would be arbitrary.

Though complicated, even this revised picture would actually be misleadingly simple, a sort of shorthand cartoon, because the fusing of branches usually would not represent the joining of whole genomes, only the transfers of single or multiple genes. The full picture would have to display simultaneously the superimposed genealogical patterns of thousands of different families of genes (the rRNA genes form just one such family).

If there had never been any lateral transfer, all these individual gene trees would have the same topology (the same branching order), and the ancestral genes at the root of each tree would have all been present in the genome of the universal last common ancestor, a single ancient cell. But extensive transfer means that neither is the case: gene trees will differ (although many will have regions of similar topology), *and* there would never have been a single cell that could be called the last universal common ancestor.

As Woese has written, "The ancestor cannot have been a particular organism, a single organismal lineage. It was communal, a loosely knit, diverse conglomeration of primitive cells that evolved as a unit, and it eventually developed to a stage where it broke into several distinct

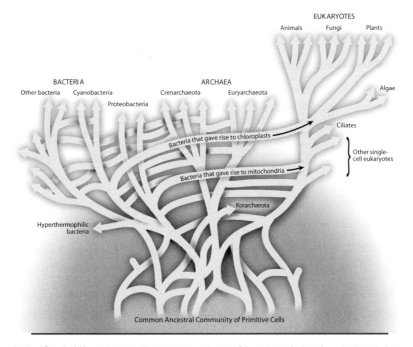

Revised "tree" of life retains a treelike structure at the top of the eukaryotic domain and acknowledges that eukaryotes obtained mitochondria and chloroplasts from bacteria. But it also includes an extensive network of untreelike links between branches. Those links have been inserted somewhat randomly to symbolize the rampant lateral gene transfer of single or multiple genes that has always occurred between unicellular organisms. This "tree" also lacks a single cell at the root; the three major domains of life probably arose from a population of primitive cells that differed in their genes.

communities, which in their turn become the three primary lines of descent [bacteria, archaea and eukaryotes]." In other words, early cells, each having relatively few genes, differed in many ways. By swapping genes freely, they shared various of their talents with their contemporaries. Eventually this collection of eclectic and changeable cells coalesced into the three basic domains known today. These domains remain recognizable because much (though by no means all) of the gene transfer that occurs these days goes on within domains.

Some biologists find these notions confusing and discouraging. It is as if we have failed at the task that Darwin set for us: delineating the unique structure of the tree of life. But in fact, our science is working just as it should. An attractive hypothesis or model (the single tree) suggested experiments, in this case the collection of gene sequences and their analysis with the methods of molecular phylogeny. The data show the model to be too simple. Now new hypotheses, having final forms we cannot yet guess, are called for.

FURTHER READING

Carl Woese. "The Universal Ancestor" in the *Proceedings of the National Academy of Sciences,* Vol. 95, No. 12, pages 6854–6859; June 9, 1998.

W. Ford Doolittle. "You Are What You Eat: A Gene Transfer Rachet Could Account for Bacterial Genes in Eukaryotic Nuclear Genomes" in *Trends in Genetics,* Vol. 14, No. 8, pages 307–311; August 1998.

W. Ford Doolittle. "Phylogenetic Classification and the Universal Tree" in *Science,* Vol. 284, pages 2124–2128; June 25, 1999.

The Birth of Complex Cells

CHRISTIAN DE DUVE

ORIGINALLY PUBLISHED IN APRIL 1996

About 3.7 billion years ago the first living organisms appeared on the earth. They were small, single-celled microbes not very different from some present-day bacteria. Cells of this kind are classified as prokaryotes because they lack a nucleus (*karyon* in Greek), a distinct compartment for their genetic machinery. Prokaryotes turned out to be enormously successful. Thanks to their remarkable ability to evolve and adapt, they spawned a wide variety of species and invaded every habitat the world had to offer.

The living mantle of our planet would still be made exclusively of prokaryotes but for an extraordinary development that gave rise to a very different kind of cell, called a eukaryote because it possesses a true nucleus. (The prefix *eu* is derived from the Greek word meaning "good.") The consequences of this event were truly epoch-making. Today all multicellular organisms consist of eukaryotic cells, which are vastly more complex than prokaryotes. Without the emergence of eukaryotic cells, the whole variegated pageantry of plant and animal life would not exist, and no human would be around to enjoy that diversity and to penetrate its secrets.

Eukaryotic cells most likely evolved from prokaryotic ancestors. But how? That question has been difficult to address because no intermediates of this momentous transition have survived or left fossils to provide direct clues. One can view only the final eukaryotic product, something strikingly different from any prokaryotic cell. Yet the problem is no longer insoluble. With the tools of modern biology, researchers have uncovered revealing kinships among a number of eukaryotic and prokaryotic features, thus throwing light on the manner in which the former may have been derived from the latter.

Appreciation of this astonishing evolutionary journey requires a basic understanding of how the two fundamental cell types differ. Eukaryotic cells are much larger than prokaryotes (typically some 10,000 times in volume), and their repository of genetic information is far more organized.

In prokaryotes the entire genetic archive consists of a single chromosome made of a circular string of DNA that is in direct contact with the rest of the cell. In eukaryotes, most DNA is contained in more highly structured chromosomes that are grouped within a well-defined central enclosure, the nucleus. The region surrounding the nucleus (the cytoplasm) is partitioned by membranes into an elaborate network of compartments that fulfill a host of functions. Skeletal elements within the cytoplasm provide eukaryotic cells with internal structural support. With the help of tiny molecular motors, these elements also enable the cells to shuffle their contents and to propel themselves from place to place.

Most eukaryotic cells further distinguish themselves from prokaryotes by having in their cytoplasm up to several thousand specialized structures, or organelles, about the size of a prokaryotic cell. The most important of such organelles are peroxisomes (which serve assorted metabolic functions), mitochondria (the power factories of cells) and, in algae and plant cells, plastids (the sites of photosynthesis). Indeed, with their many organelles and intricate internal structures, even single-celled eukaryotes, such as yeasts or amoebas, prove to be immensely complex organisms.

The organization of prokaryotic cells is much more rudimentary. Yet prokaryotes and eukaryotes are undeniably related. That much is clear from their many genetic similarities. It has even been possible to establish the approximate time when the eukaryotic branch of life's evolutionary tree began to detach from the prokaryotic trunk. This divergence started in the remote past, probably before three billion years ago. Subsequent events in the development of eukaryotes, which may have taken as long as one billion years or more, would still be shrouded in mystery were it

Prokaryotic and eukaryotic cells differ in size and complexity. Prokaryotic cells are normally about one micron across, whereas eukaryotic cells typically range from 10 to 30 microns. The latter, here represented by a hypothetical green alga, house a wide array of specialized structures—including an encapsulated nucleus containing the cell's main genetic stores.

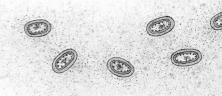

PROKARYOTIC CELLS

not for an illuminating clue that has come from the analysis of the numerous organelles that reside in the cytoplasm.

A FATEFUL MEAL

Biologists have long suspected that mitochondria and plastids descend from bacteria that were adopted by some ancestral host cell as endosymbionts (a word derived from Greek roots that means "living together

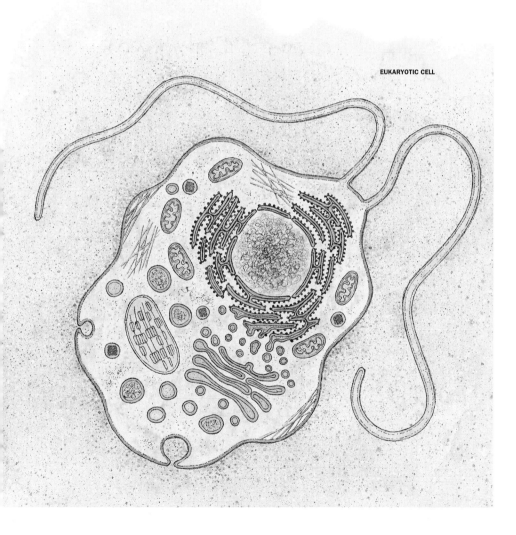

EUKARYOTIC CELL

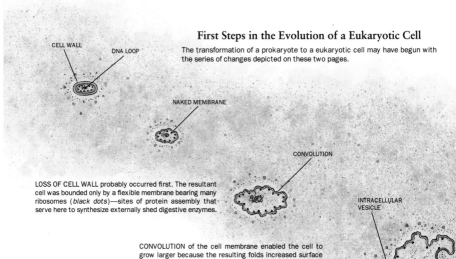

First Steps in the Evolution of a Eukaryotic Cell

The transformation of a prokaryote to a eukaryotic cell may have begun with the series of changes depicted on these two pages.

CELL WALL

DNA LOOP

NAKED MEMBRANE

CONVOLUTION

INTRACELLULAR VESICLE

LOSS OF CELL WALL probably occurred first. The resultant cell was bounded only by a flexible membrane bearing many ribosomes (*black dots*)—sites of protein assembly that serve here to synthesize externally shed digestive enzymes.

CONVOLUTION of the cell membrane enabled the cell to grow larger because the resulting folds increased surface area for the absorption of nutrients from the surrounding food supply. At this point, digestive enzymes broke down material only outside the cell.

INWARD FOLDING of the membrane allowed pockets to pinch off, forming isolated interior compartments. Digestion then occurred both outside and inside the cell. Internalization of the patch of membrane to which DNA was anchored created a sac with DNA attached—a precursor of the cell nucleus.

inside"). This theory goes back more than a century. But the notion enjoyed little favor among mainstream biologists until it was revived in 1967 by Lynn Margulis, then at Boston University, who has since tirelessly championed it, at first against strong opposition. Her persuasiveness is no longer needed. Proofs of the bacterial origin of mitochondria and plastids are overwhelming.

The most convincing evidence is the presence within these organelles of a vestigial—but still functional—genetic system. That system includes DNA-based genes, the means to replicate this DNA, and all the molecular tools needed to construct protein molecules from their DNA-encoded blueprints. A number of properties clearly characterize this genetic apparatus as prokaryotelike and distinguish it from the main eukaryotic genetic system.

Endosymbiont adoption is often presented as resulting from some kind of encounter—aggressive predation, peaceful invasion, mutually beneficial association or merger—between two typical prokaryotes. But these descriptions are troubling because modern bacteria do not exhibit such behavior. Moreover, the joining of simple prokaryotes would leave many other characteristics of eukaryotic cells unaccounted for. There is

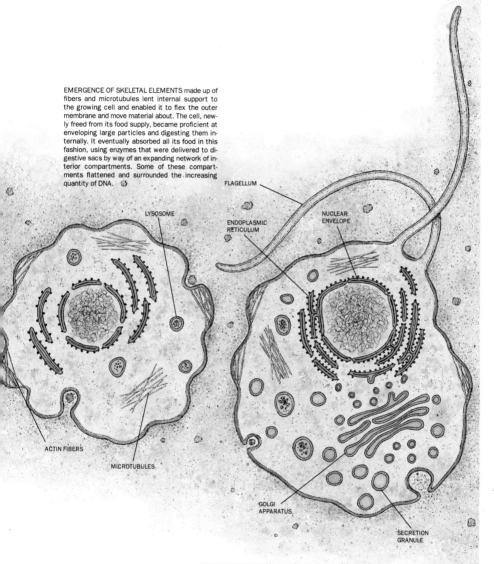

EMERGENCE OF SKELETAL ELEMENTS made up of fibers and microtubules lent internal support to the growing cell and enabled it to flex the outer membrane and move material about. The cell, newly freed from its food supply, became proficient at enveloping large particles and digesting them internally. It eventually absorbed all its food in this fashion, using enzymes that were delivered to digestive sacs by way of an expanding network of interior compartments. Some of these compartments flattened and surrounded the increasing quantity of DNA.

FLAGELLUM

LYSOSOME

ENDOPLASMIC
RETICULUM

NUCLEAR
ENVELOPE

ACTIN FIBERS

MICROTUBULES

GOLGI
APPARATUS

SECRETION
GRANULE

PRIMITIVE PHAGOCYTE, an "eating cell," ultimately developed from the sequence of incremental evolutionary advances. This cell used flagella, seen as whiplike projections, for propulsion. The phagocyte also acquired a true nucleus (as the compartments surrounding the DNA fused together), along with an increasingly complex family of cellular structures that evolved from internalized parts of the cell membrane.

a more straightforward explanation, which is directly suggested by nature itself—namely, that endosymbionts were originally taken up in the course of feeding by an unusually large host cell that had already acquired many properties now associated with eukaryotic cells.

Many modern eukaryotic cells—white blood cells, for example—entrap prokaryotes. As a rule, the ingested microorganisms are killed and broken down. Sometimes they escape destruction and go on to maim or kill their captors. On a rare occasion, both captor and victim survive in a state of mutual tolerance that can later turn into mutual assistance and, eventually, dependency. Mitochondria and plastids thus may have been a host cell's permanent guests.

If this surmise is true, it reveals a great deal about the earlier evolution of the host. The adoption of endosymbionts must have followed after some prokaryotic ancestor to eukaryotes evolved into a primitive phagocyte (from the Greek for "eating cell"), a cell capable of engulfing voluminous bodies, such as bacteria. And if this ancient cell was anything like modern phagocytes, it must have been much larger than its prey and surrounded by a flexible membrane able to envelop bulky extracellular objects. The pioneering phagocyte must also have had an internal network of compartments connected with the outer membrane and specialized in the processing of ingested materials. It would also have had an internal skeleton of sorts to provide it with structural support, and it probably contained the molecular machinery to flex the outer membrane and to move internal contents about.

The development of such cellular structures represents the essence of the prokaryote-eukaryote transition. The chief problem, then, is to devise a plausible explanation for the progressive construction of these features in a manner that can be accounted for by the operation of natural selection. Each small change in the cell must have improved its chance of surviving and reproducing (offered a selective advantage) so that the new trait would become increasingly widespread in the population.

GENESIS OF AN EATING CELL

What forces might drive a primitive prokaryote to evolve in the direction of a modern eukaryotic cell? To address this question, I will make a few assumptions. First, I shall take it that the ancestral cell fed on the debris and discharges of other organisms; it was what biologists label a heterotroph. It therefore lived in surroundings that provided it with food. An interesting possibility is that it resided in mixed prokaryotic colonies of the kind that have fossilized into layered rocks called stromatolites. Living stromatolite colonies still exist; they are formed of layers of heterotrophs topped by photosynthetic organisms that multiply with the help of sunlight and supply the lower layers with food. The fossil record indicates that such colonies already existed more than 3.5 billion years ago.

A second hypothesis, a corollary of the first, is that the ancestral organism had to digest its food. I shall assume that it did so (like most modern heterotrophic prokaryotes) by means of secreted enzymes that degraded food outside the cell. That is, digestion occurred before ingestion.

A final supposition is that the organism had lost the ability to manufacture a cell wall, the rigid shell that surrounds most prokaryotes and provides them with structural support and protection against injury. Notwithstanding their fragility, free-living naked forms of this kind exist today, even in unfavorable surroundings. In the case under consideration, the stromatolite colony would have provided the ancient organism with excellent shelter.

Accepting these three assumptions, one can now visualize the ancestral organism as a flattened, flexible blob—almost protean in its ability to change shape—in intimate contact with its food. Such a cell would thrive and grow faster than its walled-in relatives. It need not, however, automatically respond to growth by dividing, as do most cells. An alternative behavior would be expansion and folding of the surrounding membrane, thus increasing the surface available for the intake of nutrients and the excretion of waste—limiting factors on the growth of any cell. The ability to create an extensively folded surface would allow the organism to expand far beyond the size of ordinary prokaryotes. Indeed, giant prokaryotes living today have a highly convoluted outer membrane, probably a prerequisite of their enormous girth. Thus, one eukaryotic property—large size—can be accounted for simply enough.

Natural selection is likely to favor expansion over division because deep folds would increase the cell's ability to obtain food by creating partially confined areas—narrow inlets along the rugged cellular coast—within which high concentrations of digestive enzymes would break down food more efficiently. Here is where a crucial development could have taken place: given the self-sealing propensity of biological membranes (which are like soap bubbles in this respect), no great leap of imagination is required to see how folds could split off to form intracellular sacs. Once such a process was initiated, as a more or less random side effect of membrane expansion, any genetic change that would promote its further development would be greatly favored by natural selection. The inlets would have turned into confined inland ponds, within which food would now be trapped together with the enzymes that digest it. From being extracellular, digestion would have become intracellular.

Cells capable of catching and processing food in this way would have gained enormously in their ability to exploit their environment, and the resulting boost to survival and reproductive potential would have been gigantic. Such cells would have acquired the fundamental features of phagocytosis: engulfment of extracellular objects by infoldings of the cell

Final Steps in the Evolution of a Eukaryotic Cell. Adoption of prokaryotes as permanent guests within larger phagocytes marked the final phase in the evolution of eukaryotic cells. The precursors to peroxisomes may have been the first prokaryotes to develop into eukaryotic organelles. They detoxified destructive compounds created by rising oxygen levels in the atmosphere. The precursors of mitochondria proved even more adept at protecting the host cells against oxygen and offered the further ability to generate the energy-rich molecule adenosine triphosphate (ATP). The development of peroxisomes and mitochondria then allowed the adoption of the precursors of plastids, such as chloroplasts, oxygen-producing centers of photosynthesis. This final step benefited the host cells by supplying the means to manufacture materials using the energy of sunlight.

PRECURSORS OF PEROXISOMES

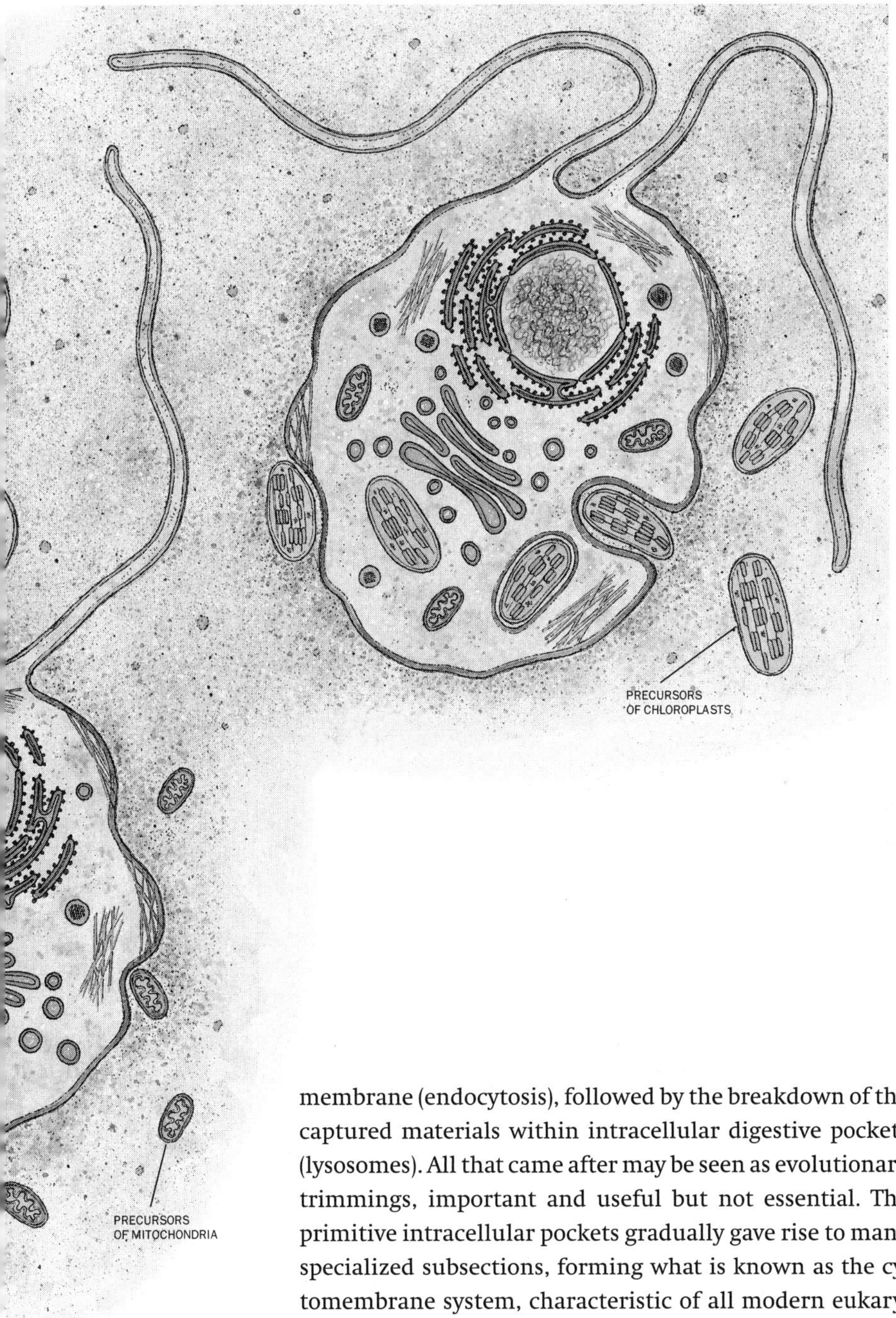

PRECURSORS
OF CHLOROPLASTS

PRECURSORS
OF MITOCHONDRIA

membrane (endocytosis), followed by the breakdown of the captured materials within intracellular digestive pockets (lysosomes). All that came after may be seen as evolutionary trimmings, important and useful but not essential. The primitive intracellular pockets gradually gave rise to many specialized subsections, forming what is known as the cytomembrane system, characteristic of all modern eukaryotic cells. Strong support for this model comes from the observation that many systems present in the cell membrane of prokaryotes are found in various parts of the eukaryotic cytomembrane system.

Interestingly, the genesis of the nucleus—the hallmark of eukaryotic cells—can also be accounted for, at least schematically, as resulting from the internalization of some of the cell's outer membrane. In prokaryotes the circular DNA chromosome is attached to the cell membrane. Infolding of this particular patch of cell membrane could create an intracellular sac bearing the chromosome on its surface. That structure could have been the seed of the eukaryotic nucleus, which is surrounded by a double membrane formed from flattened parts of the intracellular membrane system that fuse into a spherical envelope.

The proposed scenario explains how a small prokaryote could have evolved into a giant cell displaying some of the main properties of eukaryotic cells, including a fenced-off nucleus, a vast network of internal membranes and the ability to catch food and digest it internally. Such progress could have taken place by a very large number of almost imperceptible steps, each of which enhanced the cell's autonomy and provided a selective advantage. But there was a condition. Having lost the support of a rigid outer wall, the cell needed inner props for its enlarging bulk.

Modern eukaryotic cells are reinforced by fibrous and tubular structures, often associated with tiny motor systems, that allow the cells to move around and power their internal traffic. No counterpart of the many proteins that make up these systems is found in prokaryotes. Thus, the development of the cytoskeletal system must have required a large number of authentic innovations. Nothing is known about these key evolutionary events, except that they most likely went together with cell enlargement and membrane expansion, often in pacesetting fashion.

At the end of this long road lay the primitive phagocyte: a cell efficiently organized to feed on bacteria, a mighty hunter no longer condemned to reside inside its food supply but free to roam the world and pursue its prey actively, a cell ready, when the time came, to become the host of endosymbionts.

Such cells, which still lacked mitochondria and some other key organelles characteristic of modern eukaryotes, would be expected to have invaded many niches and filled them with variously adapted progeny. Yet few if any descendants of such evolutionary lines have survived to the present day. A few unicellular eukaryotes devoid of mitochondria exist, but the possibility that their forebears once possessed mitochondria and lost them cannot be excluded. Thus, all eukaryotes may well have evolved from primitive phagocytes that incorporated the precursors to mitochondria. Whether more than one such adoption took place is still being debated, but the majority opinion is that mitochondria sprang from a single stock. It would appear that the acquisition of mitochondria either saved one eukaryotic lineage from elimination or conferred such a tremendous selective advantage on its beneficiaries as to drive almost all other eukaryotes to extinction. Why then were mitochondria so overwhelmingly important?

THE OXYGEN HOLOCAUST

The primary function of mitochondria in cells today is the combustion of foodstuffs with oxygen to assemble the energy-rich molecule adenosine triphosphate (ATP). Life is vitally dependent on this process, which is the main purveyor of energy in the vast majority of oxygen-dependent (aerobic) organisms. Yet when the first cells appeared on the earth, there was no oxygen in the atmosphere. Free molecular oxygen is a product of life; it began to be generated when certain photosynthetic microorganisms, called cyanobacteria, appeared. These cells exploit the energy of sunlight to extract the hydrogen they need for self-construction from water molecules, leaving molecular oxygen as a by-product. Oxygen first entered the atmosphere in appreciable quantity some two billion years ago, progressively rising to reach a stable level about 1.5 billion years ago.

Before the appearance of atmospheric oxygen, all forms of life must have been adapted to an oxygen-free (anaerobic) environment. Presumably, like the obligatory anaerobes of today, they were extremely sensitive to oxygen. Within cells, oxygen readily generates several toxic chemical groups. These cellular poisons include the superoxide ion, the hydroxyl radical and hydrogen peroxide. As oxygen concentration rose two billion years ago, many early organisms probably fell victim to the "oxygen holocaust." Survivors included those cells that found refuge in some oxygen-free location or had developed other protection against oxygen toxicity.

These facts point to an attractive hypothesis. Perhaps the phagocytic forerunner of eukaryotes was anaerobic and was rescued from the oxygen crisis by the aerobic ancestors of mitochondria: cells that not only destroyed the dangerous oxygen (by converting it to innocuous water) but even turned it into a tremendously useful ally. This theory would neatly account for the apparent lifesaving effect of mitochondrial adoption and has enjoyed considerable favor.

Yet there is a problem with this idea. Adaptation to oxygen very likely took place gradually, starting with primitive systems of oxygen

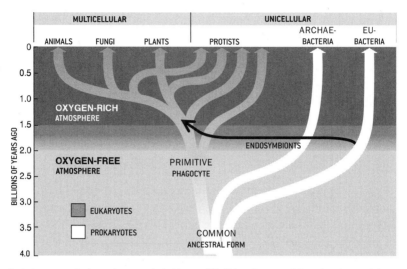

Evolutionary tree depicts major events in the history of life. This well-accepted chronology has newly been challenged by Russell F. Doolittle of the University of California at San Diego and his co-workers, who argue that the last common ancestor of all living beings existed a little more than two billion years ago.

detoxification. A considerable amount of time must have been needed to reach the ultimate sophistication of modern mitochondria. How did anaerobic phagocytes survive during all the time it took for the ancestors of mitochondria to evolve?

A solution to this puzzle is suggested by the fact that eukaryotic cells contain other oxygen-utilizing organelles, as widely distributed throughout the plant and animal world as mitochondria but much more primitive in structure and composition. These are the peroxisomes [see "Microbodies in the Living Cell," by Christian de Duve; *Scientific American,* May 1983]. Peroxisomes, like mitochondria, carry out a number of oxidizing metabolic reactions. Unlike mitochondria, however, they do not use the energy released by these reactions to assemble ATP but squander it as heat. In the process, they convert oxygen to hydrogen peroxide, but then they destroy this dangerous compound with an enzyme called catalase. Peroxisomes also contain an enzyme that removes the superoxide ion. They therefore qualify eminently as primary rescuers from oxygen toxicity.

I first made this argument in 1969, when peroxisomes were believed to be specialized parts of the cytomembrane system. I thus included peroxisomes within the general membrane expansion model I had proposed for the development of the primitive phagocyte. Afterward, experiments by the late Brian H. Poole and by Paul B. Lazarow, my associates at the

Rockefeller University, conclusively demonstrated that peroxisomes are entirely unrelated to the cytomembrane system. Instead they acquire their proteins much as mitochondria and plastids do (by a process I will explain shortly). Hence, it seemed reasonable that all three organelles began as endosymbionts. So, in 1982, I revised my original proposal and suggested that peroxisomes might stem from primitive aerobic bacteria that were adopted before mitochondria. These early oxygen detoxifiers could have protected their host cells during all the time it took for the ancestors of mitochondria to reach the high efficiency they possessed when they were adopted.

So far researchers have obtained no solid evidence to support this hypothesis or, for that matter, to disprove it. Unlike mitochondria and plastids, peroxisomes do not contain the remnants of an independent genetic system. This observation nonetheless remains compatible with the theory that peroxisomes developed from an endosymbiont. Mitochondria and plastids have lost most of their original genes to the nucleus, and the older peroxisomes could have lost all their DNA by now.

Whichever way they were acquired, peroxisomes may well have allowed early eukaryotes to weather the oxygen crisis. Their ubiquitous distribution would thereby be explained. The tremendous gain in energy retrieval provided with the coupling of the formation of ATP to oxygen utilization would account for the subsequent adoption of mitochondria, organelles that have the additional advantage of keeping the oxygen in their surroundings at a much lower level than peroxisomes can maintain.

Why then did peroxisomes not disappear after mitochondria were in place? By the time eukaryotic cells acquired mitochondria, some peroxisomal activities (for instance, the metabolism of certain fatty acids) must have become so vital that these primitive organelles could not be eliminated by natural selection. Hence, peroxisomes and mitochondria are found together in most modern eukaryotic cells.

The other major organelles of endosymbiont origin are the plastids, whose main representatives are the chloroplasts, the green photosynthetic organelles of unicellular algae and multicellular plants. Plastids are derived from cyanobacteria, the prokaryotes responsible for the oxygen crisis. Their adoption as endosymbionts quite likely followed that of mitochondria. The selective advantages that favored the adoption of photosynthetic endosymbionts are obvious. Cells that had once needed a constant food supply henceforth thrived on nothing more than air, water, a few dissolved minerals and light. In fact, there is evidence that eukaryotic cells acquired plastids at least three separate times, giving rise to green, red and

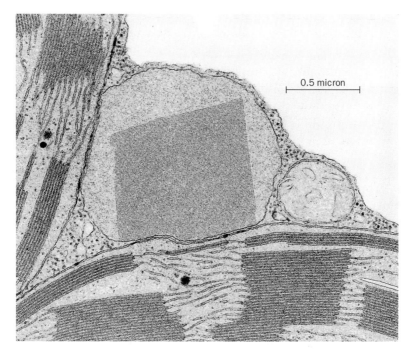

Four organelles appear in a tobacco leaf cell. The two chloroplasts (*left and bottom*) and the mitochondrion (*middle right*) evolved from prokaryotic endosymbionts. The peroxisome (*center*)—containing a prominent crystalline inclusion, most probably made up of the enzyme catalase—may have derived from an endosymbiont as well.

brown algae. Members of the first of these groups were later to form multicellular plants.

FROM PRISONER TO SLAVE

What started as an uneasy truce soon turned into the progressive enslavement of the captured endosymbiont prisoners by their phagocytic hosts. This subjugation was achieved by the piecemeal transfer of most of the endosymbionts' genes to the host cell's nucleus. In itself, the uptake of genes by the nucleus is not particularly extraordinary. When foreign genes are introduced into the cytoplasm of a cell (as in some bioengineering experiments), they can readily home to the nucleus and function there. That is, they replicate during cell division and can serve as the master templates for the production of proteins. But the migration of genes from endosymbionts to the nucleus is remarkable because it seems to have raised more difficulties than it solved. Once this transfer occurred, the proteins encoded by these genes began to be manufactured in the cytoplasm of the host cell (where the products of all nuclear genes are constructed). These molecules had then to migrate into the endosymbiont to be of use. Somehow this seemingly unpromising scheme not only withstood the hazards of evolution but also proved so successful that all endosymbionts retaining copies of transferred genes eventually disappeared.

Today mitochondria, plastids and peroxisomes acquire proteins from the surrounding cytoplasm with the aid of complex transport structures in their bounding membranes. These structures recognize parts of newly made protein molecules as "address tags" specific to each organelle. The transport apparatus then allows the appropriate molecules to travel through the membrane with the help of energy and of specialized proteins (aptly called chaperones). These systems for bringing externally made proteins into the organelles could conceivably have evolved from similar systems for protein secretion that existed in the original membranes of the endosymbionts. In their new function, however, those systems would have to operate from outside to inside.

The adoption of endosymbionts undoubtedly played a critical role in the birth of eukaryotes. But this was not the key event. More significant (and requiring a much larger number of evolutionary innovations) was the long, mysterious process that made such acquisition possible: the slow conversion, over as long as one billion years or more, of a prokaryotic ancestor into a large phagocytic microbe possessing most attributes of modern eukaryotic cells. Science is beginning to lift the veil that shrouds this momentous transformation, without which much of the living world, including humans, would not exist.

FURTHER READING

T. Cavalier-Smith. "The Origin of Eukaryote and Archaebacterial Cells" in *Annals of the New York Academy of Sciences,* Vol. 503, pages 17–54; July 1987.

Christian de Duve. *Blueprint for a Cell: The Nature and Origin of Life.* Neil Patterson Publishers/ Carolina Biological Supply Company, 1991.

Betsy D. Dyer and Robert A. Obar. *Tracing the History of Eukaryotic Cells: The Enigmatic Smile.* Columbia University Press, 1994.

Christian de Duve. *Vital Dust: Life as a Cosmic Imperative.* BasicBooks, 1995.

Viral Quasispecies

MANFRED EIGEN

ORIGINALLY PUBLISHED IN JULY 1993

According to Greek mythology, when curious Pandora opened a forbidden box she set loose all the miseries and evils known to the world. One of them was undoubtedly the virus—the very name of which is Latin for slime, poison and stench. Viruses cause a mind-boggling assortment of illnesses, ranging from the common cold to acquired immunodeficiency syndrome (AIDS), perhaps the most feared scourge of modern times.

Viruses have the ability to mystify lay-people and experts alike. Early in their studies of viruses, investigators became puzzled by the high mutation rates they observed: the magnitudes indicated that viruses must evolve more than a million times faster than cellular microorganisms. If that were true, how could viruses maintain their identities as pathogenic species over any evolutionarily significant period? Why didn't they mutate out of existence?

Those questions have generally been unanswerable within the traditional theoretical framework of biology. Borrowing ideas from both mathematics and chemistry, however, my colleagues and I have recently introduced a concept, the quasispecies, that can illuminate the problems in new ways. A viral species, we have shown, is actually a complex, self-perpetuating population of diverse, related entities that act as a whole.

The substitution of "quasispecies" for "species" is not merely semantic. It offers insights into the behavior of viruses. In the case of AIDS, for example, it helps in determining when the human immunodeficiency virus (HIV) first evolved and where it may have come from. If one were to extrapolate only from the epidemiologic data, AIDS would seem to have first appeared in 1979. Our data, in contrast, suggest that HIV is a very old virus. Moreover, the quasispecies concept points toward potential treatments for AIDS and other diseases that have so far been resistant to vaccines.

To begin to understand viral quasispecies, we must ask ourselves, What is a virus? In 1959 Nobel laureate André Lwoff's answer was "A virus is a

114

virus!"—a truism, perhaps, but one that cuts to the uniqueness of viruses in the living world. Essentially, a virus is a genetic program that carries the simple message "Reproduce me!" from one cell to another. Because a virus represents only one or a few of the messengers vying for the attention of its host, it must employ certain biochemical tricks to recruit the host's replication machinery for its selfish purpose. Often those ploys result in the host cell's death.

Viruses fall into many different categories, but one way to distinguish among them is by looking at the molecules that carry their genetic messages. Perhaps the simplest form of virus is represented by a single strand of ribonucleic acid (RNA), made up of several thousand individual nucleotide subunits. If this RNA is a so-called plus strand, it can be read directly by the host's translation apparatus, the ribosome, much as the host's own messenger RNA can. Examples of such plus strand viruses are the bacteriophage Qß, a parasite of the bacterium *Escherichia coli,* and the polio-1 virus, which causes spinomuscular paralysis. Other viruses encode their messages as minus strands of RNA. Inside a cell, minus strands must be transcribed into complementary plus strands before viral replication can begin. Influenza A, one of the most common epidemic diseases, is caused by a minus strand virus.

A third class of single-strand RNA viruses consists of retroviruses. After a retrovirus infects a host cell, a viral enzyme called reverse transcriptase changes the single strand of viral RNA into a double strand of deoxyribonucleic acid (DNA). That DNA can then incorporate itself into the host's genome, thereby making the viral message an inheritable feature of the cell. HIV belongs to the retroviral family. Its target is the immune system, which ought to provide protection against the virus.

Because viruses are so dependent on the replicative systems of their hosts, scientists generally believe viruses in their present form must have evolved after cellular life. It is even possible that viruses descended from parts of their host's genetic programs that turned their inside knowledge of cells to the goal of duplicating themselves. Whatever the case, viruses are useful models for studying how molecules may have organized themselves into self-perpetuating units at the dawn of life. They show how information can be generated and processed at the molecular level. The essence of their genetic information is self-preservation, which they achieve through mutagenesis, reproduction, proliferation and adaptation to a steadily changing environment.

The genome of a single-strand RNA virus such as HIV, which comprises only 10,000 nucleotides, is small and simple compared with that of most cells. Yet from a molecular standpoint, it is unimaginably complex. Each of those nucleotides contains one of four possible bases: adenine, uracil, guanine or cytosine. The unique sequence specified by the genome of HIV therefore represents just one choice out of $4^{10,000}$ possibilities—a number roughly equivalent to a one followed by 6,000 zeros.

Most such sequences would not qualify as viruses: they could not direct their own duplication. Nevertheless, even if only a tiny fraction of them are viruses, the number is still huge. If the entire universe were completely filled with hydrogen atoms—each about one trillionth of a trillionth of a cubic centimeter in volume—it could hold only about 10^{108} of them. Hence, an array of $10^{6,000}$ differing RNA sequences is beyond comprehension.

Fortunately, it is not beyond the analytic reach of mathematics. We can construct a theoretical framework that encompasses that vast array and reveals relations among the elements. To do so, we must first develop a geometry—a concept of space—that would allow us to represent the informational differences among the sequences as precise spatial distances. In this space, each nucleotide sequence must occupy a unique position. The positions must also be arranged to reflect the informational kinship between the sequences. In other words, each sequence should be only one unit away from all the other sequences that differ from it by only one nucleotide; it should be two units away from those differing by two nucleotides, and so on.

Sequence space proves to be an invaluable tool for interpreting what a viral species is. The term "species" is used in both biology and chemistry. In chemistry, a species is a defined chemical compound, such as trinitrotoluene or benzene. In biology, the definition is not quite as sharp: members of a given living species must show common traits and must be at least potentially able to produce offspring by recombining their genetic material. At the genetic level, a biological species is represented by a gigantic variety of differing DNA molecules.

Biologists generally speak of the wild type of a species: the form that predominates in a population and that is particularly well suited to the environment in which it lives. If one found an individual that perfectly embodied that wild type, its unique sequence of genomic DNA would specify the wild type at the genetic level and would occupy a single point in the sequence space. That view of the wild type accords with the classical model of natural selection. Although mutations occur steadily, they presumably disappear because the mutant types are less fit than the wild type. Alternatively, a mutant may have advantages, in which case it becomes the new wild type. Either outcome tends to keep all the members of a species at or very near one point in a genome sequence space.

That picture was modified by the neutral theory advanced in the 1960s by Motoo Kimura of the National Institute of Genetics in Mishima, Japan. Kimura argued that many mutations, such as those causing differences in blood types, are neither advantageous nor disadvantageous. Consequently, a small but statistically defined fraction of the neutral mutations would continuously replace the existing wild type in the population. The genome of a species would therefore drift steadily but randomly through a certain volume of sequence space.

Despite those differences, both the classical Darwinian and the neutralist theories favor the idea that wild-type populations will localize sharply in sequence space after completing an advantageous or neutral shift. Also, both theories assume that mutations appear blindly, irrespective of their selective value. No single neutral or advantageous mutation would occur more frequently than any disadvantageous one.

That view, however, is not sustained by the modern kinetic theory of molecular evolution, nor is it backed by experiments with viruses. After all, evolutionary selection is a consequence of the ability of a genome to replicate itself accurately. Imagine a case in which the process of replication is so highly error-prone that no copy resembled its parental sequence. The resulting population would behave like an ideal gas, expanding until it filled the sequence space at a very low density. Selection acting on such a population could not define it or confine it in any way. The population would lose all its integrity.

If we were to reduce the error rate of replication progressively, variation in the population would disperse less and less as the offspring came to resemble their parents more and more. At some critical error rate, the effect of selection on the population would change radically: the expansive force of mutation would strike a balance with the compressive force of selection. The diffuse gas of related sequences would suddenly condense into a finite but extended region.

This region in sequence space can be visualized as a cloud with a center of gravity at the sequence from which all the mutations arose. It is a self-sustaining population of sequences that reproduce themselves imperfectly but well enough to retain a collective identity over time. Like a real cloud, it need not be symmetric, and its protrusions can reach far from the center because some mutations are more likely than others or may have higher survival values that allow them to produce more offspring. That cloud is a quasispecies.

Biologically, the quasispecies is the true target of selection. All the members of a quasispecies—not just the consensus sequence—help to perpetuate the stable population. The fitness of the entire population is what matters, not the fitness of individual members. The wild type of a quasispecies refers to an average for all the members, not to a particularly fit individual. Chemically, the quasispecies is a multitude of distinct but related nucleic acid polymers. Its wild type is the consensus sequence that represents an average for all the mutants, weighted to reflect their individual frequency. Physically, the quasispecies is a localized distribution in sequence space that forms and dissolves cooperatively in very much the same way that molecules of water pass through phase transitions as they freeze or evaporate. Its stability is constrained by the error threshold, which may be interpreted as a kind of "melting point" for the genome information. The population density at each point of sequence space depends on the fitness value of that particular sequence. A mathematician would describe the distribution of sequences in a quasispecies with a vector that refers to the maximum growth within the set of coupled kinetic equations for all the mutants.

One might wonder why in this model an advantageous or neutral mutant would have a better chance to occur than a deleterious one. New mutants appear at the periphery of the quasispecies distribution, where they are produced by the erroneous copying of mutants already present. Because the population of a mutant in the quasispecies depends on its degree of fitness, well-adapted mutants have a better chance of producing offspring; deleterious mutants produce no offspring at all. Because the chance of finding a well-adapted or advantageous mutant is greatest in a region of sequence space associated with high fitness, there is a large bias toward producing such well-adapted mutants. Calculations show that this effect speeds up the evolutionary opportunization of viruses by many orders of magnitude, as compared with truly random, unbiased mutations.

Because the error rate directly determines the size and integrity of a quasispecies, it is the most telling characteristic of a virus. The error rate is the probability that an error will occur when one nucleotide in a sequence is being copied. It can depend both on the type of nucleotide substitution taking place and on its position in the sequence. The position is important because the ribosome interprets the nucleotides three at a time, in a group called a codon. In most codons the first two positions suffice to specify the amino acid to be incorporated into a protein. Mutations in the first two positions may therefore be more stringently maintained by selection. When researchers speak of the error rate of an entire viral sequence, they are referring to an average for all the positions.

In general, the error rate of a virus is roughly proportional to the reciprocal of its sequence length—that is, about one error per replicated sequence. If the error rate were much larger, almost every replication event would produce an unfit mutation. For an entity that produces as many offspring as a virus, an error rate reciprocal to the sequence length is highly significant. Consider a typical infection process, which starts when at least one viable virus enters a host organism. If that virus is not eradicated, it will replicate. Before an infection is detectable, the viral population must rise to around 10^9, which would take about 30 generations. If the error rate is more or less equal to the reciprocal of the sequence length, then on average one error will have been added in each generation.

Consequently, any two viruses taken from an obviously infected host are likely to differ from each other at 30 nucleotide positions or more. When researchers first noticed the sequence diversity of the HIV viruses they found in individual patients, they thought it was evidence of multiple infections by different strains. The work of Simon Wain Hobson of the Pasteur Institute in Paris has demonstrated, however, that the diverse HIV sequences in patients are usually related to one another. His work clearly confirms that viruses, and immunodeficiency viruses in particular, are quasispecies.

The proliferation of a viral quasispecies is a more complex phenomenon than the simple replication of a wild type. Viral replication takes the form of a hypercycle, a set of interlocking feedback loops that describes a regulated co-evolution within a cell of the viral genes and the viral proteins essential to replication that are encoded by those genes. Michael Gebinoga of the Max Planck Institute for Biophysical Chemistry in Göttingen has quantified the process in vivo for the Qß bacteriophage. He found evidence of two feedback cycles, one based on the enzyme replicase, which promotes replication, and the other based on the viral coat protein, which limits it. The first molecules of replicase and other proteins produced by the infectious plus strand are fairly accurate because most copies of the viral genes in the cell are similar to the originals. Errors accumulate mostly during later stages in the infection cycle. For that reason, the synthesis of replicase seems to occur primarily early after infection. Yet even viral sequences that make defective proteins are copied because the replicative machinery acts on all the strands indiscriminately. When an infected *E. coli* cell bursts after 40 minutes, it releases around 10,000 phage particles, of which only 1,000 or less are infectious.

Analyses of sequence space can reveal information about the evolution of viral quasispecies that would otherwise be inaccessible. A straightforward procedure for studying the evolution would be to follow the changes in a viral gene over time. A researcher would need to collect samples of a virus over a period of many successive years. The difficulty is that even for quickly mutating viruses, the amount of change that can accumulate in only a few years—say, the lifetime of a Ph.D. thesis—is too small to measure meaningfully. Hence, the experiment would never be done.

In the mid-1980s Peter Palese of Mount Sinai School of Medicine found a better way. He was lucky enough to obtain samples of influenza A virus that had been isolated and frozen during outbreaks of the disease over a span of about 50 years. Palese and his co-workers analyzed the gene sequence common to those samples. From that information, they plotted the evolutionary relations among the viruses from each epidemic. The "family tree" they created shows the worldwide spread of the virus from a common source in successive waves during each epidemic. The tips of the branches are the isolated virus samples; the nodes, or connections of branches, correspond to the consensus sequences of their shared ancestors. In collaboration with Walter M. Fitch of the University of California at Irvine, Palese found for influenza A an essentially linear relation between the degree of difference for any two sequences and the amount of time since their divergence. Depending on the sequences they examined, two to four mutations appeared per year. The tip-to-node distances on the tree, which reflected the spread of individual sequences, corresponded to roughly five years of evolution.

Unfortunately, the case of influenza A is as yet unique: no other collections of viruses that extend across 50 years currently exist. Nevertheless, other researchers have made progress by employing a different approach. Whereas Palese tracked the evolution of a virus over time, those workers have reconstructed evolutionary trees by making inferences from the similarities of different viruses and viral strains that abound at approximately the same time. Gerald Myers of Los Alamos National Laboratory has made such a tree for the AIDS-causing strain HIV-1, using samples collected from 1985 to 1987.

The principal difference between the tree for HIV-1 and that for influenza A virus is the length of their branches. According to the scheme

Myers developed, all the early strains of HIV-1 came from African sources. Looking at the tree, we can almost trace the journey of the virus from that continent to the rest of the world. Indeed, one can extend the tree even further back into evolution by finding the relations between HIV-1, HIV-2 and various forms of simian immunodeficiency viruses (SIVs).

For determining when these viruses diverged, it would be helpful if the separation in the sequences could be used as a measure of evolutionary time. Sadly, the problem is not that simple. If two long, originally identical sequences mutate randomly, it is at first unlikely that they will undergo the same changes at the same positions. Mutations will increase their distance from the original consensus sequence, and those changes will accumulate almost linearly with respect to time.

Eventually, however, when enough mutations have accumulated, some of them will probably reverse a previous change or duplicate a change in the other sequence. As a result, the amount of difference between the sequences will decrease or stay constant, and their distance from the original consensus sequence will finally fluctuate around a given value. Past a certain point, then, the passage of more time does not add more distance. For a genetic sequence in which any one of the four nucleotides could occupy any position, that distance is 75 percent of the total sequence length.

Moreover, the assumption of uniform substitution probabilities is usually not correct. Some positions are almost constant because of fitness constraints; some vary at a normal rate, whereas still others are hypervariable and change rapidly in response to the selection pressure imposed on them by the immune response of their host. The constant, variable and hypervariable positions would each evolve according to a different distance-time relation. Applying different relations to an interpretation of the evolutionary distances would give results for old divergences that differed by orders of magnitude. The lengths of the branches in the evolutionary trees cannot divulge when new viruses evolved.

Sequence space diagrams can, however. My colleagues Katja Nieselt-Struwe and Ruthild Winkler-Oswatitsch of Göttingen, Andreas Dress of the mathematics department of Bielefeld University and I have taken that approach. We developed a mathematical method of analyzing the relations within a quasispecies that we call statistical geometry in sequence space. That analysis allows us to determine how often on average different types of changes occur at different positions. It enables us to classify different positions in the viral sequences as constant, variable or hypervariable. From that information, we can deduce roughly how long different

viral lineages have existed and the frequency with which different types
of mutations occur.

What do the statistical geometries of the influenza A, polio-1 and immunodeficiency viruses reveal? For the tree of influenza A virus, the probability
of mutations that would parallel or reverse previous changes is small. As
Palese's study indicated, the amount of difference between strains of the
virus increases almost linearly over time. An intriguing prediction also
emerges from the data: if all the mutable positions in the virus continue to
change at the indicated rates, the influenza virus should completely lose
its identity within a few hundred years. Because some positions must be
constant, the influenza A virus will probably remain a pathogen, because
to survive, it will need to infect humans, but we cannot predict what its
pathology will be.

For polio-1 virus, the picture is entirely different. In the studied sequence segment, the nucleotides that occupy the first and second positions in each codon scarcely change at all. Mutations at those positions
must be strongly eliminated from the quasispecies by selection. Conversely, the nucleotides at the third codon positions are almost completely randomized. As a result, even though the poliovirus has about the
same error rate as the influenza virus, only mutations that do not change
the encoded amino acids appear in the quasispecies. The proteins in the
poliovirus are very highly conserved.

The immunodeficiency viruses have a third type of statistical geometry.
All three codon positions are appreciably randomized for all types of
changes. We have been able to determine the prevalence of constant, variable and hypervariable sites within the gene for an HIV surface protein
that we analyzed. From that information, we were able to estimate how
long it must have taken for the immunodeficiency viruses to have diverged to the observed degree.

About 20 percent of the positions are constant, apparently because
they are necessary for HIV to function as a retrovirus. They establish
that HIV is the descendant of an old viral family. About 70 percent of the
positions are variable and have an average lifetime of about 1,000 years
(give or take a few hundred). They seem to give HIV its specific characteristics. Many of these positions differ in HIV-1, HIV-2 and the SIV sequences,
which indicates that they must have evolutionarily diverged long ago.
My colleagues and I estimate that it was 600 to 1,200 years ago (or even
longer, because more constant positions may yet be hidden in the data).
Contrary to the evidence of the epidemiologic curves, therefore, HIV is

not a new virus, although its pathogenicity may have varied over the centuries.

About 200 positions in the studied HIV gene—about 10 percent of the total—are hypervariable and change on average within 30 years. They provide the tremendous variability that enables HIV to thwart the attempts by its host's immune system to eliminate it. They may also be directly responsible for much of the damage that the virus does to the immune system. According to a theory advanced in 1992 by Robert M. May and Martin A. Novak and their colleagues at the University of Oxford, HIV uses its capacity for variance to out-flank the immune response of its host. The number of different sequences that result from mutations at hypervariable sites outruns by far the capacity of the immune system to generate lymphocytes. If HIV can change at all its hypervariable sites in 30 years, it could exhaust the immune system in only a fraction of that time. The virus can produce mutants that evade the immunologic defenses, particularly because its infection targets are the T lymphocytes that control the immune response.

Computer simulations carried out by the Oxford group verify those predictions. That theory, based on the quasispecies nature of the virus, also satisfactorily explains the decade-long delay that usually occurs between the initial viral infection and the fatal state of the disease, when the immune system breaks down fairly suddenly. It may take that many years for HIV to exhaust the adaptive resources of the immune system. New experiments will test whether this explanation is correct.

The statistical geometry data also offer insights into ways of fighting HIV and other viruses. The most common way to rid an infected individual of a virus is to stimulate, activate or support the immune system, as a vaccine does. An awareness of the variational flexibility of viruses suggests that three additional strategies must also be explored to improve vaccines. One is to find stable immunologic features in the viral quasispecies against which highly specific monoclonal antibodies could be directed. The second is to create antibodies that can act against a broad spectrum of the likely mutant viruses that would otherwise permit a quasispecies to escape attack. The third is to spot such escape mutants during an early phase of infection and to outmaneuver them with specific agents before they can produce progeny.

The most fruitful approaches may vary with different viruses. For example, the immune system can quickly learn to recognize the almost constant protein features of the poliovirus. That virus has no chance of surviving if it encounters a vaccinated host. The real effectiveness of that protection became apparent only recently when researchers discovered that the mild strain of polio-1 virus in the Sabin vaccine differs from the pathogenic wild type at only two nucleotide positions. It is entirely possible, therefore, that a few of the polioviruses from a vaccine do mutate into a pathogenic state inside the host. Yet by the time those mutations occur, the immunologic protection of the host is already practically perfect. The success of the Sabin vaccine in saving the lives of countless children is unchallenged.

Influenza is a quite different case, as are other viruses. The targets for the immune response against influenza change steadily. Although the immune system eventually copes with the virus and quells the infection, there is no lasting protection. As a consequence, people can contract influenza repeatedly, and new vaccines must be prepared every few years. John J. Holland of the University of California at San Diego and Esteban Domingo of the Independent University of Madrid have observed that the viruses responsible for foot-and-mouth disease and vesicular stomatitis, an infection of the oral membranes in livestock, behave in a similar way. HIV, with its many variable and hypervariable positions, mutates even more rapidly and radically. Vaccines may not have any lasting value against such infections.

But vaccines are only one way to fight viruses. The administration of drugs that block viral replication is an extremely common therapy—and for AIDS it is currently the sole therapy that is in any way effective at slowing the progress of the disease. In theory, artificial chains of RNA could be administered to patients to prevent or eliminate viral infections. Those RNA molecules would hinder viral replication, either by binding to the viral RNA or by competing with it for essential enzymes. Specific factors that interfere with viral replication could also be incorporated into host cells by genetic technology. Yet all these approaches may have harmful side effects or would need to clear significant technical hurdles.

A further complication is that viruses may be able to mutate around such obstacles. In my laboratory Björn F. Lindemann has used the understanding of the replicative mechanism of the Qß bacteriophage to test one antiviral strategy. He inserted the gene for the viral coat protein into cells. The cells became resistant to infection because the coat protein, a natural regulator of the phage's replication, blocked the transcription of viral genes.

Yet this strategy did not work perpetually: given sufficient time and generations, the Qß bacteriophage adapted by mutating into a form that ignored the coat protein signal. Lindemann demonstrated that fact using one of the automated "evolution machines" developed recently in my laboratory. In these devices, viruses grow in host cells for extended periods under mild selection pressures. Evolutionary biotechnology, or applied molecular evolution, as it is often called, is a rapidly emerging field of research that may have many applications in new antiviral strategies [see "Directed Molecular Evolution," by Gerald F. Joyce; *Scientific American,* December 1992].

One strategy may be resistant to the evasive maneuvers of viruses: it would exploit their nature as quasispecies and thereby undermine the very basis of their existence. Even in a successful viral quasispecies, only a small fraction of the viral sequences in a host cell are viable. If the error rates of viruses can be increased moderately, just enough to cross the critical error threshold that defines their quasispecies, they would experience a catastrophic loss of information. The viral quasispecies would fall apart because it would be producing too many nonviable mutants.

Using drugs that produce mutations, Domingo and Holland have demonstrated that this approach works against the virus that causes foot-and-mouth disease. For such a strategy to work as a therapy, however, the drugs must change the error rate of only the viral replicase and not of enzymes essential to the host's well-being. Careful study of replicase mechanisms should bring about such a possibility of interfering with virus infection. This strategy would be precisely the opposite of immunization therapies that attempt to prevent the appearance of escape mutants.

As of today, we know little about the origin of viruses or their role in the evolution of the biosphere. Viruses come and go: some adapt; others disappear. The undeniable reality is that an estimated 13 million people worldwide are infected with HIV. Pandora's box is still open and releasing new ills. Nevertheless, our growing understanding of viruses suggests that, as in the original myth, hope has not escaped.

FURTHER READING

Manfred Eigen, John McCaskill and Peter Schuster. "Molecular Quasi-Species" in *Journal of Physical Chemistry,* Vol. 92, No. 24, pages 6881–6891; December 1, 1988.

Manfred Eigen and Christof K. Biebricher. "Role of Genome Variation in Virus Evolution" in *RNA Genetics,* Vol. 3: *Variability of RNA Genomes.* Edited by Esteban Domingo, John J. Holland and Paul Ahlguist. CRC Press, 1988.

Manfred Eigen and Katja Nieselt-Struwe. "How Old Is the Immunodeficiency Virus" in *AIDS,* Vol. 4, Supplement 1, pages S85–S93; 1990.

Manfred Eigen and Ruthild Winkler-Oswatitsch. "Stastical Geometry on Sequence Space" in *Molecular Evolution: Computer Analysis of Protein and Nucleic Acid Sequences.* Edited by Russell F. Doolittle. Academic Press, 1990.

M. Eigen, C. K. Biebricher, M. Gebinoga and W. C. Gardiner, Jr. "The Hypercycle: Coupling of RNA and Protein Biosynthesis in the Infection Cycle of an RNA Bacteriophage" in *Biochemistry,* Vol. 30, No. 46, pages 11005–11018; November 19, 1991.

How Cells Respond to Stress

WILLIAM J. WELCH

ORIGINALLY PUBLISHED IN MAY 1993

Immediately after a sudden increase in temperature, all cells—from the simplest bacterium to the most highly differentiated neuron—increase production of a certain class of molecules that buffer them from harm. When biologists first observed that phenomenon 30 years ago, they called it the heat-shock response. Subsequent studies revealed that the same response takes place when cells are subjected to a wide variety of other environmental assaults, including toxic metals, alcohols and many metabolic poisons. It occurs in traumatized cells growing in culture, in the tissues of feverish children and in the organs of heart-attack victims and cancer patients receiving chemotherapy. Because so many different stimuli elicit the same cellular defense mechanism, researchers now commonly refer to it as the stress response and to the expressed molecules as stress proteins.

In their pursuit of the structure and function of the stress proteins, biologists have learned that they are far more than just defensive molecules. Throughout the life of a cell, many of these proteins participate in essential metabolic processes, including the pathways by which all other cellular proteins are synthesized and assembled. Some stress proteins appear to orchestrate the activities of molecules that regulate cell growth and differentiation.

The understanding of stress proteins is still incomplete. Nevertheless, investigators are already beginning to find new ways to put the stress response to good use. It already shows great potential for pollution monitoring and better toxicologic testing. The promise of medical applications for fighting infection, cancer and immunologic disorders is perhaps more distant, but it is clearly on the horizon.

Such uses were far from the minds of the investigators who first discovered the stress response; as happens so often in science, it was serendipitous. In the early 1960s biologists studying the genetic basis of animal development were focusing much of their attention on the fruit fly *Drosophila melanogaster. Drosophila* is a convenient organism in which to study the maturation of an embryo into an adult, in part because it has an unusual genetic feature. Cells in its salivary glands carry four chromosomes in which the normal amount of DNA has been duplicated thousands of times; all the copies align beside one another. These so-called polytene chromosomes are so large that they can be seen through a light microscope. During each stage of the developmental process, distinct regions along the polytene chromosomes puff out, or enlarge. Each puff is the result of a specific change in gene expression.

During the course of his studies, F. M. Ritossa of the International Laboratory of Genetics and Biophysics in Naples saw that a new pattern of chromosomal puffing followed the exposure of the isolated salivary glands to temperatures slightly above those optimal for the fly's normal growth and development. The puffing pattern appeared within a minute or two after the temperature rise, and the puffs continued to increase in size for as long as 30 to 40 minutes. Over the next decade, other investigators built on Ritossa's findings.

In 1974 Alfred Tissières, a visiting scientist from the University of Geneva, and Herschel K. Mitchell of the California Institute of Technology demonstrated that the heat-induced chromosomal puffing was accompanied by the high-level expression of a unique set of "heat shock" proteins. Those new chromosomal puffs represented sites in the DNA where specific messenger RNA molecules were made; these messenger RNAs carried the genetic information for synthesizing the individual heat-shock proteins.

By the end of the 1970s evidence was accumulating that the heat-shock response was a general property of all cells. Following a sudden increase in temperature, bacteria, yeast, plants and animal cells grown in culture all increased their expression of proteins that were similar in size to the *Drosophila* heat-shock proteins. Moreover, investigators were finding that cells produced one or more heat-shock proteins whenever they were exposed to heavy metals, alcohols and various other metabolic poisons.

Because so many different toxic stimuli brought on similar changes in gene expression, researchers started referring to the heat-shock response more generally as the stress response and to the accompanying products as stress proteins. They began to suspect that this universal response to adverse changes in the environment represented a basic cellular defense mechanism. The stress proteins, which seemed to be expressed only in times of trouble, were presumably part of that response.

Mounting evidence during the next few years confirmed that stress proteins did play an active role in cellular defense. Researchers were able to identify and isolate the genes that encoded the individual stress proteins. Mutations in those genes produced interesting cellular abnormalities. For example, bacteria carrying mutations in the genes encoding several of the stress proteins exhibited defects in DNA and RNA synthesis, lost their ability to undergo normal cell division and appeared unable to degrade proteins properly. Such mutants were also incapable of growth at high temperatures.

Cell biologists soon discovered that, as in bacteria, the stress response played an important role in the ability of animal cells to withstand brief exposures to high temperatures. Animal cells given a mild heat shock—one sufficient to increase the levels of the stress proteins—were better protected against a second heat treatment that would otherwise have been lethal. Moreover, those thermotolerant cells were also less susceptible to other toxic agents. Investigators became convinced that the stress response somehow protected cells against varied environmental insults.

As scientists continued to isolate and characterize the genes encoding the stress proteins from different organisms, two unexpected results emerged. First, many of the genes that encoded the stress proteins were remarkably similar in all organisms. Elizabeth A. Craig and her colleagues at the University of Wisconsin reported that the genes for heat-shock protein (hsp) 70, the most highly induced stress protein, were more than 50 percent identical in bacteria, yeast and *Drosophila*. Apparently, the stress proteins had been conserved throughout evolution and likely served a similar and important function in all organisms.

The second unexpected finding was that many stress proteins were also expressed in normal and unstressed cells, not only in traumatized ones. Consequently, researchers subdivided the stress proteins into two groups: those constitutively expressed under normal growth conditions and those induced only in cells experiencing stress.

Investigators were still perplexed as to how so many seemingly different toxic stimuli always led to the increased expression of the same group of proteins. In 1980 Lawrence E. Hightower, working at the University of Connecticut, provided a possible answer. He noticed that many of the agents that induced the stress response were protein denaturants—that is, they caused proteins to lose their shapes. A protein consists of long chains of amino acids folded into a precise conformation. Any disturbance of the folded conformation can lead to the protein's loss of biological function.

Hightower therefore suggested that the accumulation of denatured or abnormally folded proteins in a cell initiated a stress response. The stress

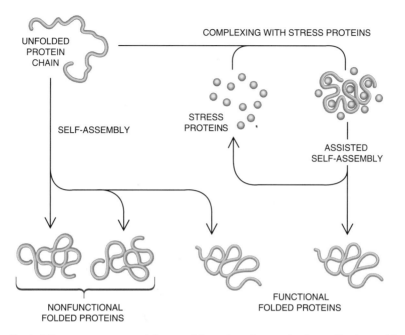

Protein folding occurs spontaneously because of thermodynamic constraints imposed by the protein's sequence of hydrophilic and hydrophobic amino acids. Although proteins can fold themselves into biologically functional configurations (self-assembly), errors in folding can occasionally occur. Stress proteins seem to help ensure that cellular proteins fold themselves rapidly and with high fidelity.

proteins, he reasoned, might somehow facilitate the identification and removal of denatured proteins from the traumatized cell. Within a few years Richard Voellmy of the University of Miami and Alfred L. Goldberg of Harvard University tested and confirmed Hightower's proposal. In a landmark study, they showed that injecting denatured proteins into living cells was sufficient to induce a stress response.

Thereafter, several laboratories set out to purify and characterize the biochemical properties of the stress proteins. The most highly inducible heat-shock protein, hsp 70, was the focus of much of this work. Using molecular probes, researchers learned that after a heat shock, much hsp 70 accumulated inside a nuclear structure called the nucleolus. The nucleolus manufactures ribosomes, the organelles that synthesize proteins. That location for hsp 70 was intriguing: previous work had demonstrated that after heat shock, cells stopped making ribosomes. Indeed, their nucleolus became awash in denatured ribosomal particles. Hugh R. B. Pelham of the Medical Research Council's Laboratory of Molecular Biology in Cambridge,

England, therefore suggested that hsp 70 might somehow recognize denatured intracellular proteins and restore them to their correctly folded, biologically active shape.

In 1986 Pelham and his colleague Sean Munro succeeded in isolating several genes, all of which encoded proteins related to hsp 70. They noticed that one form of hsp 70 was identical to immunoglobulin binding protein (BiP). Other researchers had shown that BiP was involved in the preparation of immunoglobulins, or antibodies, as well as other proteins for secretion. BiP bound to newly synthesized proteins as they were being folded or assembled into their mature form. If the proteins failed to fold or assemble properly, they remained bound to BiP and were eventually degraded. In addition, under conditions in which abnormally folded proteins accumulated, the cell synthesized more BiP.

Taken together, those observations indicated that BiP helped to orchestrate the early events associated with protein secretion. BiP seemed to act as a molecular overseer of quality control, allowing properly folded proteins to enter the secretory pathway but holding back those unable to fold correctly.

As more genes encoding proteins similar to hsp 70 and BiP came to light, it became evident that there was an entire family of hsp 70-related proteins. All of them shared certain properties, including an avid affinity for adenosine triphosphate (ATP), the molecule that serves as the universal, intracellular fuel. With only one exception, all these related proteins were present in cells growing under normal conditions (they were constitutive), yet in cells experiencing metabolic stress, they were synthesized at much higher levels. Moreover, all of them mediated the maturation of other cellular proteins, much as BiP did. For example, the cytoplasmic forms of hsp 70 interacted with many other proteins that were being synthesized by ribosomes.

In healthy or unstressed cells the interaction of the hsp 70 family member with immature proteins was transient and ATP-dependent. Under conditions of metabolic stress, however, in which newly synthesized proteins experienced problems maturing normally, the proteins remained stably bound to an hsp 70 escort.

The idea that members of the hsp 70 family participated in the early steps of protein maturation paralleled the results emerging from studies of a different family of stress proteins. Pioneering work by Costa Georgopoulos of the University of Utah and others had shown that mutations in the genes for two related stress proteins, groEL and groES, render bacteria unable to support the growth of small viruses that depend on the cellular machinery provided by their hosts. In the absence of functional groEL or groES, many viral proteins fail to assemble properly.

Proteins similar to the bacterial groEL and groES stress proteins were eventually found in plant, yeast and animal cells. Those proteins, which are known as hsp 10 and hsp 60, have been seen only in mitochondria and chloroplasts. Recent evidence suggests that more forms probably appear in other intracellular compartments.

Biochemical studies have provided compelling evidence that hsp 10 and hsp 60 are essential to protein folding and assembly. The hsp 60 molecule consists of two seven-membered rings stacked one atop the other. This large structure appears to serve as a "work-bench" onto which unfolded proteins bind and acquire their final three-dimensional structure. According to current thought, the folding process is extremely dynamic and involves a series of binding and release events. Each event requires energy, which is provided by the enzymatic splitting of ATP, and the participation of the small hsp 10 molecules. Through multiple rounds of binding and release, the protein undergoes conformational changes that take it to a stable, properly folded state.

Investigators suspect that both the hsp 60 and the hsp 70 families work together to facilitate protein maturation. As a new polypeptide emerges from a ribosome, it is likely to become bound to a form of hsp 70 in the cytoplasm or inside an organelle. Such an interaction may prevent the growing polypeptide chain from folding prematurely. Once its synthesis is complete, the new polypeptide, still bound to its hsp 70 escort, would be transferred to a form of hsp 60, on which folding of the protein and its assembly with other protein components would commence.

These new observations regarding the properties of hsp 70 and hsp 60 have forced scientists to reconsider previous models of protein folding. Work done in the 1950s and 1960s had established that a denatured protein could spontaneously refold after the denaturing agent was removed. This work led to the concept of protein self-assembly, for which Christian B. Anfinsen received a Nobel Prize in Chemistry in 1972. According to that model, the process of folding was dictated solely by the sequence of amino acids in the polypeptide. Hydrophobic amino acids (those that are not water soluble) would position themselves inside the coiling molecule, while hydrophilic amino acids (those that are water soluble) would move to the surface of the protein to ensure their exposure to the aqueous cellular environment. Folding would thus be driven entirely by thermodynamic constraints.

The principle of self-assembly is still regarded as the primary force that drives proteins into their final conformation. Now, however, many

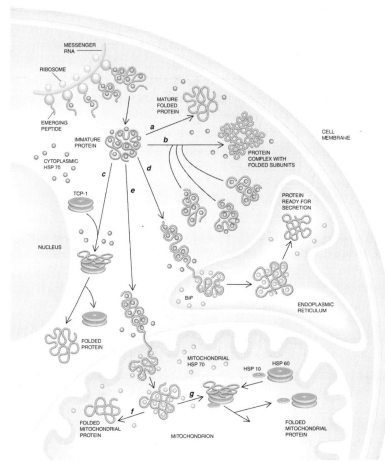

Several pathways for folding and distributing proteins inside cells are managed by stress proteins. In many cases, different stress proteins seem to work in tandem. The cytoplasmic form of hsp 70 binds to proteins being produced by the ribosomes to prevent their premature folding. The hsp 70 may dissociate from the protein and allow it to fold itself into its functional shape (a) or to associate with other proteins and thereby form larger, multimeric complexes (b). In some cases, proteins are passed from hsp 70 to another stress protein, TCP-1, before final folding and assembly occur (c). If the protein is destined for secretion, it may be carried to the endoplasmic reticulum and given to BiP or another related stress protein that directs its final folding (d). Other proteins are transferred to mitochondria or other organelles (e). Inside the mitochondrion, another specialized form of hsp 70 sometimes assists the protein in its final folding (f), but in many cases the protein is passed on to a complex of hsp 60 and hsp 10 (g). The hsp 60 molecule seems to serve as a "workbench" on which the mitochondrial protein folds.

investigators suspect that protein folding requires the activity of other cellular components, including the members of the hsp 60 and hsp 70 families of stress proteins.

Accordingly, R. John Ellis of the University of Warwick and other scientists have begun to refer to hsp 60, hsp 70 and other stress proteins as "molecular chaperones." Although the molecules do not convey information for the folding or assembly of proteins, they do ensure that those processes occur quickly and with high fidelity. They expedite self-assembly by reducing the possibility that a maturing protein will head down an inappropriate folding pathway.

Having established a role for some stress proteins as molecular chaperones in healthy and unstressed cells, investigators have turned their attention to determining why those proteins are expressed at higher levels in times of stress. One clue is the conditions that increase the expression of the stress proteins. Temperatures that are sufficient to activate the stress response may eventually denature some proteins inside cells. Heat-denatured proteins, like newly synthesized and unfolded proteins, would therefore represent targets to which hsp 70 and hsp 60 can bind. Over time, as more thermally denatured proteins become bound to hsp 60 and hsp 70, the levels of available molecular chaperones drop and begin to limit the ability of the cell to produce new proteins. The cell somehow senses this reduction and responds by increasing the synthesis of new stress proteins that serve as molecular chaperones.

Researchers suspect that a rise in the expression of stress proteins may also be a requirement for the ability of cells to recover from a metabolic insult. If heat or other metabolic insults irreversibly denature many cellular proteins, the cell will have to replace them. Raising the levels of those stress proteins that act as molecular chaperones will help facilitate the synthesis and assembly of new proteins. In addition, higher levels of stress proteins may prevent the thermal denaturation of other cellular proteins.

The repair and synthesis of proteins are vital jobs in themselves. Nevertheless, stress proteins also serve a pivotal role in the regulation of other systems of proteins and cellular responses. Another family of stress proteins, epitomized by one called hsp 90, is particularly noteworthy in this regard.

Initial interest in hsp 90 was fueled by reports of its association with some cancer-causing viruses. In the late 1970s and early 1980s cancer biologists were focusing considerable attention on the mechanism by which certain viruses infect cells and cause them to become malignant. In the

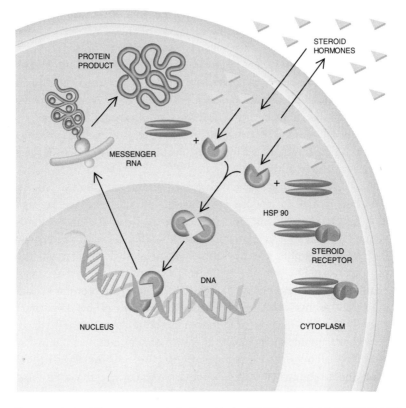

Responses to steroid hormones are controlled in part by hsp 90. This stress protein helps to maintain steroid receptors in their inactive form. When hormones are present, they bind to the receptor, and the hsp 90 is released. The activated receptor complex can then interact with DNA and initiate the expression of genes for certain proteins.

case of Rous sarcoma virus, investigators had pinpointed a viral gene that was responsible for the development of malignant properties. The enzyme it produced, pp6src, acted on other proteins that probably regulated cellular growth. Three laboratories independently reported that after its synthesis in the cytoplasm, pp60src rapidly associates with two proteins: one called p50 and the other hsp 90.

When pp60src is in the cytoplasm and is linked to its two escorts, it is enzymatically inactive. As the trio of molecules moves to the plasma membrane, the hsp 90 and the p50 fall away and allow the pp60src to deposit itself in the membrane and become active. Similar interactions between hsp 90, p50 and cancer-causing enzymes encoded by several other tumor viruses have been discovered. When bound to hsp 90 and p50, these viral enzymes seem incapable of acting on the cellular targets necessary for the development of the malignant state.

Some studies have also linked hsp 90 to another important class of molecules in mammalian cells, the steroid hormone receptors. Steroid hormones mediate several vital biological processes in animals. For example, the glucocorticoid steroids help to suppress inflammation. Other steroid hormones play important roles in sexual differentiation and development. When a steroid receptor binds to its specific hormone, the receptor becomes capable of interacting with DNA and either activating or repressing the expression of certain genes.

A crucial question concerned how steroid receptors were kept inactive inside a cell. The answer became clear following the characterization of both the active and inactive forms of the progesterone receptor. In the absence of hormone the receptor associates with several cellular proteins, among them hsp 90, which maintain it in an inactive state. After binding to progesterone, the receptor is released from the hsp 90 and experiences a series of events that allows it to bind with DNA. As with the viral enzymes, hsp 90 seems to regulate the biological activity of steroid hormone receptors.

Scientists are beginning to realize practical applications for the stress response. Medicine is one area that stands to benefit. When an individual suffers a heart attack or stroke, the delivery of blood to the heart or brain is temporarily compromised, a condition referred to as ischemia. While deprived of oxygen, the affected organ cannot maintain its normal levels of ATP, which causes essential metabolic processes to falter. When blood flow is restored, the ischemic organ is rapidly reoxygenated—yet that too can be harmful. Often the rapid reexposure to oxygen generates highly reactive molecular species, known as free radicals, that can do further damage.

In animal studies, researchers have observed the induction of stress responses in both the heart and brain after brief episodes of ischemia and reperfusion. The magnitude of the resulting stress response appears to correlate directly with the relative severity of the damage. Clinicians are therefore beginning to examine the utility of using changes in stress protein levels as markers for tissue and organ injury.

Cells that produce high levels of stress proteins appear better able to survive the ischemic damage than cells that do not. Consequently, raising the levels of stress proteins, perhaps by pharmacological means, may provide additional protection to injured tissues and organs. Such a therapeutic approach might reduce the tissue damage from ischemia incurred during surgery or help to safeguard isolated organs used for transplantation, which often suffer from ischemia and reperfusion injury.

One exciting development concerns the role of the stress response in immunology and infectious diseases. Tuberculosis, malaria, leprosy, schistosomiasis and other diseases that affect millions of people every year are a consequence of infection by bacteria or parasitic microorganisms. Immunologists have found that the stress proteins made by these organisms are often the major antigens, or protein targets, that the immune system uses to recognize and destroy the invaders. The human immune system may be constantly on the lookout for alien forms of stress proteins. The stress proteins of various pathogens, when produced in the laboratory by recombinant-DNA techniques, may therefore have potential as vaccines for preventing microbial infections. In addition, because they are so immunogenic, microbial stress proteins are being considered as adjuvants. Linked to viral proteins, they could enhance immune responses against viral infections.

Immunologists have also discovered a possible connection between stress proteins and autoimmune diseases. Most autoimmune diseases arise when the immune system turns against antigens in healthy tissues. In some of these diseases, including rheumatoid arthritis, ankylosing spondylitis and systemic lupus erythematosus, antibodies against the patient's own stress proteins are sometimes observed. If those observations are confirmed on a large number of patients, they may prove helpful in the diagnosis and perhaps the treatment of autoimmune disorders.

Because microbial stress proteins are so similar in structure to human stress proteins, the immune system may constantly be obliged to discern minor differences between the stress proteins of the body and those of invading microorganisms. The possibility that the stress proteins are uniquely positioned at the interface between tolerance to an infectious organism and autoimmunity is an intriguing idea that continues to spark debate among researchers.

The presence of antibodies against microbial stress proteins may prove useful in diagnostics. For example, the bacterium *Chlamydia trachomatis* causes a number of diseases, including trachoma, probably the world's leading cause of preventable blindness, and pelvic inflammatory disease, a major cause of infertility in women. Infection with chlamydia generally triggers the production of antibodies against chlamydial antigens, some of which are stress proteins. Often that immune response is effective and eventually eliminates the pathogen. Yet in some individuals, particularly those who have had repeated or chronic chlamydial infections, the immune response is overly aggressive and causes injury and scarring in the surrounding tissues.

Richard S. Stephens and his colleagues at the University of California at San Francisco have observed that more than 30 percent of women with pelvic inflammatory disease and more than 80 percent of women who have had ectopic pregnancies possess abnormally high levels of antibodies against the chlamydial groEL stress protein. Measurements of antibodies against chlamydial stress proteins may prove useful for identifying women at high risk for ectopic pregnancies or infertility.

The link between stress proteins, the immune response and autoimmune diseases becomes even more intriguing in light of other recent discoveries. Some members of the hsp 70 family of stress proteins are remarkably similar in structure and function to the histocompatibility antigens. The latter proteins participate in the very early stages of immune responses by presenting foreign antigens to cells of the immune system.

Researchers have wondered how any one histocompatibility protein could bind to a diverse array of different antigenic peptides. Recently Don C. Wiley and his colleagues at Harvard University helped to resolve that issue by determining the three-dimensional structure of the class I histocompatibility proteins. A pocket or groove on the class I molecule, they found, is able to bind to different antigenic peptides. Simultaneously, James E. Rothman, who was then at Princeton University, reported that members of the hsp 70 family of stress proteins were also capable of binding to short peptides. That property of hsp 70 is consistent with its role in binding to some parts of unfolded or newly made polypeptide chains.

Computer models revealed that hsp 70 probably has a peptide-binding site analogous to that of the class I histocompatibility proteins. The apparent resemblance between the two classes of proteins appears even more intriguing because several of the genes that encode hsp 70 are located very near the genes for the histocompatibility proteins. Taken together, all the observations continue to support the idea that stress proteins are integral components of the immune system.

The ability to manipulate the stress response may also prove important in developing new approaches to treating cancer. Tumors often appear to be more thermally sensitive than normal tissues. Elevating the temperature of tissues to eradicate tumors is one idea that is still at the experimental stage. Nevertheless, in early tests, the use of site-directed hyperthermia, alone or in conjunction with radiation or other conventional therapies, has brought about the regression of certain types of tumors.

The stress response is not necessarily the physician's ally in the treatment of cancer—it may also be one of the obstacles. Because stress proteins afford cells added protection, anticancer therapies that induce a stress response may make a tumor more resistant to subsequent treatments. Still, researchers may yet discover ways to inhibit the ability of a tumor to mount a stress response and thereby render it defenseless against a particular therapy.

Scientists are also beginning to explore the potential use of the stress response in toxicology. Changes in the levels of the stress proteins, particularly those produced only in traumatized cells, may prove useful for assessing the toxicity of drugs, cosmetics, food additives and other products. Such work is only at a preliminary stage of development, but several application strategies are already showing signs of success.

Employing recombinant-DNA technologies, researchers have constructed cultured lines of "stress reporter" cells that might be used to screen for biological hazards. In such cells the DNA sequences that control the activity of the stress protein genes are linked to a reporter gene that encodes an enzyme, such as β-galactosidase. When these cells experience metabolic stress and produce more stress proteins, they also make the reporter enzyme, which can be detected easily by various assays. The amount of β-galactosidase expressed in a cell can be measured by adding a chemical substrate. If the reporter enzyme is present, the cell turns blue, and the intensity of the color is directly proportional to the concentration of the enzyme in the cell.

Using such reporter cells, investigators can easily determine the extent of the stress response induced by chemical agents or treatments. If such assays prove reliable, they could ultimately reduce or even replace the use of animals in toxicology testing.

An extension of the technique could also be used to monitor the dangers of environmental pollutants, many of which evoke stress responses. Toward that end, scientists have begun developing transgenic stress reporter organisms. Eve G. Stringham and E. Peter M. Candido of the University of British Columbia, along with Stressgen Biotechnologies in Victoria, have created transgenic worms in which a reporter gene for β-galactosidase is under the control of the promoter for a heat-shock protein. When these transgenic worms are exposed to various pollutants, they express the reporter enzyme and turn blue. Candido's laboratory is currently determining whether those stress reporter worms might be useful for monitoring a wide variety of pollutants.

Voellmy and Nicole Bournias-Vardiabasis, then at City of Hope National Medical Center in Duarte, Calif., have used a similar approach to create a line of transgenic stress reporter fruit flies. The fruit flies turn blue when exposed to teratogens, agents that cause abnormal fetal development. Significantly, that bioassay is responsive to many of the teratogens that are known to cause birth defects in humans. The door appears open for the development of other stress reporter organisms that could prove useful in toxicological and environmental testing.

More than 30 years ago heat-shock and stress responses seemed like mere molecular curiosities in fruit flies. Today they are at the heart of an active and vital area of research. Studies of the structure and function of stress proteins have brought new insights into essential cellular processes, including the pathways of protein maturation. Scientists are also learning how to apply their understanding of the stress response to solve problems in the medical and environmental sciences. I suspect we have only begun to realize all the implications of this age-old response by which cells cope with stress.

FURTHER READING

M. Ashburner and J. J. Bonner. "The Induction of Gene Activity in Drosophila by Heat Shock" in *Cell*, Vol. 17, No. 2, pages 241–254; June 1979.

Richard I. Morimoto, Alfred Tissières and Costa Georgopoulos. *Stress Proteins in Biology and Medicine*. Cold Spring Harbor Laboratory Press, 1990.

R. John Ellis and S. M. Van der Vies. "Molecular Chaperones" in *Annual Reviews of Biochemistry*, Vol. 60, pages 321–347; 1991.

Thomas Langer, Chi Lu, Harrison Echols, John Flanagan, Manajit K. Hayer and F. Ulrich Hartl. "Successive Action of DNAK, DNAJ and GroEL along the Pathway of Chaperone-Mediated Protein Folding" in *Nature*, Vol. 356, No. 6371, pages 683–689; April 23, 1992.

William J. Welch. "Mammalian Stress Response: Cell Physiology, Structure/Function of Stress Proteins, and Implications for Medicine and Disease" in *Physiological Reviews*, Vol. 72, pages 1063–1081; October 1992.

Cell Communication: The Inside Story

JOHN D. SCOTT AND TONY PAWSON

ORIGINALLY PUBLISHED IN JUNE 2000

As anyone familiar with the party game "telephone" knows, when people try to pass a message from one individual to another in a line, they usually garble the words beyond recognition. It might seem surprising, then, that mere molecules inside our cells constantly enact their own version of telephone without distorting the relayed information in the least.

Actually, no one could survive without such precise signaling in cells. The body functions properly only because cells communicate with one another constantly. Pancreatic cells, for instance, release insulin to tell muscle cells to take up sugar from the blood for energy. Cells of the immune system instruct their cousins to attack invaders, and cells of the nervous system rapidly fire messages to and from the brain. Those messages elicit the right responses only because they are transmitted accurately far into a recipient cell and to the exact molecules able to carry out the directives.

But how do circuits within cells achieve this high-fidelity transmission? For a long time, biologists had only rudimentary explanations. In the past 15 years, though, they have made great progress in unlocking the code that cells use for their internal communications. The ongoing advances are suggesting radically new strategies for attacking diseases that are caused or exacerbated by faulty signaling in cells—among them cancer, diabetes and disorders of the immune system.

REFINING THE QUESTION

The earliest insights into information transfer in cells emerged in the late 1950s, when Edwin G. Krebs and Edmond H. Fischer of the University of Washington and the late Earl W. Sutherland, Jr., of Vanderbilt University identified the first known signal-relaying molecules in the cytoplasm (the material between the nucleus and a cell's outer membrane). All three received Nobel Prizes for their discoveries.

By the early 1980s researchers had gathered many details of how signal transmission occurs. For instance, it usually begins after a messenger responsible for carrying information between cells (often a hormone) docks temporarily, in lock-and-key fashion, with a specific receptor on a recipient cell. Such receptors, the functional equivalent of antennae, are able to relay a messenger's command into a cell because they are physically connected to the cytoplasm. The typical receptor is a protein, a folded chain of amino acids. It includes at least three domains: an external docking region for a hormone or other messenger, a component that spans the cell's outer membrane, and a "tail" that extends a distance into the cytoplasm. When a messenger binds to the external site, this linkage induces a change in the shape of the cytoplasmic tail, thereby facilitating the tail's interaction with one or more information-relaying molecules in the cytoplasm. These interactions in turn initiate cascades of further intracellular signaling.

Yet no one had a good explanation for how communiqués reached their destinations without being diverted along the way. At that time, cells were viewed as balloonlike bags filled with a soupy cytoplasm containing floating proteins and organelles (membrane-bound compartments, such as the nucleus and mitochondria). It was hard to see how, in such an unstructured milieu, any given internal messenger molecule could consistently and quickly find exactly the right tag team needed to convey a directive to the laborers deep within the cell that could execute the order.

ON THE IMPORTANCE OF LEGO BLOCKS

Today's fuller understanding grew in part from efforts to identify the first cytoplasmic proteins that are contacted by activated (messenger-bound) receptors in a large and important family: the receptor tyrosine kinases. These vital receptors transmit the commands of many hormones that regulate cellular replication, specialization or metabolism. They are so named because they are kinases—enzymes that add phosphate groups to ("phosphorylate") selected amino acids in a protein chain. And, as Tony R. Hunter of the Salk Institute for Biological Studies in La Jolla, Calif., demonstrated, they specifically put phosphates onto the amino acid tyrosine.

In the 1980s work by Joseph Schlessinger of New York University and others indicated that the binding of hormones to receptor tyrosine kinases at the cell surface causes the individual receptor molecules to

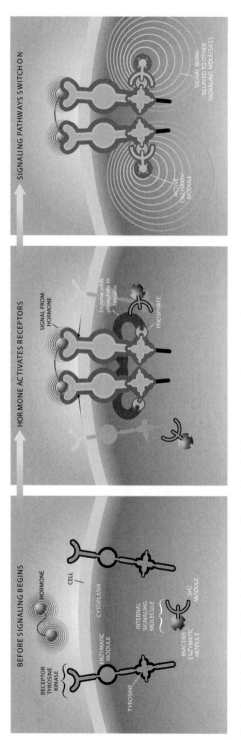

BEFORE SIGNALING BEGINS

HORMONE

CELL

CYTOPLASM

RECEPTOR
TYROSINE
KINASE

ENZYMATIC
MODULE

TYROSINE

INACTIVE
ENZYMATIC
MODULE

INTERNAL
SIGNALING
MOLECULE

SH2
MODULE

HORMONE ACTIVATES RECEPTORS

SIGNAL FROM
HORMONE

Enzyme adds
phosphate to
tyrosine

PHOSPHATE

SIGNALING PATHWAYS SWITCH ON

ACTIVE
ENZYMATIC
MODULE

SIGNAL BEING
RELAYED TO OTHER
SIGNALING MOLECULES

The molecules that form signaling circuits in cells are often modular—built from components that carry out distinct tasks. This discovery emerged in part from studies of molecules known as receptor tyrosine kinases (*pogo-stick shape in first panel*). When a hormone docks with those molecules at the surface of a cell (*second panel*), the receptors pair up and add phosphates to tyrosine, an amino acid, on each other's cytoplasmic tails. Then so-called SH2 modules in certain proteins hook onto the altered tyrosines (*last panel*). This linkage enables "talkative," enzymatic modules in the proteins to pick up the messenger's order and pass it along.

cluster into pairs and to attach phosphates to the tyrosines on each other's cytoplasmic tails. In trying to figure out what happens next, one of us (Pawson) and his colleagues found that the altered receptors interact directly with proteins that contain a module they called an SH2 domain. The term "domain" or "module" refers to a relatively short sequence of about 100 amino acids that adopts a defined three-dimensional structure within a protein.

At the time, prevailing wisdom held that messages were transmitted within cells primarily through enzymatic reactions, in which one molecule alters a second without tightly binding to it and without itself being altered. Surprisingly, though, the phosphorylated receptors did not necessarily alter the chemistry of the SH2-containing proteins. Instead many simply induced the SH2 domains to latch onto the phosphate-decorated tyrosines, as if the SH2 domains and the tyrosines were Lego blocks being snapped together.

By the mid-1990s groups led by Pawson, Hidesaburo Hanafusa of the Rockefeller University and others had revealed that many of the proteins involved in internal communications consist of strings of modules, some of which serve primarily to connect one protein to another. At times, whole proteins in signaling pathways contain nothing but linker modules.

But how did those nonenzymatic modules contribute to swift and specific communication in cells? One answer is that they help enzymatic domains transmit information efficiently. When a protein that bears a linker also includes an enzymatic module, attachment of the linker region to another protein can position the enzymatic module where it most needs to be. For example, the act of binding can simultaneously bring the enzymatic region close to factors that switch it on and into immediate contact with the enzyme's intended target. In the case of certain SH2-containing proteins, the linker module may originally be folded around the enzymatic domain in a way that blocks the enzyme's activity. When the SH2 domain unfurls to engage an activated receptor, the move liberates the enzyme to work on its target.

Even when a full protein is formed from nothing but protein-binding modules, it can function as an indispensable adapter, akin to a power strip plugged into a single socket. One module in the adapter plugs into a developing signaling complex, and the other modules allow still more proteins to join the network. An important benefit of these molecular adapters is that they enable cells to make use of enzymes that otherwise might not fit into a particular signaling circuit.

Nonenzymatic modules can support communication in other ways, too. Certain molecules in signaling pathways feature a protein-binding

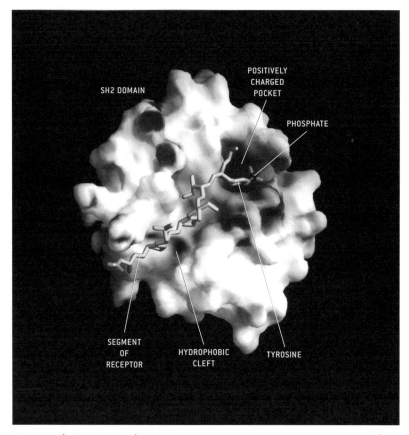

SH2 domain (*globular structure*) in a signaling molecule is bound to a segment of a receptor (*stick model*). The two fit together in part because a positively charged pocket in SH2 is attracted to a negatively charged phosphate that has been added to the amino acid tyrosine in the receptor. Also, the nearby amino acids in the receptor fit snugly into a hydrophobic (water-hating) groove on SH2. All SH2 domains can bind to phosphate-bearing tyrosines, but they differ in their binding partners because they vary in their ability to lock onto the amino acids that lie next to tyrosine in a protein.

module and a DNA-binding module that meshes with, or "recognizes," a specific sequence of DNA nucleotides in a gene. (Nucleotides are the building blocks of genes, which specify the amino acid sequences of proteins.) James E. Darnell, Jr., of the Rockefeller University showed that when one of these proteins attaches, through its linker module, to an activated receptor kinase, the interaction spurs the bound protein to detach, move to the nucleus and bind to a particular gene, thus inducing the synthesis of a protein. In this instance, the only enzyme in the signaling chain is the receptor itself; everything that happens after the receptor becomes activated occurs through proteins' recognition of other proteins or DNA.

As these various discoveries were being made, work in other areas demonstrated that the cytoplasm is not really amorphous after all. It is packed densely with organelles and proteins. Together such findings indicate that high-fidelity signaling within cells depends profoundly on the Lego-like interlocking of selected proteins through dedicated linker modules and adapter proteins. These complexes assure that enzymes or DNA-binding modules and their targets are brought together promptly and in the correct sequence as soon as a receptor at the cell surface is activated.

FAIL-SAFE FEATURES AID SPECIFICITY

Studies of receptor tyrosine kinases and of SH2 domains have also helped clarify how cells guarantee that only the right proteins combine to form any chosen signaling pathway. Soon after SH2 domains were identified, investigators realized that these modules are present in well over 100 separate proteins. What prevented different activated receptors from attracting the same SH2-containing proteins and thereby producing identical effects in cells? For the body to operate properly, it is crucial that diverse hormones and receptors produce distinct effects on cells. To achieve such specificity, receptors must engage somewhat different communication pathways.

The answer turns out to be quite simple. Every SH2 domain includes a region that fits snugly over a phosphate-bearing tyrosine (a phosphotyrosine). But each also includes a second region, which differs from one SH2 domain to another. That second region—as Lewis C. Cantley of Harvard University revealed—recognizes a particular sequence of three or so amino acids next to the phosphotyrosine. Hence, all SH2 domains can bind to phosphorylated tyrosine, but these modules vary in their preference for the adjacent amino acids in a receptor. The amino acids around the tyrosine thereby serve as a code to specify which version of the SH2 domain can attach to a given phosphotyrosine-bearing receptor. Because each SH2 domain is itself attached to a different enzymatic domain or linker module, this code also dictates which pathways will be activated downstream of the receptor. Other kinds of linker modules operate analogously.

A pathway activated by a protein called platelet-derived growth factor illustrates the principles we have described. This factor is often released after a blood vessel is injured. Its attachment to a unique receptor tyrosine kinase on a smooth muscle cell in the blood vessel wall causes such receptors to cluster and become phosphorylated on tyrosine. This change draws to the receptor a protein called Grb2, which consists of a specific SH2 domain flanked on either side by another linker domain,

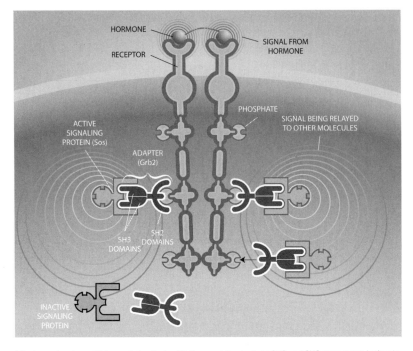

Adapter molecules, which consist entirely of linker modules such as SH2 and SH3, turn out to be important players in many signaling pathways. They enable cells to make use of proteins that would otherwise be unable to hook into a given communication circuit. Here, for instance, the adapter protein Grb2 draws an enzymatic protein—Sos—into a pathway headed by a receptor that itself has no means of interlocking with Sos.

SH3. Grb2 is a classic adapter; it has no enzymatic power at all, but its SH3 domains (which like to bind to the amino acid proline) hook an enzyme-containing protein called Sos to the receptor. There Sos activates a membrane-associated protein known as Ras, which triggers a series of enzymatic events. These reactions ultimately stimulate proteins in the nucleus to activate genes that cause the cells to divide, an action that promotes wound healing.

The signaling networks headed by receptor tyrosine kinases seem to rely on relatively small adapter proteins. Analyses of communication circuits in nerve cells (neurons) of the brain show that some proteins in neuronal pathways have an incredibly large number of linker domains. These proteins are often called scaffolding molecules, as they permanently hold groups of signaling proteins together in one place. The existence of such scaffolds means that certain signaling networks are hardwired into cells. That hardwiring can enhance the speed and accuracy of information transfer.

SCAFFOLDS ABOUND

One well-studied scaffolding protein goes by the name PSD-95. It operates primarily in neurons involved in learning. In nerve tissue, signals pass from one neuron to another at contact points called synapses. The first neuron releases a chemical messenger—a neurotransmitter—into a narrow cleft between the cells. Receptors on the second cell grab the neurotransmitter and then cause ion channels in the membrane to open. This influx of ions activates enzymes that are needed to propagate an electrical impulse. Once generated, the impulse travels down the axon, a long projection, to the axon's abundant tiny branches, inducing them to release more neurotransmitter. For the impulse to be produced, many components of the signaling system must jump into action virtually simultaneously.

Among the multiple linker modules in PSD-95 are three so-called PDZ domains. One binds to the cytoplasmic tail of the receptor for the neurotransmitter glutamate. A second grabs onto a membrane-spanning ion channel (which controls the inflow of potassium), and a third clasps proteins in the cytoplasm (as does an additional module in the scaffold). PSD-95 thus yokes together several signaling components at once, enabling them to coordinate their activities. The eye of a fruit fly also relies on a PDZ-containing scaffolding protein—InaD—for the efficient relay of visual information from the eye to the brain.

Yet another preformed signaling complex has been found only recently, in mammalian neurons. The core is a scaffolding protein named yotiao. As one of us (Scott) and his colleagues have shown, this molecule grasps a dual-purpose, membrane-spanning protein that is both a glutamate receptor and an ion channel. It also clasps a kinase that adds phosphate to, and thereby opens, the ion channel when the receptor is activated by glutamate. And it anchors a phosphatase, an enzyme that removes phosphates from proteins. The bound phosphatase closes the ion channel whenever glutamate is absent from the receptor. This elegant arrangement ensures that ions flow through the channel only when glutamate is docked with the receptor.

Kinases and phosphatases control most activities in cells. If one kinase activates a protein, some phosphatase will be charged with inactivating that protein, or vice versa. Yet human cells manufacture hundreds of different kinases and phosphatases. Scaffolding proteins, it appears, are a common strategy for preventing the wrong kinases and phosphatases from acting on a target; they facilitate the proper reactions by holding selected kinases and phosphatases near the precise proteins they are supposed to regulate.

MANY PAYOFFS

From an evolutionary perspective, the advent of a modular signaling system would be very useful to cells. By mixing and matching existing modules, a cell can generate many molecules and combinations of molecules and can build an array of interconnected pathways without having to invent a huge repertoire of building blocks. What is more, when a new module does arise, its combination with existing modules can increase versatility tremendously—just as adding a new area code to a city turns already assigned phone numbers into entirely new ones for added customers.

For cell biologists, merely chipping away at the mystery of how cells carry out their myriad tasks is often reward enough for their efforts. But the new findings have a significance far beyond intellectual satisfaction.

The much publicized Human Genome Project will soon reveal the nucleotide sequence of every gene in the human body. To translate that information into improved understanding of human diseases, those of us who study the functioning of cells will have to discern the biological roles of any newly discovered genes. That is, we will need to find out what the corresponding proteins do and what happens when they are overproduced, underproduced or made incorrectly.

We already know the amino acid sequences and the functions of many modules in signaling proteins. Hence, we have something of a key for determining whether the nucleotide sequence of a previously unknown gene codes for a signaling protein and, if it does, which molecules the protein interacts with. When we have enough of those interactions plotted, we may be able to draw a wiring diagram of every cell type in the body. Even with only a partial diagram, we may uncover ways to "rewire" cells when something goes wrong—halting aberrant signals or rerouting them to targets of our own choosing. We might, for instance, funnel proliferative commands in cancer cells into pathways that instruct the cells to kill themselves instead of dividing.

By learning the language that cells use to speak to one another and to their internal "workers," we will be able to listen in on their conversations and, ideally, find ways to intervene when the communications go awry and cause disease. We may yet reduce "body language" to a precise science.

FURTHER READING

Bruce Alberts et al. *Molecular Biology of the Cell.* Garland Books, 1994.

Tony Pawson. "Protein Modules and Signalling Networks" in *Nature,* Vol. 373, pages 573–580; February 16, 1995.

Tony Pawson and John D. Scott. "Signaling through Scaffold, Anchoring and Adaptor Proteins" in *Science,* Vol. 278, pages 2075–2080; December 19, 1997.

Tony Hunter. "Signaling: 2000 and Beyond" in *Cell,* Vol. 100, No. 1, pages 113–127; January 7, 2000.

The European Molecular Biology Laboratory SMART database Web site is at http://smart .embl-heidelberg.de/

The Howard Hughes Medical Institute News Web site is at www.hhmi.org/news/scott.htm

The Oregon Health Sciences Vollum Institute Web site is at www.ohsu.edu/vollum/faculty/ scott/index.htm

The Samuel Lunenfeld Research Institute at the Mount Sinai Hospital, Toronto, Web site is at www.mshri.on.ca/pawson/research.html

Life, Death and the Immune System

SIR GUSTAV J. V. NOSSAL

ORIGINALLY PUBLISHED IN SEPTEMBER 1993

What did Franz Schubert, John Keats and Elizabeth Barrett Browning have in common? Each was a creative genius, but each also had his or her life tragically shortened by a communicable disease that today could have been prevented or cured. Progress in the treatment of such diseases undoubtedly ranks as one of the greatest achievements of modern science. Smallpox has been completely eradicated, and poliomyelitis and measles may be problems of the past by the end of the century. So great has been the headway against infectious diseases that until the current AIDS pandemic, industrialized countries had placed them on the back burner among major national concerns.

Such staggering improvements in public health alone would justify tremendous efforts to understand the human immune system. Yet the field of immunology embraces more than just the nature and prevention of infections. Immunologic research is pointing toward new approaches for treating cancer and diseases that result from lapses or malfunctions in the immune response. This work also provides a scientific framework for examining the chemical organization of living systems and integrating that information into an understanding of how the organism functions as a whole.

I am a little ashamed to admit that I did not immediately recognize the underlying importance of immunology. As a medical student in the 1950s, I became interested in viruses, hoping that the analysis of their growth might reveal the most profound details of the life process. I aspired to study under Sir Frank Macfarlane Burnet, the prominent Australian virologist, at the Walter and Eliza Hall Institute of Medical Research in Melbourne.

After my graduation and hospital training, I was lucky enough to be accepted. Burnet wrote, however, that he had become interested less in viruses than in exploring the human immune system. I was utterly dismayed. To my thinking, the early giants—Louis Pasteur, Paul Ehrlich and Emil A. von Behring—had already discovered the fundamental truths about immunity. Public health, the major application of immunology research, seemed the dullest of the subjects in the medical curriculum.

Since then I have learned how wrong I was. Just as I began my graduate work, a series of immune-related discoveries began ushering in an extraordinary chapter in the history of biomedicine. Researchers observed that the white blood cells called lymphocytes, which destroy pathogenic microbes that enter the body, can attack cancer cells and hold them in check, at least temporarily. Other experiments showed that those same lymphocytes can also behave in less desirable ways. For example, they can act against the foreign cells in transplanted organs and cause graft rejection. If the regulation of the immune system breaks down, lymphocytes can attack cells belonging to the very body that they should be protecting, leading to a potentially fatal autoimmune disease.

All these findings intensified interest in one of the most central and baffling mysteries of the immune system: how it is able to recognize the seemingly infinite number of viruses, bacteria and other foreign elements that threaten the health of the organism. In most biochemical interactions, such as the binding of a hormone to a receptor or the adhesion of a virus to its host cell, eons of evolution have refined the chemistry involved so that each molecule unites with its partner in a precise, predetermined way. The immune system, in contrast, cannot anticipate what foreign molecule it will confront next.

One of the crucial elements that helps the immune system meet that challenge is antibody, a large protein molecule discovered in 1890 by von Behring and Shibasaburo Kitasato. Antibodies latch onto and neutralize foreign invaders such as bacteria and viruses; they also coat microbes in a way that makes them palatable to scavenger cells, such as macrophages. Each type of antibody acts on only a very specific target molecule, known as an antigen. Consequently, antibodies that attack anthrax bacilli have no effect against typhoid. For decades, biologists thought of the antigen as a kind of template around which the antibody molecule molded itself to assume a complementary form. This theory, first clearly articulated by Felix Haurowitz in the 1930s and later espoused by Linus Pauling, held sway until about 1960.

By the mid-1960s the template model was in trouble. Gordon L. Ada of the Hall Institute and I demonstrated that antibody-making cells did not contain any antigen around which to shape an antibody. Studies of enzymes showed that the structure of a protein depends only on the particular sequence of its amino acid subunits. Furthermore, Francis Crick deduced that, in biological systems, information flows from DNA to RNA to protein. For this reason, antigen proteins could not define new antibody proteins: the information for the antibody structures had to be encoded in the genes. Those findings raised a puzzling question: If genes dictate the manufacture of antibodies, how can there be specific genes for each of the millions of different antibodies that the body can fabricate?

In 1955 Niels K. Jerne, then at the California Institute of Technology, had already hit on a possible explanation for the incredible diversity of antibodies. He suggested that the immune response is selective rather than instructive—that is, mammals have an inherent capacity to synthesize billions of different antibodies and that the arrival of an antigen only accelerates the formation of the antibody that makes the best fit.

Two years later Burnet and David W. Talmage of the University of Colorado independently hypothesized that antibodies sit on the surface of lymphocytes and that each lymphocyte bears only one kind of antibody. When a foreign antigen enters the body, it eventually encounters a lymphocyte having a matching receptor and chemically stimulates it to divide and to mass-produce the relevant antibody. In 1958 Joshua Lederberg, then visiting the Hall Institute, and I demonstrated that when an animal is immunized with two different antigens, any given cell does in fact make just one type of antibody.

Soon thereafter Gerald M. Edelman of the Rockefeller University and Rodney R. Porter of the University of Oxford discovered that antibodies are composed of four small proteins called chains. Each antibody possesses two identical heavy chains and two identical light chains. An intertwining light chain and heavy chain form an active site capable of recognizing an antigen, so each antibody molecule has two identical recognition sites. Knowing that two chains contribute to the binding site helps to explain the great diversity of antibodies because of the large number of possible pair combinations.

A set of experiments initiated by Susumu Tonegawa of the Basel Institute for Immunology led to the definitive description of how the immune system can produce so many different antibody types. He found that, unlike nearly all other genes in the body, those that contain the code for the heavy chains do not preexist in the fertilized egg. Instead the code resides in four sets of mini-genes located in widely separated parts of the nucleus. Antibody diversity springs from the size of these mini-gene families: there are more than 100 kinds of V (variable) genes, 12 D (diversity) genes and four J (joining) genes. The C, or constant, genes vary in ways that affect only the function of the antibody, not its antigen affinity.

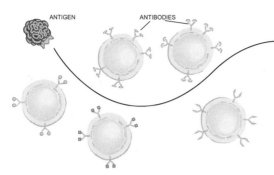

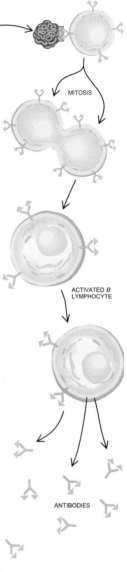

Clonal selection enables the immune system to react to a myriad of possible pathogens. Lymphocytes having any one of millions of different surface antibodies constantly roam the body. When the antigen on the surface of a foreign entity meets a lymphocyte having a matching antibody (*top*), the lymphocyte swells and begins to divide rapidly (*right*). Once they reach maturity, *B* cells secrete antibodies that attack the invader (*bottom*); *T* cells generate lymphokines, chemicals that boost the activity of other cells in the immune system.

During the development of an antibody-forming cell, one member from each set of mini-genes jumps out of its original position and links with the other jumpers to form a complete V-D-J-C gene. This genetic rearrangement allows for 4,800 different varieties ($100 \times 12 \times 4 \times 1$) of heavy chains. The same process occurs in the assembly of the light-chain genes, except that they have only V, J and C segments, so there are about 400 basic combinations for them. The diversity of heavy and light chains allows for the existence of $4,800 \times 400$, or 1,920,000, antibody genes. Moreover, special enzymes can insert a few extra DNA coding units at the junctions between the V and D or D and J segments when they interlink, which further increases the number of possible antibody constructions.

Despite their enormous versatility, antibodies alone cannot provide full protection from infectious attack. Some diseases, such as tuberculosis, slip inside their host cells so quickly that they can hide from antibody molecules. In these cases, a second form of immune response comes into play. When the infected cells become inflamed, lymphocytes attack them so as to confine the infection. This defense mechanism is known as cell-mediated immunity, in contrast with the so-called humoral immunity mediated by antibodies.

In the early 1960s Jacques F.A.P. Miller, then at the Chester Beatty Research Institute in London, and Noel L. Warner and Aleksander Szenberg of the Hall Institute determined that lymphocytes fall into two different classes, each of which controls one of the two types of immune response. Cell-mediated immunity involves a type of lymphocyte that originates in the thymus and is thus called a *T* cell. Humoral immunity occurs through the action of antibodies, which are produced by the lymphocytes known as *B* cells that form in the bone marrow.

   T cells and *B* cells differ not only in their function but also in the way they locate a foreign invader. As Talmage and Burnet hypothesized, *B* cells can recognize antigens because they carry antibodies on their surface. Each *T* cell also has a unique receptor, but unlike *B* cells, *T* cells cannot "see" the entire antigen. Instead the receptors on *T* cells recognize protein fragments of antigens, or peptides, linear sequences of eight to 15 amino acids. *T* cells spot foreign peptide sequences on the surface of body cells, including bits of virus, mutated molecules in cancer cells or even sections of the inner part of a microbe. A molecule known as a major histocompatibility complex (MHC) protein brings the peptide to the cell surface, where the *T* cell can bind to it.

   T cells and antibodies make perfect partners. Antibodies respond swiftly to toxin molecules and to the outer surfaces of microbes; *T* cells discover the antigens of hidden inner pathogens, which makes them particularly effective at tracking down infectious agents. For instance, a virus might be able, through mutation, to change its outer envelope rapidly and in this way frustrate neutralization by antibodies. That same virus might contain within its core several proteins that are so essential to its life process that mutations are not permitted. When that virus replicates inside cells, short peptide chains from those viral proteins break off and travel to the cell surface. They serve as ripe targets for the *T* cell, which can then attack the infected cell and inhibit the spread of the virus.

   So far I have described *T* and *B* lymphocytes as though they operate independently, but in actuality they form a tightly interwoven system. *T* cells make close contact with *B* cells, stimulate them into an active state and secrete lymphokines, molecules that promote antibody formation.

T cells also can suppress antibody formation by releasing inhibitory lymphokines.

B cells, in turn, process antigens into the form to which *T* cells most readily respond, attach the antigens to MHC molecules and display them on the cell surface. In this way, *B* cells help to stimulate *T* cells into an active state. Researchers have observed that *B* cells can also inhibit *T* cell responses under experimental conditions. Such highly regulated positive and negative feedback loops are a hallmark of the organization of the immune system.

The specialization of the immune system does not end with its division into *B* and *T* cells. *T* cells themselves comprise two subpopulations, CD4 (helper) and CD8 (killer) *T* cells. CD4 cells recognize peptides from proteins that have been taken up by macrophages and other specialized antigen-capturing cells. CD8 cells react to samples of peptides originating within a cell itself, such as a segment of a virus in an infected cell or mutant proteins in a cancer cell. Each variety of *T* cell utilizes its own form of MHC to make the peptides noticeable.

When CD4 *T* cells encounter the proper chemical signal, they produce large amounts of lymphokines to accelerate the division of other *T* cells and to promote inflammation. Some CD4 cells specialize in helping *B* cells, others in causing inflammation. Activated CD8 cells produce much smaller amounts of lymphokines but develop the capacity to punch holes into target cells and to secrete chemicals that kill infected cells, limiting the spread of a virus. Because of their murderous nature, CD8 *T* cells are also referred to as cytotoxic *T* cells.

B cells undergo an especially stunning transformation once activated. Before it meets antigen, the *B* cell is a small cell having a compact nucleus and very little cytoplasm—a head office without much happening on the factory floor. When the cell springs into action, it divides repeatedly and builds up thousands of assembly points in its cytoplasm for the manufacture of antibodies, as well as an extensive channeling system for packaging and exporting the antibodies. One *B* cell can pump out more than 10 million antibody molecules an hour.

My co-workers and I routinely cultivate a single *B* cell to grow a "clone" comprising hundreds of daughter cells. After one week, those clones can generate 100 billion identical antibody molecules to study. Such clonal cultures have enabled us to witness another of the *B* cell's remarkable talents. *B* cells can switch from making one isotype, or functional variety, of antibody to another without changing the antigen to which the antibody binds. Each isotype of an antibody derives from a different form of the C mini-gene.

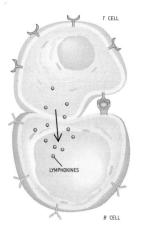

T CELL

LYMPHOKINES

B CELL

B lymphocyte in its resting state is little more than a nucleus surrounded by a thin enclosure of cytoplasm (*left*). Once a *B* cell meets a matching antigen, it develops an extended body (*center*) containing polyribosomes, which make antibodies, and an elaborate channel system for exporting those antibodies. *T* lymphocytes can regulate the behavior of *B* cells by administering lymphokines through an intimate junction somewhat like a nerve synapse (*right*). During these interactions, the *B* cell can also influence the activity of the *T* cell.

Each antibody isotype has its own peculiar advantage. One isotype serves as a first line of defense; another specializes in neutralizing toxins; a third suffuses mucus and so helps to create a barrier against infectious agents attempting to enter through the nose, throat or intestines. In response to lymphokines from *T* cells, *B* cells can switch from one isotype of antibody to another within a day or so.

Both *B* and *T* lymphocytes get a helping hand from various other cells and molecules. When antibodies attach to a bacterium, they can activate complement, a class of enzymes that kill bacteria by destroying their outer membranes. Some lymphokines send out a chemical call to macrophages, granulocytes and other white blood cells that clean up the mess at an infected site by gobbling up germs and dead cells. Such tidiness is enormously important: a patient having no granulocytes faces grave risk of death from the infectious bacteria that feed on cellular corpses. Clearly, all the white blood cells work together as a well-orchestrated team.

Amid all the complex operations of the immune defenses, it is utterly crucial that lymphocytes remain consistently benign toward the body's own cells, commonly referred to as self, while reacting aggressively to those that it recognizes as foreign, or nonself. Burnet postulated that self-recognition is not genetically determined but rather is learned by the immune system during the organism's embryonic stage. He suggested that a foreign antigen introduced into an embryo before the immune system had developed would trick the lymphocytes into regarding the foreign molecule as self. Burnet's attempts to prove his theory by injecting an influenza vaccine into chick embryos did not elicit the expected null response, however.

In 1953 Rupert E. Billingham, Leslie Brent and Sir Peter B. Medawar, working at University College, London, succeeded where Burnet had failed. The three men were exploring ways to transplant skin from one individual to another—in order, for instance, to treat a burn victim. Medawar had previously discovered that the body rejected such skin grafts because of an immune response. When he came across Burnet's theoretical writings, Medawar and his colleagues set about injecting in-bred mouse embryos with spleen-derived cells from a different mouse strain. Some embryos died as a result of this insult, but those that survived to adulthood accepted skin grafts from the donor strain. A patch of black fur growing on a white mouse dramatically showcased the discovery of actively acquired immunologic tolerance; for the first time, lymphocytes were fooled into recognizing nonself as self. Burnet and Medawar shared a Nobel Prize for their work.

Subsequent research clarified why Burnet's experiment had gone awry. Medawar's group used living cells as an antigen source—specifically, cells that could move into critical locations such as the thymus and the bone marrow. As long as those donor cells lived, they continued to make antigens that influenced the emerging lymphocytes. Burnet's influenza vaccine, on the other hand, had been rapidly consumed and broken down by scavenger cells; not enough antigen reached the immune system to induce a significant degree of tolerance.

The realization that immune response depends heavily on the vast diversity of antibodies on the body's innumerable B cells suggested the mechanism by which lymphocytes learn to ignore cells of the self. An immune reaction represents the activation of specific lymphocytes selected from the body's varied repertoire. It seemed logical that tolerance of self could be seen as the mirror image of immunity: the systematic deletion of those cells that respond to self-antigen.

Genetic influences and environmental triggers can cause the usual immunologic rules to break down. In those instances, B cells or T cells, or both, may respond to self-antigens, attacking the body's own cells and leading to a devastating autoimmune disease. Some such disorders result from misdirected antibodies: in hemolytic anemia, antibodies attack red blood cells, and in myasthenia gravis, antibodies turn on a vital protein on muscle cells that receives signals from nerves. T cells play the villain's role in other autoimmune diseases: in insulin-dependent diabetes, T lymphocytes destroy insulin-producing cells in the pancreas, and in multiple sclerosis, they direct their fury against the insulation surrounding nerve fibers in the brain and spinal cord.

Treating autoimmune diseases necessitates abolishing or at least restraining the immune system. Immunosuppressive and anti-inflammatory drugs can achieve the desired effect, but such a blunderbuss approach suppresses not only the bad, antiself response but also all the desirable immune reactions. Fortunately, researchers are making some progress toward the ideal goal of reestablishing specific immunologic tolerance to the beleaguered self-antigen.

One kind of therapy involves feeding the patient large quantities of the attacked self-antigen; surprisingly enough, such an approach can selectively restrain future responses to that antigen. Researchers have achieved similar results by administering antigens intravenously while the T cells are temporarily blindfolded by monoclonal antibodies that block their antigen receptors. Some treatments for autoimmune diseases based on these approaches have reached the stage of clinical trials.

Successful organ transplantation also requires shutting down an undesired aspect of immune response. In principle, the surgeon can begin supplying immunosuppressive drugs at the time of surgery, preempting a lymphocyte attack. Most organ transplants provoke such a strong T cell response that the doses of drugs needed to prevent organ rejection are even higher than those used to treat autoimmune diseases. Fortunately, those dosages can be reduced after a few months. Newer, more powerful immunosuppressive drugs are leading to good success rates for transplants of the kidney, heart, liver, bone marrow, heart-lung and pancreas; recently a few small-bowel transplants have taken. Researchers are also striving to develop targeted drugs that dampen the organ rejection response while still allowing the body to react to infectious diseases.

Transplantation has become so successful that doctors often confront a shortage of organs from recently deceased donors. Workers therefore are renewing their efforts to perform xenotransplantation, the transplantation of organs from animal donors. Tissue from endocrine glands can be cultured so that it loses some of its antigenic punch, raising the possibility that insulin-secreting cells from pigs will one day be grafted into diabetics. Chemical treatments may be able to "humanize" crucial molecules in animal organs so as to ameliorate the ferocity of immune rejection. Nevertheless, xenotransplantation faces formidable technical and ethical obstacles.

Immunologic attacks on tissues in the body need not be horrific; they could actually be beneficial if directed against cancers. Indeed, one controversial theory—the immune surveillance theory, first articulated by Lewis Thomas when he was at New York University—holds that eliminating precancerous cells is one of the prime duties of the constantly patrolling lymphocytes.

People whose immune system has been suppressed by drugs—mostly recipients of organ transplants—do in fact experience a higher incidence of leukemias, lymphomas and skin cancers fairly soon after transplantation than do similar individuals in the general population. After three decades of observing kidney transplant patients, physicians find that those individuals also experience a somewhat elevated susceptibility to many common cancers, such as those of the lung, breast, colon, uterus and prostate. These findings hint that immune surveillance may act to hold at least certain cancers in check. Alternatively, drug-associated cancers may be the result of some mechanism other than immunosuppression.

Further evidence of the immune system's role in preventing cancer comes from studies of mouse cancers induced by viruses or by chemical carcinogens. Those cancers often provoke strong immune responses when transplanted into genetically identical mice, which proves that the cancerous cells bear antigens that mark them as abnormal. Spontaneously arising cancers in mice, which are likely to be more akin to human cancer, provoke little or no immune response, however.

Yet even spontaneous cancers may carry some tumor-specific antigens that could arouse a reaction from the immune system if other chemical signals are present. One highly potent trigger molecule is known as B7. When inserted into the cells of a tumor, B7 can convert deadly, uncontrollable cancer cells into ones that T cells attack and destroy. B7 is not itself an antigen, but it evidently helps antigenic molecules in the tumor cell to activate T cells.

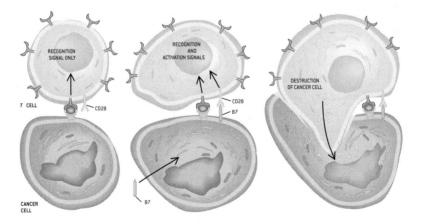

Cancer cells can elude attack by lymphocytes even if they bear distinctive antigens. That absence of immune response may occur because cancerous cells lack the proper costimulatory molecules (*left*). Researchers are attempting to induce the body to fight tumors by inserting the molecule B7 into cancer cells (*center*). When B7 engages CD28, a complementary molecule on the surface of T cells, it generates a signal that instigates an assault on the cancer cells (*right*).

The discovery of immunostimulating molecules such as B7 has renewed interest in the possibility of developing anticancer vaccines. Such treatments might be effective against malignant melanoma, the cancer arising from pigmented moles. These cancers contain a family of proteins collectively called MAGE, which has been extensively studied by Thierry Boon at the Ludwig Institute for Cancer Research in Brussels. In laboratory experiments, a peptide derived from MAGE can provoke a strong attack from cytotoxic T cells. If researchers could learn how to manipulate the antigen properly—perhaps by injecting a patient with MAGE or its constituent peptides, along with molecules designed to strengthen immunity—they might be able to create an effective therapy for melanoma.

Another way to fight cancers involves boosting the immune response to aberrant forms of a class of proteins known as mucins. Normal mucins consist of a protein core almost completely enveloped by a shell of sugar molecules. Many cancerous cells, most notably those associated with tumors of the gastrointestinal tract, lung or ovary, contain altered mucins whose cores are exposed. Workers have identified peptides from the core proteins in mucins to which T cells strongly respond. Vaccines constructed from those peptides may be able to induce cytotoxic T cells to attack the naked core proteins and thereby kill the cancerous cells.

Devising cancer vaccines presents a difficult challenge. Tumor cells have a great capacity to mutate, which allows them to avoid destruction by discarding or changing their distinctive antigens. Killing every single tumor cell, as must be done to cure cancer, will not be easy in advanced cases of cancer. And yet experimental vaccines have shown tantalizing signs of success. In tests on patients who had several forms of widespread cancer, such as melanoma, kidney cancer and certain forms of leukemia, roughly one fifth of them experienced a dramatic regression of their tumors in response to these vaccines. Little is known about why those people responded and others did not.

Many workers believe cancer vaccines will come into their own as weapons against the few mutant cells that persist in the body after cancer surgery, chemotherapy or radiation therapy. These surviving cells can cause a recurrence of the cancer even after an apparently successful primary therapy. In principle, killing the few million cancer cells that remain after a primary treatment should be easier than eliminating the hundreds of billions that exist beforehand.

Despite the promise of such innovative techniques, new and improved vaccines against infectious disease continue to be the most urgent and immediate application of immunologic research. In this arena, the World

Health Organization's Expanded Program on Immunization (EPI) has stood out as a laudable triumph amid the generally troubled global public health scene. With wonderful help from UNICEF, the World Bank, Rotary International and the developing countries' health authorities, EPI provides protection against six major diseases—diphtheria, whooping cough, tetanus, poliomyelitis, measles and tuberculosis—to over 80 percent of the more than 100 million children born every year in the Third World.

Last year EPI added hepatitis B vaccine to its list, although cost considerations have limited the number of doses available. In many Asian and African countries, 5 to 10 percent of the population become chronic carriers of the hepatitis B virus; a significant proportion of these acquire severe liver disease and finally liver cancer. An infant who receives the vaccine at birth does not become a carrier and is protected from the virus. Mass vaccination against hepatitis B is worthwhile even in Western countries, not only because of the risk faced by homosexual men but also because many of these countries now include significant Asian or African-derived populations.

Encouraging though the trends are, an enormous amount remains to be done in the realm of immunization. Effective vaccines against several forms of meningitis are not yet in widespread use. The available vaccines against typhoid, cholera, tuberculosis and influenza are only partially effective. No generally available vaccine exists for many common diseases, such as pneumonia, diarrhea, malaria and cancers caused by human papillomavirus and glandular fever virus. Furthermore, rich and poor countries alike face the practical problems of delivering the vaccine to those who need it and making sure that it is used. The World Health Organization badly needs extra funds to sustain its marvelous thrust in research and deployment.

Devising a vaccine against AIDS is one of the most urgent and daunting tasks facing immunology researchers. There are now at least 10 million people around the world infected with the human immunodeficiency virus (HIV), which causes AIDS; most of these people live in developing countries. HIV manifests a dizzying capacity to mutate, and it can hide from the immune system inside lymphocytes and scavenger cells. Still, there are some encouraging signs that the virus can be defeated. HIV often lies dormant in humans for years, which suggests that immune processes hold the virus in check for long periods. Antibodies can neutralize HIV, and cytotoxic T cells can kill at least some of the virus-carrying cells. Vaccines have prevented AIDS-like infections in monkeys. It will take several years, however, to determine whether any of the present clinical trials hold real promise.

The AIDS crisis has so enhanced public awareness of immunology that when I attend social or business functions and reveal that I am an immunologist, people commonly respond, "Oh, then you must be working on AIDS!" They are often surprised to hear that immunology is a vast science that predates the identification of AIDS by many decades.

And yet the interdisciplinary nature of immunology has had, I believe, a significant salutary effect on all the biological sciences. When I was young, many researchers worried that as the specialties and subspecialties bloomed, scientists would discover more and more about less and less, so that the research enterprise would splinter into myriad fragments, each bright and shiny in its own right but having little connection to the others.

Rather a new, integrated biology has arisen, built on the foundation of molecular biology, protein chemistry and cell biology, encompassing fields as diverse as neurobiology, developmental biology, endocrinology, cancer research and cardiovascular physiology. A fundamental finding made within one discipline spreads like wildfire through the others.

Immunology sits at the center of the action. The cells of the immune system constitute ideal tools for basic biological research. They grow readily in the test tube, display a rich diversity of chemical receptors and manufacture molecules of great specificity and power; consequently, the lymphocyte is perhaps the best understood of all living cells. Moreover, immunology embraces many interlinked molecular and cellular systems and considers how they affect the organism as a whole. As a result, the immune system has become an instructive model of the life process. Enough of the master plan has been revealed to provide a sturdy springboard for future research, but enough remains hidden to challenge the most intrepid explorer.

FURTHER READING

Arthur M. Silverstein. *A History of Immunology.* Academic Press, 1989.

G. J. V. Nossal. "Immunologic Tolerance: Collaboration Between Antigen and Lymphokines" in *Science,* Vol. 245, pages 147–153; July 14, 1989.

I. M. Roitt. *Essential Immunology.* Seventh edition. Blackwell Scientific Publications, 1991.

Noel R. Rose and Ian R. Mackay. *The Autoimmune Diseases.* Academic Press, 1992.

Cybernetic Cells

W. WAYT GIBBS

ORIGINALLY PUBLISHED IN AUGUST 2001

Three centuries of reductionism in biology recently culminated in its ultimate triumph. Dissecting life into ever smaller pieces— organisms to organs, tissues to cells, chromosomes to DNA to genes— scientists at last hit the limit. They identified each molecular rung on the chemical ladders of the majority of the human genome. Even before the draft sequence was in hand this past February, some researchers with a philosophical bent began looking ahead to the next major phase of biology—the era of integrationism. It is clear that computer models will be the main tools with which all the biochemical pieces will be placed into a complete theory. But if the variety of "virtual cells" under development is any indication, there is no consensus yet on how best to use those tools.

"People are imagining that this is the final step," observes Drew Endy of the Molecular Sciences Institute at the University of California at Berkeley. "We have the complete parts list for a human being. Now it seems just a matter of assembling the parts in a computer and flipping the switch" to untie all the knotted mysteries of medicine. In fact, he says, "Nothing could be further from the truth."

Endy speaks as one who learned the hard way. In 1994 he and John Yin of the University of Wisconsin–Madison began programming a computer model that would incorporate virtually everything known about the way that a certain virus, T7 bacteriophage, infects *Escherichia coli* bacteria that live in the human gut. The virus looks like a lunar lander. It uses clawlike appendages to grasp the outer wall of a bacterium as the phage injects its DNA into the cell. The genetic material hijacks the cell's own reproductive apparatus, forcing it to churn out bacteriophage clones until it bursts.

Endy and Yin's model simulated mathematically how all 56 of the virus's genes were translated into 59 proteins, how those proteins subverted the host cell and even how the viruses would evolve resistance to various RNA-based drugs. That seems impressive. But peek inside the equations, Endy says, and you'll find that despite including measurements from 15 years of laborious experiments, "there are still a tremendous number of degrees of freedom." The equations can be tweaked to produce almost any behavior. "A useful model must suggest a hypothesis that forces the model builder to do an experiment," Endy says. This one didn't.

Many early attempts to re-create life in silico suffered the same problem. And so most biologists still use computers as little more than receptacles for the surge of data gushing from their robotic sequencers and gene chip analyzers. The "models" they publish in their journal articles are sketchy caricatures based on the best theory they have: the central dogma that a gene in DNA is converted to an RNA that is translated to a protein that performs a particular biochemical function.

But the past few years have seen a growing movement among mathematically minded biologists to challenge the central dogma as simplistic and to use computer simulation to search for a more powerful theory. "We're witnessing a grand-scale Kuhnian revolution in biology," avers Bernhard Ø. Palsson, head of the genetic circuits research group at the University of California at San Diego. Two years ago Palsson co-founded Genomatica, one of several companies that are creating computer models of cells to try to avoid some of the mistakes that make drug development so costly and slow.

Indeed, reports James E. Bailey of the Institute of Biotechnology at the Swiss Federal Institute of Technology in Zurich, "the cost to discover drugs is actually going up," despite billions of dollars invested in monoclonal antibodies, cloning, sequencing, combinatorial chemistry and robotics. One reason those technologies haven't paid off as hoped, he says, is that they are "based on the naive idea that you can redirect the cell in a way that you want it to go by sending in a drug that inhibits only one protein." The central dogma says that that should usually work. But nine times out of 10 it doesn't.

Consider, too, Bailey urges, that geneticists have engineered hundreds of "knockout" strains of bacteria and mice to disable a particular gene. And yet in many of those mutants, the broken gene causes no apparent abnormality. The central dogma also cannot readily explain how the complex behavior of myriad human cell types emerges from a mere 30,000 or so genes.

"I could draw you a map of all the components in a cell and put all the proper arrows connecting them," says Alfred G. Gilman, a Nobel Prize–winning biochemist at the University of Texas Southwestern Medical Center at Dallas. But for even the simplest single-celled microorganism,

"I or anybody else would look at that map and have absolutely no ability to predict anything."

Bailey compares the confused state of microbiology with astronomy in the 16th century. "The astronomers had large archives detailing the movement and positions of celestial objects," he says. "But they couldn't predict the planetary motions with accuracy. They would never have believed that all the orbits are elliptic and described by a simple equation. Nevertheless, Kepler proved it. Now, I don't pretend there is any simple equation for the biology of a cell. But we should be looking for unifying principles that will order our facts into some understanding."

One early candidate to emerge from the more sophisticated cell simulations now under construction is the principle of robustness. Life of every kind has to cope with dramatic swings in temperature, changes in food supply, assaults by toxic chemicals, and attacks from without and within. To survive and prosper, cells must have backup systems and biological networks that tolerate interference.

Masaru Tomita saw this property emerge in virtual experiments he ran on his E-Cell model. With teammates at the Laboratory for Bioinformatics at Keio University in Fujisawa, Japan, Tomita built the virtual cell from 127 genes, most borrowed from *Mycoplasma genitalium,* a single-celled microbe that has the smallest genome yet discovered in a self-reproducing life-form. The team's ultimate goal is to find the minimal number of genes needed to create a self-sufficient organism and then synthesize it—an eminently reductionist strategy. But Tomita was surprised when he changed by several orders of magnitude the strength at which various genes in the model were expressed: the E-Cell's behavior hardly budged at all.

"That was an interesting revelation for us as well," says Jeff K. Trimmer, a life scientist at Entelos. The Menlo Park, Calif.–based firm has built a functional model of a human fat cell, as well as whole-body models that attempt to mimic the physiological response of obese and diabetic patients to diet and drug treatments. Pharmaceutical firms such as Eli Lilly, Bristol-Myers Squibb, and Johnson & Johnson have hired Entelos to help them prioritize their drug candidates. But when Entelos scientists adjust the virtual cell to reflect the activity of the drug, "we're often quite surprised at how little efficacy a dramatic change in cellular state has on the disease condition," Trimmer says.

Several model-building biologists suspect that what most strongly affects how a cell behaves in response to a drug or disease is not whether any particular gene is turned up or down, and not whether any single protein is blocked, but how all the genes and proteins interact dynamically. Like a connect-the-dots flip book, the story emerges from the links, which shift over time. If that is so, modelers could face a big problem: for most biochemical systems, scientists don't know what reacts with what, and when.

John R. Koza, a computer scientist at Stanford University, recently conducted an experiment that may help biologists connect their genetic dots. Koza is a pioneer in genetic programming, a technique for evolving software by instructing the computer to generate random programs, mutate them repeatedly and then screen them to identify the ones that perform the desired task best. Nicely closing a circle of metaphor, Koza used genetic programming to re-create a small but complicated part of the E-Cell model, itself built from software to mimic genes.

Koza rigged his system to evolve programs that piece together known enzymes into chemical machinery that can convert fatty acid and glycerol to diacylglycerol. Each variant program was converted, for the sake of convenience, to an equivalent electrical circuit, whose behavior was calculated on a commercial circuit simulator. The biological "circuits" that most closely matched the input-output patterns of E-Cell were retained for further evolution; the rest were killed.

After a day, Koza's 1,000-processor custom-made Beowulf supercomputer [see "The Do-It-Yourself Supercomputer," by William W. Hargrove et al., *Scientific American,* August 2001, p. 72] spit out a program that matched the actual reaction network. It had four enzymes, five intermediate chemicals and all the right feedback loops. It even found the correct reaction rates for each enzyme. There was a definite "right" answer; no alternative arrangements worked nearly as well.

Koza believes genetic programming can handle larger problems, perhaps one day even deducing the convoluted paths by which cells turn food into energy, growth and waste—but only in cases where biochemists have measured how cells process chemicals over time. Such data are still scarce.

The observation that many biochemical problems most likely have an optimal answer is exploited by Palsson and his colleagues in the models they have built of *E. coli, Hemophilus influenzae* and *Helicobacter pylori,* the germ found in stomach ulcers. They comb the literature to reconstruct as much of the biochemical networks as they can. "Then we subject them to constraints that they must abide," Palsson explains. Mass must be conserved, for example. Electrical charges must balance. Thermodynamics makes many reactions irreversible. "We try to home in on the range of solutions that are physically possible."

Markus W. Covert, a graduate student in Palsson's lab, says the goal is not perfect prediction but reliable approximation: "Engineers can design an airplane in a computer and test it virtually without ever building a prototype, even though they can't compute exactly how the air will flow." In February, Palsson's team reported that their simulation successfully predicted that *E. coli* is optimized for growth, not energy production.

This top-down approach to simulating cells has caught on. Gilman notes that an academic consortium called the Alliance for Cellular Signaling, which he chairs, has secured federal funding to build such models of the internal lives of heart muscle cells and B cells, key players in the immune system. He figures the effort will take a decade to complete, at $10 million a year. "But when we have these sorts of models," Gilman predicts, "it will be the most incredible drug discovery engine there ever was. You could model disease in that cell and then see what drug manipulation could do. Ultimately—though maybe not in 10 years—I have no doubt that there will be quantitative models of cell function, organ function and eventually whole-animal function."

"I would approach such a goal with a fair amount of humility," Bailey cautions. "History teaches us that simulations can help explore particular questions, but there won't be any master model that answers all questions. Eventually the models will become as complicated as the cell itself and as difficult to understand." Unless, perhaps, the next Kepler happens to be a computer wizard.

FURTHER READING

Drew Endy and Roger Brent. "Modelling Cellular Behaviour" in *Nature*, Vol. 409, pages 391–395; January 18, 2001.

Masaru Tomita. "Whole-Cell Simulation: A Grand Challenge of the 21st Century" in *Trends in Biotechnology*, Vol. 19, No. 6, pages 205–210; June 2001.

Details of John R. Koza's genetic programming approach can be found in the proceedings of the 2001 Pacific Symposium on Biocomputing at psb.stanford.edu/psb-online/#PATH

DINOSAURS AND OTHER MONSTERS

Rulers of the Jurassic Seas

RYOSUKE MOTANI

ORIGINALLY PUBLISHED IN DECEMBER 2000

Picture a late autumn evening some 160 million years ago, during the Jurassic time period, when dinosaurs inhabited the continents. The setting sun hardly penetrates the shimmering surface of a vast blue-green ocean, where a shadow glides silently among the dark crags of a submerged volcanic ridge. When the animal comes up for a gulp of evening air, it calls to mind a small whale—but it cannot be. The first whale will not evolve for another 100 million years. The shadow turns suddenly and now stretches more than twice the height of a human being. That realization becomes particularly chilling when its long, tooth-filled snout tears through a school of squidlike creatures.

The remarkable animal is *Ophthalmosaurus,* one of more than 80 species now known to have constituted a group of sea monsters called the ichthyosaurs, or fish-lizards. The smallest of these animals was no longer than a human arm; the largest exceeded 15 meters. *Ophthalmosaurus* fell into the medium-size group and was by no means the most aggressive of the lot. Its company would have been considerably more pleasant than that of a ferocious *Temnodontosaurus,* or "cutting-tooth lizard," which sometimes dined on large vertebrates.

When paleontologists uncovered the first ichthyosaur fossils in the early 1800s, visions of these long-vanished beasts left them awestruck. Dinosaurs had not yet been discovered, so every unusual feature of ichthyosaurs seemed intriguing and mysterious. Examinations of the fossils revealed that ichthyosaurs evolved not from fish but from land-dwelling animals, which themselves had descended from an ancient fish. How, then, did ichthyosaurs make the transition back to life in the water? To which other animals were they most related? And why did they evolve bizarre characteristics, such as backbones that look like a stack of hockey pucks and eyes as big around as bowling balls?

Despite these compelling questions, the opportunity to unravel the enigmatic transformation from landlubbing reptiles to denizens of the

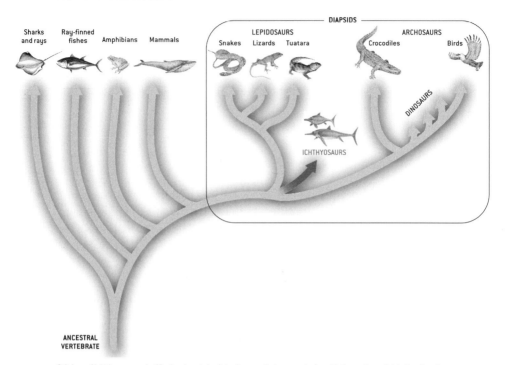

Origins of ichthyosaurs baffled paleontologists for nearly two centuries. At times thought to be closely related to everything from fish to salamanders to mammals, ichthyosaurs are now known to belong to the group called diapsids. New analyses indicate that they branched off from other diapsids at about the time lepidosaurs and archosaurs diverged from each other—but no one yet knows whether ichthyosaurs appeared shortly before that divergence or shortly after.

open sea would have to wait almost two centuries. When dinosaurs such as *Iguanodan* grabbed the attention of paleontologists in the 1830s, the novelty of the fish-lizards faded away. Intense interest in the rulers of the Jurassic seas resurfaced only a few years ago, thanks to newly available fossils from Japan and China. Since then, fresh insights have come quickly.

MURKY ORIGINS

Although most people forgot about ichthyosaurs in the early 1800s, a few paleontologists did continue to think about them throughout the 19th century and beyond. What has been evident since their discovery is that the ichthyosaurs' adaptations for life in water made them quite successful. The widespread ages of the fossils revealed that these beasts ruled the ocean from about 245 million until about 90 million years ago—roughly the entire era that dinosaurs dominated the continents. Ichthyosaur fossils were found all over the world, a sign that they migrated extensively, just as whales do today. And despite their fishy appearance, ichthyosaurs were obviously air-breathing reptiles. They did not have gills, and the configurations of their skull and jawbones were undeniably reptilian. What is more, they had two pairs of limbs (fish have none), which implied that their ancestors once lived on land.

Paleontologists drew these conclusions based solely on the exquisite skeletons of relatively late, fish-shaped ichthyosaurs. Bone fragments of the first ichthyosaurs were not found until 1927. Somewhere along the line, those early animals went on to acquire a decidedly fishy body: stocky legs morphed into flippers, and a boneless tail fluke and dorsal fin appeared. Not only were the advanced, fish-shaped ichthyosaurs made for aquatic life, they were made for life in the open ocean, far from shore. These extreme adaptations to living in water meant that most of them had lost key features—such as particular wrist and ankle bones—that would have made it possible to recognize their distant cousins on land. Without complete skeletons of the very first ichthyosaurs, paleontologists could merely speculate that they must have looked like lizards with flippers.

The early lack of evidence so confused scientists that they proposed almost every major vertebrate group—not only reptiles such as lizards and crocodiles but also amphibians and mammals—as close relatives of ichthyosaurs. As the 20th century progressed, scientists learned better how to decipher the relationships among various animal species. On applying the new skills, paleontologists started to agree that ichthyosaurs were indeed reptiles of the group Diapsida, which includes snakes, lizards, crocodiles and dinosaurs. But exactly when ichthyosaurs branched off the family tree remained uncertain—until paleontologists in Asia recently unearthed new fossils of the world's oldest ichthyosaurs.

The first big discovery occurred on the northeastern coast of Honshu, the main island of Japan. The beach is dominated by outcrops of slate, the layered black rock that is often used for the expensive ink plates of Japanese calligraphy and that also harbors bones of the oldest ichthyosaur, *Utatsusaurus*. Most *Utatsusaurus* specimens turn up fragmented and incomplete, but a group of geologists from Hokkaido University excavated two nearly complete skeletons in 1982. These specimens eventually became available for scientific study, thanks to the devotion of Nachio Minoura and his colleagues, who spent much of the next 15 years painstakingly cleaning the slate-encrusted bones. Because the bones are so fragile, they had to chip away the rock carefully with fine carbide needles as they peered through a microscope.

As the preparation neared its end in 1995, Minoura, who knew of my interest in ancient reptiles, invited me to join the research team. When I saw the skeleton for the first time, I knew that *Utatsusaurus* was exactly what paleontologists had been expecting to find for years: an ichthyosaur that looked like a lizard with flippers. Later that same year my colleague You Hailu, then at the Institute for Vertebrate Paleontology and Paleoanthropology in Beijing, showed me a second, newly discovered fossil—the world's most complete skeleton of *Chaohusaurus,* another early ichthyosaur. *Chaohusaurus* occurs in rocks the same age as those harboring remains of *Utatsusaurus,* and it, too, had been found before only in bits and pieces. The new specimen clearly revealed the outline of a slender, lizard-like body.

Utatsusaurus and *Chaohusaurus* illuminated at long last where ichthyosaurs belonged on the vertebrate family tree, because they still retained some key features of their land-dwelling ancestors. Given the configurations of the skull and limbs, my colleagues and I think that ichthyosaurs branched off from the rest of the diapsids near the separation of two major groups of living reptiles, lepidosaurs (such as snakes and lizards) and archosaurs (such as crocodiles and birds). Advancing the family-tree debate was a great achievement, but the mystery of the ichthyosaurs' evolution remained unsolved.

FROM FEET TO FLIPPERS

Perhaps the most exciting outcome of the discovery of these two Asian ichthyosaurs is that scientists can now paint a vivid picture of the elaborate adaptations that allowed their descendants to thrive in the open ocean. The most obvious transformation for aquatic life is the one from feet to flippers. In contrast to the slender bones in the front feet of most reptiles, all bones in the front "feet" of the fish-shaped ichthyosaurs are wider than they are long. What is more, they are all a similar shape. In most other four-limbed creatures it is easy to distinguish bones in the wrist (irregularly rounded) from those in the palm (long and cylindrical). Most important, the bones of fish-shaped ichthyosaurs are closely packed—without skin in between—to form a solid panel. Having all the toes enclosed in a single envelope of soft tissues would have enhanced the rigidity of the flippers, as it does in living whales, dolphins, seals and sea turtles. Such soft tissues also improve the hydrodynamic efficiency of the flippers because they are streamlined in cross section—a shape impossible to maintain if the digits are separated.

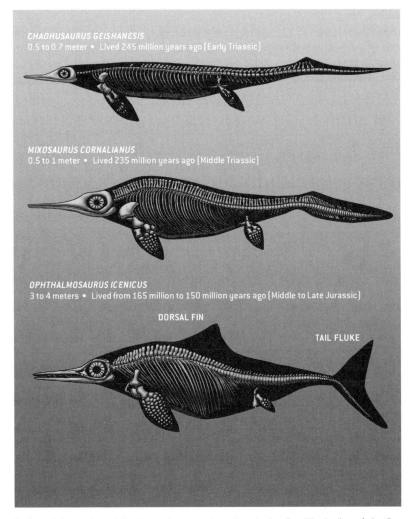

CHAOHUSAURUS GEISHANESIS
0.5 to 0.7 meter • Lived 245 million years ago (Early Triassic)

MIXOSAURUS CORNALIANUS
0.5 to 1 meter • Lived 235 million years ago (Middle Triassic)

OPHTHALMOSAURUS ICENICUS
3 to 4 meters • Lived from 165 million to 150 million years ago (Middle to Late Jurassic)

DORSAL FIN

TAIL FLUKE

Ancient skeletons have helped scientists trace how the slender, lizardlike bodies of the first ichthyosaurs (*top*) thickened into a fish shape with a dorsal fin and a tail fluke.

But examination of fossils ranging from lizard- to fish-shaped—especially those of intermediate forms—revealed that the evolution from fins to feet was not a simple modification of the foot's five digits. Indeed, analyses of ichthyosaur limbs reveal a complex evolutionary process in which digits were lost, added and divided. Plotting the shape of fin skeletons along the family tree of ichthyosaurs, for example, indicates that fish-shaped ichthyosaurs lost the thumb bones present in the earliest ichthyosaurs. Additional evidence comes from studying the order in which digits became bony, or ossified, during the growth of the fish-shaped ichthyosaur *Stenopterygius,* for which we have specimens representing various growth stages. Later, additional fingers appeared on both sides of the preexisting ones, and some of them occupied the position of the lost thumb. Needless to say, evolution does not always follow a continuous, directional path from one trait to another.

BACKBONES BUILT FOR SWIMMING

The new lizard-shaped fossils have also helped resolve the origin of the skeletal structure of their fish-shaped descendants. The descendants have backbones built from concave vertebrae the shape of hockey pucks. This shape, though rare among diapsids, was always assumed to be typical of all ichthyosaurs. But the new creatures from Asia surprised paleontologists by having a much narrower backbone, composed of vertebrae shaped more like canisters of 35-millimeter film than hockey pucks. It appeared that the vertebrae grew dramatically in diameter and shortened slightly as ichthyosaurs evolved from lizard- to fish-shaped. But why?

My colleagues and I found the answer in the swimming styles of living sharks. Sharks, like ichthyosaurs, come in various shapes and sizes. Cat sharks are slender and lack a tall tail fluke, also known as a caudal fin, on their lower backs, as did early ichthyosaurs. In contrast, mackerel sharks such as the great white have thick bodies and a crescent-shaped caudal fin similar to the later fish-shaped ichthyosaurs. Mackerel sharks swim by swinging only their tails, whereas cat sharks undulate their entire bodies. Undulatory swimming requires a flexible body, which cat sharks achieve by having a large number of backbone segments. They have about 40 vertebrae in the front part of their bodies—the same number scientists find in the first ichthyosaurs, represented by *Utatsusaurus* and *Chaohusaurus.* (Modern reptiles and mammals have only about 20.)

Undulatory swimmers, such as cat sharks, can maneuver and accelerate sufficiently to catch prey in the relatively shallow water above the continental shelf. Living lizards also undulate to swim, though not as efficiently as creatures that spend all their time at sea. It is logical to conclude, then, that the first ichthyosaurs—which looked like cat sharks and descended from a lizardlike ancestor—swam in the same fashion and lived in the environment above the continental shelf.

Undulatory swimming enables predators to thrive near shore, where food is abundant, but it is not the best choice for an animal that has to travel long distances to find a meal. Offshore predators, which hunt in

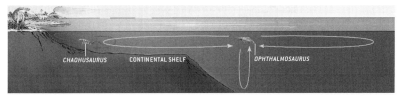

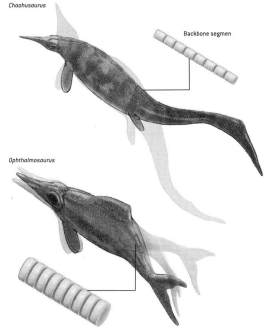

Chaohusaurus

Backbone segmen

Ophthalmosaurus

Swimming styles—and thus the habitats (*top*)—of ichthyosaurs changed as the shape of their vertebrae evolved. The narrow backbone of the first ichthyosaurs suggests that they undulated their bodies like eels (*middle*). This motion allowed for the quickness and maneuverability needed for shallow-water hunting. As the backbone thickened in later ichthyosaurs, the body stiffened and so could remain still as the tail swung back and forth (*bottom*). This stillness facilitated the energy-efficient cruising needed to hunt in the open ocean.

the open ocean where food is less concentrated, need a more energy-efficient swimming style. Mackerel sharks solve this problem by having stiff bodies that do not undulate as their tails swing back and forth. A crescent-shaped caudal fin, which acts as an oscillating hydrofoil, also improves their cruising efficiency. Fish-shaped ichthyosaurs had such a caudal fin, and their thick body profile implies that they probably swam like mackerel sharks.

Inspecting a variety of shark species reveals that the thicker the body from top to bottom, the larger the diameter of the vertebrae in the animal's trunk. It seems that sharks and ichthyosaurs solved the flexibility problem resulting from having high numbers of body segments in similar ways. As the bodies of ichthyosaurs thickened over time, the number of vertebrae stayed about the same. To add support to the more voluminous body, the backbone became at least one and a half times thicker than those of the first ichthyosaurs. As a consequence of this thickening, the body became less flexible, and the individual vertebrae acquired their hockey-puck appearance.

DRAWN TO THE DEEP

The ichthyosaurs' invasion of open water meant not only a wider coverage of surface waters but also a deeper exploration of the marine environment. We know from the fossilized stomach contents of fish-shaped ichthyosaurs that they mostly ate squidlike creatures known as dibranchiate cephalopods. Squid-eating whales hunt anywhere from about 100 to 1,000 meters deep and sometimes down to 3,000 meters. The great range in depth is hardly surprising considering that food resources are widely scattered below about 200 meters. But to hunt down deep, whales and other air-breathing divers have to go there and get back to the surface in one breath—no easy task. Reducing energy use during swimming is one of the best ways to conserve precious oxygen stored in their bodies. Consequently, deep divers today have streamlined shapes that reduce drag—and so did fish-shaped ichthyosaurs.

Characteristics apart from diet and body shape also indicate that at least some fish-shaped ichthyosaurs were deep divers. The ability of an air-breathing diver to stay submerged depends roughly on its body size: the heavier the diver, the more oxygen it can store in its muscles, blood and certain organs—and the slower the consumption of oxygen per unit of body mass. The evolution of a thick, stiff body increased the volume and mass of fish-shaped ichthyosaurs relative to their predecessors. Indeed, a fish-shaped ichthyosaur would have been up to six times heavier than a lizard-shaped ichthyosaur of the same body length. Fish-shaped ichthyosaurs also grew longer, further augmenting their bulk. Calculations based on the aerobic capacities of today's air-breathing divers (mostly mammals and birds) indicate that an animal the weight of fish-shaped *Ophthalmosaurus,* which was about 950 kilograms, could hold its breath for at least 20 minutes. A conservative estimate suggests, then, that *Ophthalmosaurus* could easily have dived to 600 meters—possibly even 1,500 meters—and returned to the surface in that time span.

Bone studies also indicate that fish-shaped ichthyosaurs were deep divers. Limb bones and ribs of four-limbed terrestrial animals include a dense outer shell that enhances the strength needed to support a body on land. But that dense layer is heavy. Because aquatic vertebrates are fairly buoyant in water, they do not need the extra strength it provides. In fact, heavy bones (which are little help for oxygen storage) can impede the ability of deep divers to return to the surface. A group of French biologists has established that modern deep-diving mammals solve that problem by making the outer shell of their bones spongy and less dense. The same type of spongy layer also encases the bones of fish-shaped ichthyosaurs, which implies that they, too, benefited from lighter skeletons.

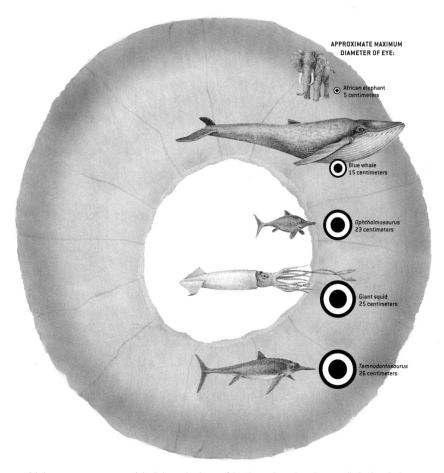

Ichthyosaur eyes were surprisingly large. Analyses of doughnut-shaped eye bones called sclerotic rings reveal that *Ophthalmosaurus* had the largest eyes relative to body size of any adult vertebrate, living or extinct, and that *Temnodontosaurus* had the biggest eyes, period.

Perhaps the best evidence for the deep-diving habits of later ichthyo-saurs is their remarkably large eyes, up to 23 centimeters across in the case of *Ophthalmosaurus*. Relative to body size, that fish-shaped ichthyosaur had the biggest eyes of any animal ever known.

The size of their eyes also suggests that visual capacity improved as ichthyosaurs moved up the family tree. These estimates are based on mea-surements of the sclerotic ring, a doughnut-shaped bone that was embed-ded in their eyes. (Humans do not have such a ring—it was lost in mam-malian ancestors—but most other vertebrates have bones in their eyes.) In the case of ichthyosaurs, the ring presumably helped to maintain the shape of the eye against the forces of water passing by as the animals swam, regardless of depth.

The diameter of the sclerotic ring makes it possible to calculate the eye's minimum f-number—an index, used to rate camera lenses, for the relative brightness of an optical system. The lower the number, the brighter the im-age and therefore the shorter the exposure time required. Low-quality lenses have a value of f/3.5 and higher; high-quality lenses have values as low as f/1.0. The f-number for the human eye is about 2.1, whereas the number for the eye of a nocturnal cat is about 0.9. Calculations suggest that a cat would be capable of seeing at depths of 500 meters or greater in most oceans. *Ophthalmosaurus* also had a minimum f-number of about 0.9, but with its much larger eyes, it probably could outperform a cat.

GONE FOR GOOD

Many characteristics of ichthyosaurs—including the shape of their bodies and backbones, the size of their eyes, their aerobic capacity, and their habitat and diet—seem to have changed in a connected way during their evolution, although it is not possible to judge what is the cause and what is the effect. Such adaptations enabled ichthyosaurs to reign for 155 mil-lion years. New fossils of the earliest of these sea dwellers are now making it clear just how they evolved so successfully for aquatic life, but still no one knows why ichthyosaurs went extinct.

Loss of habitat may have clinched the final demise of lizard-shaped ichthyosaurs, whose inefficient, undulatory swimming style limited them to near-shore environments. A large-scale drop in sea level could have snuffed out these creatures along with many others by eliminating their shallow-water niche. Fish-shaped ichthyosaurs, on the other hand, could make a living in the open ocean, where they would have had a better chance of survival. Because their habitat never disappeared, something else must have eliminated them. The period of their disappearance roughly corresponds to the appearance of advanced sharks, but no one has found direct evidence of competition between the two groups.

Scientists may never fully explain the extinction of ichthyosaurs. But as paleontologists and other investigators continue to explore their evolutionary history, we are sure to learn a great deal more about how these fascinating creatures lived.

FURTHER READING

R. L. Carroll. *Vertebrate Paleontology and Evolution.* Freeman, San Francisco, 1987.

Christopher McGowan. *Dinosaurs, Spitfires, and Sea Dragons.* Harvard University Press, 1991.

Ryosuke Motani, You Hailu and Christopher McGowan. "Eel-like Swimming in the Earliest Ichthyosaurs" in *Nature,* Vol. 382, pages 347–348; July 25, 1996.

Ryosuke Motani, Nachio Minoura and Tatsuro Ando. "Ichthyosaurian Relationships Illuminated by New Primitive Skeletons from Japan" in *Nature,* Vol. 393, pages 255–257; May 21, 1998.

Ryosuke Motani, Bruce M. Rothschild and William Wahl, Jr. "Large Eyeballs in Diving Ichthyosaurs" in *Nature,* Vol. 402, page 747; December 16, 1999.

Ryosuke Motani's Web site: www.ucmp.berkeley.edu/people/motani/ichthyo/

The Mammals That Conquered the Seas

KATE WONG

ORIGINALLY PUBLISHED IN MAY 2002

Dawn breaks over the Tethys Sea, 48 million years ago, and the blue-green water sparkles with the day's first light. But for one small mammal, this new day will end almost as soon as it has started. Tapir-like *Eotitanops* has wandered perilously close to the water's edge, ignoring its mother's warning call. For the brute lurking motionless among the mangroves, the opportunity is simply too good to pass up. It lunges landward, propelled by powerful hind limbs, and sinks its formidable teeth into the calf, dragging it back into the surf. The victim's frantic struggling subsides as it drowns, trapped in the viselike jaws of its captor. Victorious, the beast shambles out of the water to devour its kill on terra firma. At first glance, this fearsome predator resembles a crocodile, with its squat legs, stout tail, long snout and eyes that sit high on its skull. But on closer inspection, it has not armor but fur, not claws but hooves. And the cusps on its teeth clearly identify it not as a reptile but as a mammal. In fact, this improbable creature is *Ambulocetus,* an early whale, and one of a series of intermediates linking the land-dwelling ancestors of cetaceans to the 80 or so species of whales, dolphins and porpoises that rule the oceans today.

Until recently, the emergence of whales was one of the most intractable mysteries facing evolutionary biologists. Lacking fur and hind limbs and unable to go ashore for so much as a sip of freshwater, living cetaceans represent a dramatic departure from the mammalian norm. Indeed, their piscine form led Herman Melville in 1851 to describe Moby Dick and his fellow whales as fishes. But to 19th-century naturalists such as Charles Darwin, these air-breathing, warm-blooded animals that nurse their young with milk distinctly grouped with mammals. And because ancestral mammals lived on land, it stood to reason that whales ultimately descended from a terrestrial ancestor. Exactly how that might have happened, however, eluded scholars. For his part, Darwin noted in *On the Origin of Species* that a bear swimming with its mouth agape to catch insects was a plausible evolutionary starting point for whales. But the proposition attracted so much ridicule that in later editions of the book he said just that such a bear was "almost like a whale."

The fossil record of cetaceans did little to advance the study of whale origins. Of the few remains known, none were sufficiently complete or primitive to throw much light on the matter. And further analyses of the bizarre anatomy of living whales led only to more scientific head scratching. Thus, even a century after Darwin, these aquatic mammals remained an evolutionary enigma. In fact, in his 1945 classification of mammals, famed paleontologist George Gaylord Simpson noted that whales had evolved in the oceans for so long that nothing informative about their ancestry remained. Calling them "on the whole, the most peculiar and aberrant of mammals," he inserted cetaceans arbitrarily among the other orders. Where whales belonged in the mammalian family tree and how they took to the seas defied explanation, it seemed.

Over the past two decades, however, many of the pieces of this once imponderable puzzle have fallen into place. Paleontologists have uncovered a wealth of whale fossils spanning the Eocene epoch, the time between 55 million and 34 million years ago when archaic whales, or archaeocetes, made their transition from land to sea. They have also unearthed some clues from the ensuing Oligocene, when the modern suborders of cetaceans—the mysticetes (baleen whales) and the odontocetes (toothed whales)—arose. That fossil material, along with analyses of DNA from living animals, has enabled scientists to paint a detailed picture of when, where and how whales evolved from their terrestrial forebears. Today their transformation—from landlubbers to Leviathans—stands as one of the most profound evolutionary metamorphoses on record.

EVOLVING IDEAS

At around the same time that Simpson declared the relationship of whales to other mammals undecipherable on the basis of anatomy, a new comparative approach emerged, one that looked at antibody-antigen reactions in living animals. In response to Simpson's assertion, Alan Boyden of Rutgers University and a colleague applied the technique to the whale question. Their results showed convincingly that among living animals, whales are most closely related to the even-toed hoofed mammals, or artiodactyls, a group whose members include camels, hippopotamuses, pigs and ruminants such as cows. Still, the exact nature of that relationship remained unclear. Were whales themselves artiodactyls? Or did they occupy their own branch of the mammalian family tree, linked to the artiodactyl branch via an ancient common ancestor?

Support for the latter interpretation came in the 1960s, from studies of primitive hoofed mammals known as condylarths that had not yet evolved the specialized characteristics of artiodactyls or the other mammalian orders. Paleontologist Leigh Van Valen, then at the American Museum of Natural History in New York City, discovered striking resemblances between the three-cusped teeth of the few known fossil whales and those of a group of meat-eating condylarths called mesonychids. Likewise, he found shared dental characteristics between artiodactyls and another group of condylarths, the arctocyonids, close relatives of the mesonychids. Van Valen concluded that whales descended from the carnivorous, wolflike mesonychids and thus were linked to artiodactyls through the condylarths.

WALKING WHALES

A decade or so passed before paleontologists finally began unearthing fossils close enough to the evolutionary branching point of whales to address Van Valen's mesonychid hypothesis. Even then the significance of these finds took a while to sink in. It started when University of Michigan paleontologist Philip Gingerich went to Pakistan in 1977 in search of Eocene land mammals, visiting an area previously reported to shelter such remains. The expedition proved disappointing because the spot turned out to contain only marine fossils. Finding traces of ancient ocean life in Pakistan, far from the country's modern coast, is not surprising: during the Eocene, the vast Tethys Sea periodically covered great swaths of what is now the Indian subcontinent. Intriguingly, though, the team discovered among those ancient fish and snail remnants two pelvis fragments that appeared to have come from relatively large, walking beasts. "We joked about walking whales," Gingerich recalls with a chuckle. "It was unthinkable." Curious as the pelvis pieces were, the only fossil collected during that field season that seemed important at the time was a primitive artiodactyl jaw that had turned up in another part of the country.

Two years later, in the Himalayan foothills of northern Pakistan, Gingerich's team found another weird whale clue: a partial braincase from a wolf-size creature—found in the company of 50-million-year-old land mammal remains—that bore some distinctive cetacean characteristics. All modern whales have features in their ears that do not appear in any other vertebrates. Although the fossil skull lacked the anatomy necessary for hearing directionally in water (a critical skill for living whales), it clearly had the diagnostic cetacean ear traits. The team had discovered the oldest and most primitive whale then known—one that must have spent some, if not most, of its time on land. Gingerich christened the creature *Pakicetus* for its place of origin and, thus hooked, began hunting for ancient whales in earnest.

At around the same time, another group recovered additional remains of *Pakicetus*—a lower jaw fragment and some isolated teeth—that bolstered the link to mesonychids through strong dental similarities. With *Pakicetus* showing up around 50 million years ago and mesonychids known from around the same time in the same part of the world, it looked increasingly likely that cetaceans had indeed descended from the mesonychids or something closely related to them. Still, what the earliest whales looked like from the neck down was a mystery.

Further insights from Pakistan would have to wait, however. By 1983 Gingerich was no longer able to work there because of the Soviet Union's invasion of Afghanistan. He decided to cast his net in Egypt instead, journeying some 95 miles southwest of Cairo to the Western Desert's Zeuglodon Valley, so named for early 20th-century reports of fossils of archaic whales—or zeuglodons, as they were then known—in the area. Like Pakistan, much of Egypt once lay submerged under Tethys. Today the skeletons of creatures that swam in that ancient sea lie entombed in sandstone. After several field seasons, Gingerich and his crew hit pay dirt: tiny hind limbs belonging to a 60-foot-long sea snake of a whale known as *Basilosaurus* and the first evidence of cetacean feet.

Earlier finds of *Basilosaurus,* a fully aquatic monster that slithered through the seas between about 40 million and 37 million years ago, preserved only a partial femur, which its discoverers interpreted as vestigial. But the well-formed legs and feet revealed by this discovery hinted at functionality. Although at less than half a meter in length the diminutive limbs probably would not have assisted *Basilosaurus* in swimming and certainly would not have enabled it to walk on land, they may well have helped guide the beast's serpentine body during the difficult activity of aquatic mating. Whatever their purpose, if any, the little legs had big implications. "I immediately thought, we're 10 *million* years after *Pakicetus*," Gingerich recounts excitedly. "If these things still have feet and toes, we've got 10 million years of history to look at." Suddenly, the walking whales they had scoffed at in Pakistan seemed entirely plausible.

Just such a remarkable creature came to light in 1992. A team led by J. G. M. (Hans) Thewissen of the Northeastern Ohio Universities College of Medicine recovered from 48-million-year-old marine rocks in northern

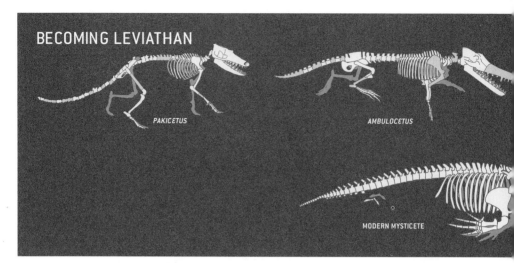

BECOMING LEVIATHAN

PAKICETUS

AMBULOCETUS

MODERN MYSTICETE

Representative Archaeocetes in the lineage leading to modern odontocetes and mysticetes trace some of the anatomical changes that enabled these animals to take to the seas. In just 15 million years, whales shed their terrestrial trappings and became fully adapted to aquatic life. Notably, the hind limbs diminished, the forelimbs transformed into flippers, and the vertebral column evolved to permit tail-powered swimming. Meanwhile the skull changed to enable underwater hearing, the nasal opening moved backward to the top of the skull, and the teeth simplified into pegs for grasping instead of grinding. Later in whale evolution, the mysticetes' teeth were replaced with baleen.

Pakistan a nearly complete skeleton of a perfect intermediate between modern whales and their terrestrial ancestors. Its large feet and powerful tail bespoke strong swimming skills, while its sturdy leg bones and mobile elbow and wrist joints suggested an ability to locomote on land. He dubbed the animal *Ambulocetus natans,* the walking and swimming whale.

SHAPE SHIFTERS

Since then, Thewissen, Gingerich and others have unearthed a plethora of fossils documenting subsequent stages of the whale's transition from land to sea. The picture emerging from those specimens is one in which *Ambulocetus* and its kin—themselves descended from the more terrestrial pakicetids—spawned needle-nosed beasts known as remingtonocetids and the intrepid protocetids—the first whales seaworthy enough to fan out from Indo-Pakistan across the globe. From the protocetids arose the dolphinlike dorudontines, the probable progenitors of the snakelike basilosaurines and modern whales.

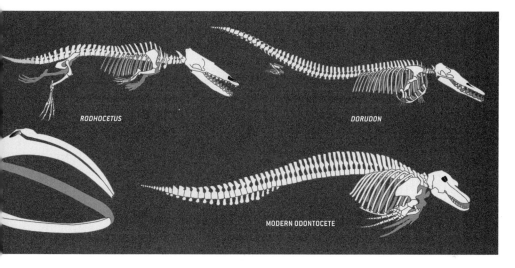

In addition to furnishing supporting branches for the whale family tree, these discoveries have enabled researchers to chart many of the spectacular anatomical and physiological changes that allowed cetaceans to establish permanent residency in the ocean realm. Some of the earliest of these adaptations to emerge, as *Pakicetus* shows, are those related to hearing. Sound travels differently in water than it does in air. Whereas the ears of humans and other land-dwelling animals have delicate, flat eardrums, or tympanic membranes, for receiving airborne sound, modern whales have thick, elongate tympanic ligaments that cannot receive sound. Instead a bone called the bulla, which in whales has become quite dense and is therefore capable of transmitting sound coming from a denser medium to deeper parts of the ear, takes on that function. The *Pakicetus* bulla shows some modification in that direction, but the animal retained a land mammal–like eardrum that could not work in water.

What, then, might *Pakicetus* have used its thickened bullae for? Thewissen suspects that much as turtles hear by picking up vibrations from the ground through their shields, *Pakicetus* may have employed its bullae to pick up ground-borne sounds. Taking new postcranial evidence into consideration along with the ear morphology, he envisions *Pakicetus* as an ambush predator that may have lurked around shallow rivers, head to the ground, preying on animals that came to drink. *Ambulocetus* is even more likely to have used such inertial hearing, Thewissen says, because it had the beginnings of a channel linking jaw and ear. By resting its jaw on the ground—a strategy seen in modern crocodiles—*Ambulocetus* could have listened for approaching prey. The same features that allowed early whales to receive sounds from soil, he surmises, preadapted them to hearing in the water.

Zhe-Xi Luo of the Carnegie Museum of Natural History in Pittsburgh has shown that by the time of the basilosaurines and dorudontines, the first fully aquatic whales, the ropelike tympanic ligament had probably already evolved. Additionally, air sinuses, presumably filled with sponge-like tissues, had formed around the middle ear, offering better sound resolution and directional cues for underwater hearing. Meanwhile, with the external ear canal closed off (a prerequisite for deep-sea diving), he adds, the lower jaw was taking on an increasingly important auditory role, developing a fat-filled canal capable of conducting sound back to the middle ear.

Later in the evolution of whale hearing, the toothed and baleen whales parted ways. Whereas the toothed whales evolved the features necessary to produce and receive high-frequency sounds, enabling echolocation for hunting, the baleen whales developed the ability to produce and receive very low frequency sounds, allowing them to communicate with one another over vast distances. Fossil whale ear bones, Luo says, show that by around 28 million years ago early odontocetes already had some of the bony structures necessary for hearing high-pitched sound and were thus capable of at least modest echolocation. The origin of the mysticete's low-frequency hearing is far murkier, even though the fossil evidence of that group now dates back to as early as 34 million years ago.

Other notable skull changes include movement of the eye sockets from a crocodilelike placement atop the head in *Pakicetus* and *Ambulocetus* to a lateral position in the more aquatic protocetids and later whales. And the nasal opening migrated back from the tip of the snout in *Pakicetus* to the top of the head in modern cetaceans, forming the blowhole. Whale dentition morphed, too, turning the complexly cusped, grinding molars of primitive mammalian ancestors into the simple, pronglike teeth of modern odontocetes, which grasp and swallow their food without chewing. Mysticetes lost their teeth altogether and developed comblike plates of baleen that hang from their upper jaws and strain plankton from the seawater.

The most obvious adaptations making up the whale's protean shift are those that produced its streamlined shape and unmatched swimming abilities. Not surprisingly, some bizarre amphibious forms resulted along the way. *Ambulocetus,* for one, retained the flexible shoulder, elbow, wrist and finger joints of its terrestrial ancestors and had a pelvis capable of supporting its weight on land. Yet the creature's disproportionately large hind limbs and paddlelike feet would have made walking somewhat awkward. These same features were perfect for paddling around in the fish-filled shallows of Tethys, however.

Moving farther out to sea required additional modifications, many of which appear in the protocetid whales. Studies of one member of this group, *Rodhocetus,* indicate that the lower arm bones were compressed and already on their way to becoming hydrodynamically efficient, says University of Michigan paleontologist Bill Sanders. The animal's long, delicate feet were probably webbed, like the fins used by scuba divers. *Rodhocetus* also exhibits aquatic adaptations in its pelvis, where fusion between the vertebrae that form the sacrum is reduced, loosening up the lower spine to power tail movement. These features, says Gingerich, whose team discovered the creature, suggest that *Rodhocetus* performed a leisurely dog paddle at the sea surface and a swift combination of otterlike hind-limb paddling and tail propulsion underwater. When it went ashore to breed or perhaps to bask in the sun, he proposes, *Rodhocetus* probably hitched itself around somewhat like a modern eared seal or sea lion.

By the time of the basilosaurines and dorudontines, whales were fully aquatic. As in modern cetaceans, the shoulder remained mobile while the elbow and wrist stiffened, forming flippers for steering and balance. Farther back on the skeleton, only tiny legs remained, and the pelvis had dwindled accordingly. Analyses of the vertebrae of *Dorudon,* conducted by Mark D. Uhen of the Cranbrook Institute of Science in Bloomfield Hills, Mich., have revealed one tail vertebra with a rounded profile. Modern whales have a similarly shaped bone, the ball vertebra, at the base of their fluke, the flat, horizontal structure capping the tail. Uhen thus suspects that basilosaurines and dorudontines had tail flukes and swam much as modern whales do, using so-called caudal oscillation. In this energetically efficient mode of locomotion, motion generated at a single point in the vertebral column powers the tail's vertical movement through the water, and the fluke generates lift.

Exactly when whales lost their legs altogether remains unknown. In fact, a recent discovery made by Lawrence G. Barnes of the Natural History Museum of Los Angeles County hints at surprisingly well developed hind limbs in a 27-million-year-old baleen whale from Washington State, suggesting that whale legs persisted far longer than originally thought. Today, however, some 50 million years after their quadrupedal ancestors first waded into the warm waters of Tethys, whales are singularly sleek. Their hind limbs have shrunk to externally invisible vestiges, and the pelvis has diminished to the point of serving merely as an anchor for a few tiny muscles unrelated to locomotion.

MAKING WAVES

The fossils uncovered during the 1980s and 1990s advanced researchers' understanding of whale evolution by leaps and bounds, but all morphological signs still pointed to a mesonychid origin. An alternative view of cetacean roots was taking wing in genetics laboratories in the U.S., Belgium and Japan, however. Molecular biologists, having developed sophisticated techniques for analyzing the DNA of living creatures, took Boyden's 1960s immunology-based conclusions a step further. Not only were whales more closely related to artiodactyls than to any other living mammals, they asserted, but in fact whales were themselves artiodactyls, one of many twigs on that branch of the mammalian family tree. Moreover, a number of these studies pointed to an especially close relationship between whales and hippopotamuses. Particularly strong evidence for this idea came in 1999 from analyses of snippets of noncoding DNA called SINES (short interspersed elements), conducted by Norihiro Okada and his colleagues at the Tokyo Institute of Technology.

The whale-hippo connection did not sit well with paleontologists. "I thought they were nuts," Gingerich recollects. "Everything we'd found was consistent with a mesonychid origin. I was happy with that and happy with a connection through mesonychids to artiodactyls." Whereas mesonychids appeared at the right time, in the right place and in the right form to be considered whale progenitors, the fossil record did not seem to contain a temporally, geographically and morphologically plausible artiodactyl ancestor for whales, never mind one linking whales and hippos specifically. Thewissen, too, had largely dismissed the DNA findings. But "I stopped rejecting it when Okada's SINE work came out," he says.

It seemed the only way to resolve the controversy was to find, of all things, an ancient whale anklebone. Morphologists have traditionally defined artiodactyls on the basis of certain features in one of their anklebones, the astragalus, that enhance mobility. Specifically, the unique artiodactyl astragalus has two grooved, pulleylike joint surfaces. One connects to the tibia, or shinbone; the other articulates with more distal anklebones. If whales descended from artiodactyls, researchers reasoned, those that had not yet fully adapted to life in the seas should exhibit this double-pulleyed astragalus.

That piece of the puzzle fell into place last fall, when Gingerich and Thewissen both announced discoveries of new primitive whale fossils. In the eastern part of Baluchistan Province, Gingerich's team had found partially articulated skeletons of *Rodhocetus balochistanensis* and a new protocetid genus, *Artiocetus*. Thewissen and his colleagues recovered from a bone bed in the Kala Chitta Hills of Punjab, Pakistan, much of the long-sought postcranial skeleton of *Pakicetus,* as well as that of a smaller member of the pakicetid family, *Ichthyolestes*. Each came with an astragalus bearing the distinctive artiodactyl characteristics.

The anklebones convinced both longtime proponents of the mesony-chid hypothesis that whales instead evolved from artiodactyls. Gingerich has even embraced the hippo idea. Although hippos themselves arose long after whales, their purported ancestors—dog- to horse-size, swamp-dwelling beasts called anthracotheres—date back to at least the middle Eocene and may thus have a forebear in common with the cetaceans. In fact, Gingerich notes that *Rodhocetus* and anthracotheres share features in their hands and wrists not seen in any other later artiodactyls. Thewissen agrees that the hippo hypothesis holds much more appeal than it once did. But he cautions that the morphological data do not yet point to a particular artiodactyl, such as the hippo, being the whale's closest relative, or sister group. "We don't have the resolution yet to get them there," he remarks, "but I think that will come."

What of the evidence that seemed to tie early whales to mesonychids? In light of the new ankle data, most workers now suspect that those similarities probably reflect convergent evolution rather than shared ancestry and that mesonychids represent an evolutionary dead end. But not everyone is convinced. Maureen O'Leary of the State University of New York at Stony Brook argues that until all the available evidence—both morphological and molecular—is incorporated into a single phylogenetic analysis, the possibility remains that mesonychids belong at the base of the whale pedigree. It is conceivable, she says, that mesonychids are actually ancient artiodactyls but ones that reversed the ankle trend. If so, mesonychids could still be the whales' closest relative, and hippos could be their closest living relative. Critics of that idea, however, point out that although folding the mesonychids into the artiodactyl order offers an escape hatch of sorts to supporters of the mesonychid hypothesis, it would upset the long-standing notion that the ankle makes the artiodactyl.

Investigators agree that figuring out the exact relationship between whales and artiodactyls will most likely require finding additional fossils—particularly those that can illuminate the beginnings of artiodactyls in general and hippos in particular. Yet even with those details still unresolved, "we're really getting a handle on whales from their origin to the end of archaeocetes," Uhen reflects. The next step, he says, will be to figure out how the mysticetes and odontocetes arose from the archaeocetes and when their modern features emerged. Researchers may never unravel all the mysteries of whale origins. But if the extraordinary advances made over the past two decades are any indication, with continued probing, answers to many of these lingering questions will surface from the sands of time.

FURTHER READING

J. G. M. Thewissen, ed. *The Emergence of Whales: Evolutionary Patterns in the Origin of Cetacea.* Plenum Publishing, 1998.

J. G. M. Thewissen, E. M. Williams, L. J. Roe and S. T. Hussain. "Skeletons of Terrestrial Cetaceans and the Relationship of Whales to Artiodactyls" in *Nature,* Vol. 413, pages 277–281; September 20, 2001.

Philip D. Gingerich, Munir ul Haq, Iyad S. Zalmout, Intizar Hussain Khan and M. Sadiq Malkani. "Origin of Whales from Early Artiodactyls: Hands and Feet of Eocene Protocetidae from Pakistan" in *Science,* Vol. 293, pages 2239–2242; September 21, 2001.

The Encyclopedia of Marine Mammals. Edited by W. F. Perrin, Bernd G. Würsig and J. G. M. Thewissen. Academic Press, 2002.

Breathing Life into *Tyrannosaurus rex*

GREGORY M. ERICKSON

ORIGINALLY PUBLISHED IN SEPTEMBER 1999; UPDATED IN 2004

Dinosaurs ceased to walk the earth 65 million years ago, yet they still live among us. Velociraptors star in movies, and *Triceratops* toys clutter toddlers' bedrooms. Of these charismatic animals, however, one species has always ruled our fantasies. Children, filmmaker Steven Spielberg and professional paleontologists agree that the superstar was and is *Tyrannosaurus rex.*

The late Harvard University paleontologist Stephen Jay Gould said that every species designation represents a theory about that animal. The very name *Tyrannosaurus rex*—"tyrant lizard king"—evokes a powerful image of this species. John R. Horner of Montana State University and science writer Don Lessem wrote in their book *The Complete* T. Rex, "We're lucky to have the opportunity to know *T. rex,* study it, imagine it, and let it scare us. Most of all, we're lucky *T. rex* is dead." And paleontologist Robert T. Bakker of the Glenrock Paleontological Museum in Wyoming described *T. rex* as a "10,000-pound roadrunner from hell," a tribute to its obvious size and power.

In Spielberg's *Jurassic Park,* which boasted the most accurate popular depiction of dinosaurs ever, *T. rex* was, as usual, presented as a killing machine whose sole purpose was aggressive, bloodthirsty attacks on helpless prey. *T. rex*'s popular persona, however, is as much a function of artistic license as of concrete scientific evidence. A century of study and the existence of 30 fairly complete *T. rex* specimens have generated substantial information about its anatomy. But inferring behavior from anatomy alone is perilous, and the true nature of *T. rex* continues to be largely shrouded in mystery. Whether it was even primarily a predator or a scavenger is still the subject of debate.

Over the past decade or so, a new breed of scientists has begun to unravel some of *T. rex*'s better-kept secrets. These paleobiologists try to put a creature's remains in a living context—they attempt to animate the silent and still skeleton of the museum display. *T. rex* is thus changing before our

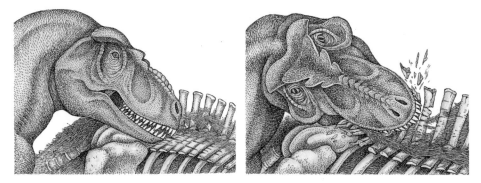

Nipping strategy (*left*) enabled *T. rex* to remove strips of flesh in tight spots, such as between vertebrae, using only the front teeth. Massive force generated by *T. rex* in the "puncture and pull" biting technique (*right*) was sufficient to have created the huge furrows on the surface of the section of a fossil *Triceratops* pelvis shown in the inset at the right. The enormous body of the *T. rex* and its powerful neck musculature enabled the "pull" in "puncture and pull."

eyes as paleobiologists use fossil clues, some new and some previously overlooked, to develop fresh ideas about the nature of these magnificent animals.

Rather than draw conclusions about behavior based solely on anatomy, paleobiologists demand proof of actual activities. Skeletal assemblages of multiple individuals shine a light on the interactions among *T. rex* and between them and other species. In addition, so-called trace fossils reveal activities through physical evidence, such as bite marks on bones and wear patterns on teeth. Also of great value as trace fossils are coprolites, fossilized feces. (Remains of a herbivore, such as *Triceratops* or *Edmontosaurus,* in *T. rex* coprolites certainly provide smoking-gun proof of species interactions!)

One assumption that paleobiologists are willing to make is that closely related species may have behaved in similar ways. *T. rex* data are therefore being corroborated by comparisons with those of earlier members of the family Tyrannosauridae, including their cousins *Albertosaurus, Gorgosaurus* and *Daspletosaurus,* collectively known as albertosaurs.

SOLO OR SOCIAL?

Tyrannosaurs are usually depicted as solitary, as was the case in *Jurassic Park.* (An alternative excuse for that film's loner is that the movie's genetic wizards wisely created only one.) Mounting evidence, however, points to gregarious *T. rex* behavior, at least for part of the animals' lives. Two *T. rex* excavations in the Hell Creek Formation of eastern Montana are most compelling.

In 1966 Los Angeles County Museum researchers attempting to exhume a Hell Creek adult were elated to find another, smaller individual resting atop the *T. rex* they had originally sought. This second fossil was identified at first as a more petite species of tyrannosaur. My examination of the histological evidence—the microstructure of the bones—now suggests that the second animal was actually a subadult *T. rex*. A similar discovery was made during the excavation of "Sue," the largest and most complete fossil *T. rex* ever found. Sue is perhaps as famous for its $8.36-million auction price following ownership haggling as for its paleontological status [see "No Bones about It," by Karin Vergoth; News and Analysis, *Scientific American,* December 1997]. Remains of a subadult and a juvenile *T. rex* were later found in Sue's quarry by researchers from the Black Hills Institute of Geological Research in Hill City, S.D. Experts who have worked the Hell Creek Formation, myself included, generally agree that long odds argue against multiple, loner *T. rex* finding their way to the same burial. The more parsimonious explanation is that the animals were part of a group.

An even more spectacular find from 1910 further suggests gregarious behavior among the Tyrannosauridae. Researchers from the American Museum of Natural History in New York City working in Alberta, Canada, found a bone bed—a deposit with fossils of many individuals—holding at least nine of *T. rex*'s close relatives, albertosaurs.

Philip J. Currie and his team from the Royal Tyrrell Museum of Paleontology in Alberta have relocated the 1910 find and are conducting the first detailed study of the assemblage. Such aggregations of carnivorous animals can occur when one after another gets caught in a trap, such as a mud hole or soft sediment at a river's edge, in which a prey animal that has attracted them is already ensnared. Under those circumstances, however, the collection of fossils should also contain those of the hunted herbivore. The lack of such herbivore remains among the albertosaurs (and among the three–*T. rex* assemblage that included Sue) indicates that the herd most likely associated with one another naturally and perished together from drought, disease or drowning.

From examination of the remains collected so far, Currie estimates that the animals ranged from four to almost nine meters (13 to 29 feet) in length. This variation in size hints at a group composed of juveniles and adults. One individual is considerably larger and more robust than the others. Although it might have been a different species of albertosaur, a mixed bunch seems unlikely. I believe that if *T. rex* relatives did indeed have a social structure, this largest individual may have been the patriarch or matriarch of the herd.

Tyrannosaurs in herds, with complex interrelationships, are in many ways an entirely new species to contemplate. But science has not morphed them into a benign and tender collection of Cretaceous Care Bears: some of the very testimony for *T. rex* group interaction is partially healed bite marks that reveal nasty interpersonal skills.

A paper published by Currie and Darren Tanke, also at the Royal Tyrrell Museum, highlights this evidence. Tanke is a leading authority on pale-opathology—the study of ancient injuries and disease. He has detected a unique pattern of bite marks among theropods, the group of carnivorous dinosaurs that encompasses *T. rex* and other tyrannosaurs. These bite marks consist of gouges and punctures on the sides of the snout, on the sides and bottom of the jaws, and occasionally on the top and back of the skull.

Interpreting these wounds, Tanke and Currie reconstructed how these dinosaurs fought. They believe that the animals faced off but primarily gnawed at one another with one side of their complement of massive teeth rather than snapping from the front. The workers also surmise that the jaw-gripping behavior accounts for peculiar bite marks found on the sides of tyrannosaur teeth. The bite patterns imply that the combatants maintained their heads at the same level throughout a confrontation. Based on the magnitude of some of the fossil wounds, *T. rex* clearly showed little reserve in battle and sometimes inflicted severe damage to its conspecific foe. One tyrannosaur studied by Tanke and Currie sports a souvenir tooth embedded in its own jaw, perhaps left by a fellow combatant.

The usual subjects—food, mates and territory—may have prompted the vigorous clashes among tyrannosaurs. Whatever the motivation behind the fighting, the fossil record demonstrates that the behavior was repeated throughout a tyrannosaur's life. Injuries among younger individuals seem to have been more common, possibly because a juvenile was subject to attack by members of its own age group as well as by large adults. (Nevertheless, the fossil record may also be slightly misleading and simply contain more evidence of injuries in young *T. rex*. Nonlethal injuries to adults would have eventually healed, destroying the evidence. Juveniles were more likely to die from adult-inflicted injuries, and they carried those wounds to the grave.)

BITES AND BITS

Imagine the large canine teeth of a baboon or lion. Now imagine a mouthful of much larger canine-type teeth, the size of railroad spikes and with serrated edges. Kevin Padian of the University of California at Berkeley has summed up the appearance of the huge daggers that were *T. rex* teeth: "lethal bananas."

Despite the obvious potential of such weapons, the general opinion among paleontologists had been that dinosaur bite marks were rare. The few published reports before 1990 consisted of brief comments buried in articles describing more sweeping new finds, and the clues in the marred remains concerning behavior escaped contemplation.

Some researchers have nonetheless speculated about the teeth. As early as 1973, Ralph E. Molnar, now at the Museum of Northern Arizona in Flagstaff, began musing about the strength of the teeth, based on their shape. Later, James O. Farlow of Indiana University–Purdue University Fort Wayne and Daniel L. Brinkman of Yale University performed elaborate morphological studies of tyrannosaur dentition, which made them confident that the "lethal bananas" were robust, thanks to their rounded cross-sectional configuration, and would endure bone-shattering impacts during feeding.

In 1992 I was able to provide material support for such speculation. Kenneth H. Olson, a Lutheran pastor and superb amateur fossil collector for the Museum of the Rockies in Bozeman, Mont., came to me with several specimens. One was a one-meter-wide, 1.5-meter-long partial pelvis from an adult *Triceratops*. The other was a toe bone from an adult *Edmontosaurus* (duck-billed dinosaur). I examined Olson's specimens and found that both bones were riddled with gouges and punctures up to 12 centimeters long and several centimeters deep. The *Triceratops* pelvis had nearly 80 such indentations. I documented the size and shape of the marks and used orthodontic dental putty to make casts of some of the deeper holes. The teeth that had made the holes were spaced some 10 centimeters apart. They left punctures with eye-shaped cross sections. They clearly included carinae, elevated cutting edges, on their anterior and posterior faces. And those edges were serrated. The totality of the evidence pointed to these indentations being the first definitive bite marks from a *T. rex*.

This finding had considerable behavioral implications. It confirmed for the first time the assumption that *T. rex* fed on its two most common contemporaries, *Triceratops* and *Edmontosaurus*. Furthermore, the bite patterns opened a window into *T. rex*'s actual feeding techniques, which apparently involved two distinct biting behaviors. *T. rex* usually used the "puncture and pull" strategy, in which biting deeply with enormous force was followed by drawing the teeth through the penetrated flesh and bone, which typically produced long gashes. In this way, a *T. rex* appears to have de-tached the pelvis found by Olson from the rest of the *Triceratops* torso. *T. rex* also employed a nipping approach in which the front (incisiform) teeth grasped and stripped the flesh in tight spots between vertebrae, where only the muzzle of the beast could fit. This method left vertically aligned, parallel furrows in the bone.

Many of the bites on the *Triceratops* pelvis were spaced only a few cen-timeters apart, as if the *T. rex* had methodically worked its way across the hunk of meat as we would nibble an ear of corn. With each bite, *T. rex* ap-pears also to have removed a small section of bone. We presumed that the missing bone had been consumed, confirmation for which shortly came, and from an unusual source.

In 1997 Karen Chin, now at the University of Colorado, received a pe-culiar, tapered mass that had been unearthed by a crew from the Royal Saskatchewan Museum. The object, which weighed 7.1 kilograms and measured 44 by 16 by 13 centimeters, proved to be a *T. rex* coprolite. The specimen, the first ever confirmed from a theropod and more than twice as large as any previously reported meat eater's coprolite, was chock-full of pulverized bone. Once again making use of histological methods, Chin and I determined that the shattered bone came from a young herbivorous dinosaur. *T. rex* did ingest parts of the bones of its food sources and, fur-thermore, partially digested these items with strong enzymes or stomach acids.

Following the lead of Farlow and Molnar, Olson and I have argued ve-hemently that *T. rex* probably left multitudinous bite marks, despite the paucity of known specimens. Absence of evidence is not evidence of ab-sence, and we believe two factors account for this toothy gap in the fossil record. First, researchers have never systematically searched for bite marks. Even more important, collectors have had a natural bias against finds that might display bite marks. Historically, museums desire complete skeletons rather than single, isolated parts. But whole skeletons tend to be the remains of animals that died from causes other than predation and were rapidly buried before being dismembered by scavengers. The shred-ded bits of bodies eschewed by museums, such as the *Triceratops* pelvis, are precisely those specimens most likely to carry the evidence of feeding.

Indeed, Aase Roland Jacobsen of the University of Århus in Denmark recently surveyed isolated partial skeletal remains and compared them with nearly complete skeletons in Alberta. She found that 3.5 times as many of the individual bones (14 percent) bore theropod bite marks as did the less disrupted remains (4 percent). Paleobiologists therefore view the majority of the world's natural history museums as deserts of behavioral evidence when compared with fossils still lying in the field waiting to be discovered and interpreted.

HAWK OR VULTURE?

Some features of tyrannosaur biology, such as coloration, vocalizations or mating displays, may remain mysteries. But their feeding behavior is accessible through the fossil record. The collection of more trace fossils may finally settle a great debate in paleontology—the 80-year controversy over whether *T. rex* was a predator or a scavenger.

When *T. rex* was first found a century ago, scientists immediately labeled it a predator. But sharp claws and powerful jaws do not necessarily a predator make. For example, most bears are omnivorous and kill only a small proportion of their food. In 1917 Canadian paleontologist Lawrence Lambe examined a partial albertosaur skull and inferred that tyrannosaurs fed on soft, rotting carrion. He came to this conclusion after noticing that the teeth were relatively free of wear. (Future research would show that 40 percent of shed tyrannosaur teeth are severely worn and broken, damage that occurs in a mere two to three years, based on my estimates of their rates of tooth replacement.) Lambe thus established the minority view that the beasts were in fact giant terrestrial "vultures." The ensuing arguments in the predator-versus-scavenger dispute have centered on the anatomy and physical capabilities of *T. rex,* leading to a tiresome game of point-counterpoint.

Scavenger advocates adopted the "weak tooth theory," which maintained that *T. rex*'s elongate teeth would have failed in predatory struggles or in bone impacts. They also contended that its diminutive arms precluded lethal attacks and that *T. rex* would have been too slow to run down prey.

Predator supporters answered with biomechanical data. They cited my own bite-force studies that demonstrate that *T. rex* teeth were actually quite robust. (I personally will remain uncommitted in this argument until the discovery of direct physical proof.) They also noted that Kenneth Carpenter of the Denver Museum of Natural History and Matthew Smith, then at the Museum of the Rockies, estimated that the "puny" arms of a *T. rex* could curl nearly 180 kilograms. And they pointed to the work of Per Christiansen of the University of Copenhagen, who believes, based on limb proportion, that *T. rex* may have been able to sprint at 47 kilometers an hour (29 miles an hour). Such speed would be faster than that of any of

T. rex's contemporaries, although endurance and agility, which are difficult to quantify, are equally important in such considerations.

Even these biomechanical studies fail to resolve the predator-scavenger debate—and they never will. The critical determinant of *T. rex*'s ecological niche is discovering how and to what degree it utilized the animals living and dying in its environment, rather than establishing its presumed adeptness for killing. Both sides concede that predaceous animals, such as lions and spotted hyenas, will scavenge and that classic scavengers, such as vultures, will sometimes kill. And mounting physical evidence leads to the conclusion that tyrannosaurs both hunted and scavenged.

Within *T. rex*'s former range exist bone beds consisting of hundreds and sometimes thousands of edmontosaurs that died from floods, droughts and causes other than predation. Bite marks and shed tooth crowns in these edmontosaur assemblages attest to scavenging behavior by *T. rex*. Jacobsen has found comparable evidence for albertosaur scavenging. Carpenter, on the other hand, has provided evidence that he considers solid proof of predaceous behavior in the form of an unsuccessful attack by a *T. rex* on an adult *Edmontosaurus*. The intended prey purportedly escaped with several broken tailbones that later healed. The only animal with the stature, proper dentition and biting force to account for this injury is *T. rex*.

Quantification of such discoveries could help determine the degree to which *T. rex* undertook each method of obtaining food, and paleontologists could avoid future arguments by adopting standard definitions of predator and scavenger. Such a convention is necessary, because a wide range of views pervades vertebrate paleontology as to what exactly makes for each kind of feeder. For example, some extremists contend that if a carnivorous animal consumes any carrion at all, it should be called a scavenger. But such a constrained definition negates a meaningful ecological distinction, because it would include nearly all the world's carnivorous birds and mammals.

In a definition more consistent with most paleontologists' common-sense categorization, a predatory species would be one in which most individuals acquire most of their meals from animals they or their peers killed. Most individuals in a scavenging species, on the other hand, would not be responsible for the deaths of most of their food.

Trace fossils could open the door to a systematic approach to the predator-scavenger controversy, and the resolution could come from testing hypotheses about entire patterns of tyrannosaur feeding preferences. For instance, Jacobsen has pointed out that evidence of a preference for less dangerous or easily caught animals supports a predator niche. Conversely, scavengers would be expected to consume all species equally.

Within this logical framework, Jacobsen has compelling data supporting predation. She surveyed thousands of dinosaur bones from Alberta and learned that unarmored hadrosaurs are twice as likely to bear tyrannosaur bite marks as are the more dangerous horned ceratopsians. Tanke, who participated in the collection of these bones, relates that no bite marks have been found on the heavily armored, tanklike ankylosaurs.

Jacobsen cautions, though, that other factors confuse this set of findings. Most of the hadrosaur bones are from isolated individuals, but most ceratopsians in her study are from bone beds. Again, these beds contain more whole animals that have been fossilized unscathed, creating the kind of tooth-mark bias discussed earlier. A survey of isolated ceratopsians would be enlightening. And analysis of more bite marks that reveal failed predatory attempts, such as those reported by Carpenter, could also turn up preferences, or the lack thereof, for less dangerous prey.

Jacobsen's finding that cannibalism among tyrannosaurs was rare—only 2 percent of albertosaur bones had albertosaur bite marks, whereas 14 percent of herbivore bones did—might also support predatory preferences instead of a scavenging niche for *T. rex*, particularly if these animals were in fact gregarious. Assuming that they had no aversion to consuming flesh of their own kind, it would be expected that at least as many *T. rex* bones would exhibit signs of *T. rex* dining as do herbivore bones. A scavenging *T. rex* would have had to stumble on herbivore remains, but if *T. rex* traveled in herds, freshly dead conspecifics would seem to have been a guaranteed meal.

Coprolites may also provide valuable evidence about whether *T. rex* had any finicky eating habits. Because histological examination of bone found in coprolites can give the approximate stage of life of the consumed animal, Chin and I have suggested that coprolites may reveal a *T. rex* preference for feeding on vulnerable members of herds, such as the very young. Such a bias would point to predation, whereas a more impartial feeding pattern, matching the normal patterns of attrition, would indicate scavenging. Meaningful questions may lead to meaningful answers.

In the past century, paleontologists have recovered enough physical remains of *Tyrannosaurus rex* to give the world an excellent idea of what these monsters looked like. The attempt to discover what *T. rex* actually *was* like relies on those fossils that carry precious clues about the daily activities of dinosaurs. Paleontologists now appreciate the need for reanalysis of finds that were formerly ignored and have recognized the biases in collection practices, which have clouded perceptions of dinosaurs. The intentional pursuit of behavioral data should accelerate discoveries of dinosaur paleobiology. And new technologies may tease information out of fossils that we currently deem of little value. The *T. rex*, still alive in the imagination, continues to evolve.

FURTHER READING

Ralph E. Molnar and James O. Farlow. "Carnosaur Paleobiology" in *Dinosauria*. Edited by David B. Weishampel, Peter Dodson and Halszka Osmólska. University of California Press, 1990.

John R. Horner and Don Lessem. *The Complete T. Rex*. Simon & Schuster, 1993.

Gregory M. Erickson, Samuel D. Van Kirk, Jinntung Su, Marc E. Levenston, William E. Caler and Dennis R. Carter. "Bite-Force Estimation for *Tyrannosaurus rex* from Tooth-Marked Bones" in *Nature*, Vol. 382, pages 706–708; August 22, 1996.

Gregory M. Erickson. "Incremental Lines of von Ebner in Dinosaurs and the Assessment of Tooth Replacement Rates Using Growth Line Counts" in *Proceedings of the National Academy of Sciences USA*, Vol. 93, No. 25, pages 14623–14627; December 10, 1996.

Karen Chin, Timothy T. Tokaryk, Gregory M. Erickson and Lewis C. Calk. "A King-Sized Theropod Coprolite" in *Nature*, Vol. 393, pages 680–682; June 18, 1998.

Madagascar's Mesozoic Secrets

JOHN J. FLYNN AND ANDRÉ R. WYSS

ORIGINALLY PUBLISHED IN FEBRUARY 2002

Three weeks into our first fossil-hunting expedition in Madagascar in 1996, we were beginning to worry that dust-choked laundry might be all we would have to show for our efforts. We had turned up only a few random teeth and bones—rough terrain and other logistical difficulties had encumbered our search. With our field season drawing rapidly to a close, we finally stumbled on an encouraging clue in the southwestern part of the island. A tourist map hanging in the visitor center of Isalo National Park marked a local site called "the place of animal bones." We asked two young men from a neighboring village to take us there right away.

Our high hopes faded quickly as we realized the bleached scraps of skeletons eroding out of the hillside belonged to cattle and other modern-day animals. This site, though potentially interesting to archaeologists, held no promise of harboring the much more ancient quarry we were after. Later that day another guide, accompanied by two dozen curious children from the village, led us to a second embankment similarly strewn with bones. With great excitement we spotted two thumb-size jaw fragments that were undoubtedly ancient. They belonged to long-extinct, parrot-beaked cousins of the dinosaurs called rhynchosaurs.

The rhynchosaur bones turned out to be a harbinger of a spectacular slew of prehistoric discoveries yet to come. Since then, the world's fourth-largest island has become a prolific source of new information about animals that walked the land during the Mesozoic era, the interval of the earth's history (from 250 million to 65 million years ago) when both dinosaurs and mammals were making their debut. We have unearthed the bones of primitive dinosaurs that we suspect are older than any found previously. We have also stirred up controversy with the discovery of a shrewlike creature that seems to defy a prominent theory of mammalian history by being in the "wrong" hemisphere. These exquisite specimens, among numerous others collected over five field seasons, have enabled us to begin painting a picture of ancient Madagascar and to shape our strategy for a sixth expedition this summer.

Much of our research over the past two decades has been aimed at unraveling the history of land-dwelling animals on the southern continents. Such questions have driven other paleontologists to fossil-rich locales in South Africa, Brazil, Antarctica and India. Rather than probing those established sites for additional finds, we were lured to Madagascar: the island embraces vast swaths of Mesozoic age rocks, but until recently only a handful of terrestrial vertebrate fossils from that time had been discovered there. Why? We had a hunch that no one had looked persistently enough to find them.

Persistence became our motto as we launched our 1996 expedition. Our team consisted of a dozen scientists and students from the U.S. and the University of Antananarivo in Madagascar. Among other benefits, our partnership with the country's leading university facilitated the acquisition of collecting and exporting permits—requisite components of all paleontological fieldwork. Before long, however, we ran headlong into logistical obstacles that surely contributed to earlier failures to find ancient fossils on the island. Mesozoic deposits in western Madagascar are spread over an area roughly the size of California. Generations of oxcarts and foot travel have carved the only trails into more remote areas, and most of them are impassable by even the brawniest four-wheel-drive vehicles. We had to haul most of our food, including hundreds of pounds of rice, beans and canned meats, from the capital. Fuel shortages sometimes seriously restricted mobility, and our work was even thwarted by wildfires, which occur frequently and rage unchecked. New challenges often arose unexpectedly, requiring us to adjust our plans on the spot.

ANCIENT LUCK

Perhaps the most daunting obstacle we faced in prospecting such a large region was deciding where to begin. Fortunately, we were not planning our search blindly. The pioneering fieldwork of geologists such as Henri Besairie, who directed Madagascar's ministry of mines during the mid-1900s, provided us with large-scale maps of the island's Mesozoic rocks. From those studies we knew that a fortuitous combination of geologic factors had led to the accumulation of a thick blanket of sediments over most of Madagascar's western lowlands—and gave us good reason to believe that ancient bones and teeth might have been trapped and preserved there.

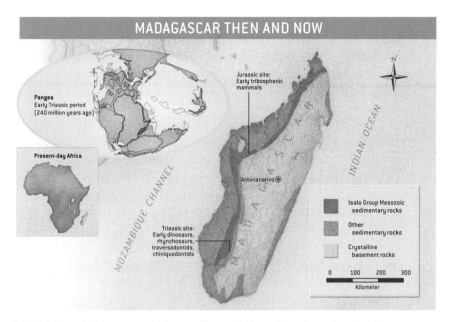

Fossil-bearing rocks drape western Madagascar. These rocks formed from the sand, mud, and occasional remnants of dead animals that accumulated in valleys when the island began to separate from Africa.

At the dawn of the Mesozoic era 250 million years ago, it would have been possible to walk from Madagascar to almost anywhere else in the world. All of the planet's landmasses were united in the supercontinent Pangea, and Madagascar was nestled between the west coast of what is now India and the east coast of present-day Africa. The world was a good deal warmer than at present—even the poles were free of ice. In the supercontinent's southern region, called Gondwana, enormous rivers coursed into lowland basins that would eventually become the Mozambique Channel, which today spans the 250 miles between Madagascar and eastern Africa.

These giant basins represent the edge of the geologic gash created as Madagascar began pulling away from Africa more than 240 million years ago. This seemingly destructive process, called rifting, is an extremely effective way to accumulate fossils. (Indeed, many of the world's most important fossil vertebrate localities occur in ancient rift settings—including the famous record of early human evolution in the much younger rift basins of east Africa.) The rivers flowing into the basins carried with them mud, sand, and occasionally the carcasses or bones of dead animals. Over time the rivers deposited this material as a sequence of vast layers. Continued rifting and the growing mass of sediment caused the floors of the basins to sink ever deeper. This depositional process persisted for nearly 100 million years, until the basin floors thinned to the breaking point and molten rock ascended from the planet's interior to fill the gap as new ocean crust.

Up to that point nature had afforded Madagascar three crucial ingredients required for fossil preservation: dead organisms, holes in which to bury them (rift basins), and material to cover them (sand and mud). But special conditions were also needed to ensure that the fossils were not destroyed during the subsequent 160 million years. Again, geologic circumstances proved fortuitous. As the newly separated landmasses of Africa and Madagascar drifted farther apart, their sediment-laden coastlines rarely experienced volcanic eruptions or other events that could have destroyed buried fossils. Also key for fossil preservation is that the ancient rift basins ended up on the western side of the island, which today is dotted with dry forests, grasslands and desert scrub. In a more humid environment, such deposits would have eroded away or would be hidden under dense vegetation like the kind that hugs much of the island's eastern coast.

Initially Madagascar remained attached to the other Gondwanan landmasses: India, Australia, Antarctica and South America. It did not attain islandhood until it split from India about 90 million years ago. Sometime since then, the island acquired its suite of bizarre modern creatures, of which lemurs are the best known. For more than a century, researchers have wondered how long these modern creatures and their ancestors have inhabited the island. Illuminating discoveries by another team of paleontologists indicate that almost all major groups of living vertebrates arrived on Madagascar since sometime near the end of the Mesozoic era 65 million years ago. Our own probing has focused on a more ancient interval of Madagascar's history—the first two periods of the Mesozoic era.

PAY DIRT

One of the joys of working in little-charted terrain has been that if we manage to find anything, its scientific significance is virtually assured. That's why our first discoveries near Isalo National Park were so exciting. The same evening in 1996 that we found the rhynchosaur jaw fragments, University of Antananarivo student Léon Razafimanantsoa spotted the six-inch-long skull of another interesting creature. We immediately identified the animal as a peculiar plant eater, neither mammal nor reptile, called a traversodontid cynodont.

The rhynchosaur jaws and the exquisite traversodontid skull—the first significant discoveries of our ongoing U.S.-Malagasy project—invigorated our expedition. The first fossil is always the hardest one to find; now we could hunker down and do the detailed collecting work necessary to begin piecing together an image of the past. The white sandstones we were excavating had formed from the sand carried by the rivers that poured into lowlands as Madagascar unhinged from Africa. Within these prehistoric valleys rhynchosaurs and traversodontids, both four-legged creatures ranging from three to 10 feet in length, probably grazed together much the same way zebras and wildebeests do in Africa today. The presence of rhynchosaurs, which are relatively common in coeval rocks around the world, narrowed the date of this picture to sometime within the Triassic period (the first of three Mesozoic time intervals), which spans from 250 million to 205 million years ago. And because traversodontids were much more diverse and abundant during the first half of the Triassic than during the second, we thought initially that this scene played out sometime before about 230 million years ago.

During our second expedition, in 1997, a third type of animal challenged our sense of where we were in time. Shortly after we arrived in southwestern Madagascar, one of our field assistants, a local resident named Mena, showed us some bones that he had found across the river from our previous localities. We were struck by the fine-grained red rock adhering to the bones—everything we had found until that point was buried in the coarse white sandstone. Mena led us about half a mile north of the rhynchosaur and traversodontid site to the bottom of a deep gully. Within a few minutes we spotted the bone-producing layer from which his unusual specimens had rolled. A rich concentration of fossils was entombed within the three-foot-thick layer of red mudstones, which had formed in the floodplains of the same ancient rivers that deposited the white sands. Excavation yielded about two dozen specimens of what appeared to be dinosaurs. Our team found jaws, strings of vertebrae, hips, claws, an articulated forearm with some wrist bones, and other assorted skeletal elements. When we examined these and other bones more closely, we realized that we had uncovered remains of two different species of prosauropods (not yet formally named), one of which appears to resemble a species from Morocco called *Azendohsaurus*. These prosauropods, which typically appear in rocks between 225 million and 190 million years old, are smaller-bodied precursors of the long-necked sauropod dinosaurs, including such behemoths as *Brachiosaurus*.

When we discovered that dinosaurs were foraging among rhynchosaurs and traversodontids, it became clear that we had unearthed a collection of fossils not known to coexist anywhere else. In Africa, South America and other parts of the world, traversodontids are much less abundant and less diverse once dinosaurs appear. Similarly, the most common type of rhynchosaur we found, *Isalorhynchus,* lacks advanced characteristics and thus is inferred to be more ancient than the group of rhynchosaurs that is found with other early dinosaurs. What is more, the Malagasy fossil assemblage lacks remains of several younger reptile groups usually found with the earliest dinosaurs, including the heavily armored, crocodilelike phytosaurs and aetosaurs. The occurrence of dinosaurs with more ancient kinds of animals, plus the lack of younger groups, suggests that the Malagasy prosauropods are as old as any dinosaur ever discovered, if not older.

Only one early dinosaur site—at Ischigualasto, Argentina—contains a rock layer that has been dated directly; all other early dinosaur sites with similar fossils are thus estimated to be no older than its radioisotopic age of about 228 million years. (Reliable radioisotopic ages for fossils are obtainable only from rock layers produced by contemporaneous volcanoes. The Malagasy sediments accumulated in a rift basin with no volcanoes nearby.) Based on the fossils present, we have tentatively concluded that our dinosaur-bearing rocks slightly predate the Ischigualasto time span. And because prosauropods represent one of the major branches of the dinosaur evolutionary tree, we know that the common ancestor of all dinosaurs must be older still. Rocks from before about 245 million years ago have been moderately well sampled around the world, but none of them has yielded dinosaurs. That means the search for the common ancestor of all dinosaurs must focus on a relatively poorly known and ever narrowing interval of Middle Triassic rocks, between about 240 million and 230 million years old.

MOSTLY MAMMALS

Dinosaurs naturally attract considerable attention, being the most conspicuous land animals of the Mesozoic. Less widely appreciated is the fact that mammals and dinosaurs sprang onto the evolutionary stage at nearly the same time. At least two factors account for the popular misconception that mammals arose only after dinosaurs became extinct: Early mammals all were chipmunk-size or smaller, so they don't grab the popular imagination in the way their giant Mesozoic contemporaries do. In addition, the fossil record of early mammals is quite sparse, apart from very late in the Mesozoic. To our delight, Madagascar has once again filled in two mysterious gaps in the fossil record. The traversodontid cynodonts from the Isalo deposits reveal new details about close mammalian relatives, and a younger fossil from the northwest side of the island poses some controversial questions about where and when a key advanced group of mammals got its start.

The Malagasy traversodontids, the first known from the island, include some of the best-preserved representatives of early cynodonts ever discovered. ("Cynodontia" is the name applied to a broad group of land animals that includes mammals and their nearest relatives.) Accordingly, these bones provide a wealth of anatomical information previously undocumented for these creatures. These cynodonts are identified by, among other diagnostic features, a simplified lower jaw that is dominated by a single bone, the dentary. Some specimens include both skulls and skeletons. Understanding the complete morphology of these animals is crucial for resolving the complex evolutionary transition from the large cold-blooded, scale-covered animals with sprawling limbs (which dominated the continents prior to the Mesozoic) to the much smaller warm-blooded, furry animals with an erect posture that are so plentiful today.

Many kinds of mammals, with many anatomical variations, now inhabit the planet. But they all share a common ancestor marked by a single, distinctive suite of features. To determine what these first mammals looked like, paleontologists must examine their closest evolutionary relatives within the Cynodontia, which include the traversodontids and their much rarer cousins, the chiniquodontids (also known as probainognathians), both of which we have found in southwestern Madagascar. Traversodontids almost certainly were herbivorous, because their wide cheek teeth are designed for grinding. One of our four new Malagasy traversodontid species also has large, stout, forward-projecting incisors for grasping vegetation. The chiniquodontids, in contrast, were undoubtedly carnivorous, with sharp, pointed teeth. Most paleontologists agree that some chiniquodontids share a more recent common ancestor with mammals than the herbivorous traversodontids do. The chiniquodontid skulls and skeletons we found in Madagascar will help reconstruct the bridge between early cynodonts and true mammals.

Not only are Madagascar's Triassic cynodonts among the best preserved in the world, they also sample a time period that is poorly known elsewhere. The same is true for the youngest fossils our expeditions have uncovered—those from a region of the northwest where the sediments are about 165 million years in age. (That date falls within the middle of the Jurassic, the second of the Mesozoic's three periods.) Because these sediments were considerably younger than our Triassic rocks, we allowed ourselves the hope that we might find remains of an ancient mammal. Not a single mammal had been recorded from Jurassic rocks of a southern landmass at that point, but this did nothing to thwart our motivation.

Once again, persistence paid off. During our 1996 field season, we had visited the village of Ambondromahabo after hearing local reports of abundant large fossils of the sauropod dinosaur *Lapperentosaurus.* Sometimes where large animals are preserved, the remains of smaller animals can also be found—though not as easily. We crawled over the landscape, eyes held a few inches from the ground. This uncomfortable but time-tested strategy turned up a few small theropod dinosaur teeth, fish scales and other bone fragments, which had accumulated at the surface of a small mound of sediment near the village.

These unprepossessing fossils hinted that more significant items might be buried in the sediment beneath. We bagged about 200 pounds of sediment and washed it through mosquito-net hats back in the capital, Antananarivo, while waiting to be granted permits for the second leg of our trip—the leg to the southwest that turned up our first rhynchosaur jaws and traversodontid skull.

During the subsequent years back in the U.S., while our studies focused on the exceptional Triassic material, the tedious process of sorting the Jurassic sediment took place. A dedicated team of volunteers at the Field Museum in Chicago—Dennis Kinzig, Ross Chisholm and Warren Valsa—spent many a weekend sifting through the concentrated sediment under a microscope in search of valuable flecks of bone or teeth. We didn't think much about that sediment again until 1998, when Kinzig relayed the news that they had uncovered the partial jawbone of a tiny mammal with three grinding teeth still in place. We were startled not only by the jaw's existence but also by its remarkably advanced cheek teeth. The shapes of the teeth document the earliest occurrence of Tribosphenida, a group encompassing the vast majority of living mammals. We named the new species *Ambondro mahabo,* after its place of origin.

The discovery pushes back the geologic range of this group of mammals by more than 25 million years and offers the first glimpse of mammalian evolution on the Southern continents during the last half of the Jurassic period. It shows that this subgroup of mammals may have evolved in the Southern Hemisphere rather than the Northern, as is commonly supposed. Although the available information does not conclusively resolve the debate, this important addition to the record of early fossil mammals does point out the precarious nature of long-standing assumptions rooted in a fossil record historically biased toward the Northern Hemisphere.

PLANNING PERSISTENTLY

Although our team has recovered a broad spectrum of fossils in Madagascar, scientists are only beginning to describe the Mesozoic history of the Southern continents. The number of species of Mesozoic land vertebrates known from Australia, Antarctica, Africa and South America is probably an order of magnitude smaller than the number of contemporaneous findings from the Northern Hemisphere. Clearly, Madagascar now ranks as one of the world's top prospects for adding important insights to paleontologists' knowledge of the creatures that once roamed Gondwana.

Often the most significant hypotheses about ancient life on the earth can be suggested only after these kinds of new fossil discoveries are made. Our team's explorations provide two cases in point: the fossils found alongside the Triassic prosauropods indicate that dinosaurs debuted earlier than previously recorded, and the existence of the tiny mammal at our Jurassic site implies that tribosphenic mammals may have originated in the Southern, rather than Northern, Hemisphere. The best way to bolster these proposals (or to prove them wrong) is to go out and uncover more bones. That is why our primary goal this summer will be the same as it has been for our past five expeditions: find as many fossils as possible.

Our agenda includes digging deeper into known sites and surveying new regions, blending risky efforts with sure bets. No matter how carefully formulated, however, our plans will be subject to last-minute changes, dictated by such things as road closures and our most daunting challenge to date, the appearance of frenzied boomtowns.

During our first three expeditions, we never gave a second thought to the gravels that overlay the Triassic rock outcrops in the southwestern part of the island. Little did we know that those gravels contain sapphires. By 1999 tens of thousands of people were scouring the landscape in search of these gems. The next year all our Triassic sites fell within sapphire-mining claims. Those areas are now off limits to everyone, including paleontologists, unless they get permission from both the claim holder and the government. Leaping that extra set of hurdles will be one of our foremost tasks this year.

Even without such logistical obstacles slowing our progress, it would require uncountable lifetimes to carefully survey all the island's untouched rock exposures. But now that we have seen a few of Madagascar's treasures, we are inspired to keep digging—and to reveal new secrets.

FURTHER READING

Ken Preston-Mafham. *Madagascar: A Natural History.* Foreword by Sir David Attenborough. Facts on File, 1991.

Steven M. Goodman and Bruce D. Patterson, eds. *Natural Change and Human Impact in Madagascar.* Smithsonian Institution Press, 1997.

John J. Flynn, J. Michael Parrish, Berthe Rakotosaminimanana, William F. Simpson and André R. Wyss. "A Middle Jurassic Mammal from Madagascar" in *Nature,* Vol. 401, pages 57–60; September 2, 1999.

John J. Flynn, J. Michael Parrish, Berthe Rakotosaminimanana, William F. Simpson, Robin L. Whatley and André R. Wyss. "A Triassic Fauna from Madagascar, Including Early Dinosaurs" in *Science,* Vol. 286, pages 763–765; October 22, 1999.

Which Came First, the Feather or the Bird?

RICHARD O. PRUM AND ALAN H. BRUSH

ORIGINALLY PUBLISHED IN FEBRUARY 2002

Hair, scales, fur, feathers. Of all the body coverings nature has designed, feathers are the most various and the most mysterious. How did these incredibly strong, wonderfully lightweight, amazingly intricate appendages evolve? Where did they come from? Only in the past five years have we begun to answer this question. Several lines of research have recently converged on a remarkable conclusion: the feather evolved in dinosaurs before the appearance of birds.

The origin of feathers is a specific instance of the much more general question of the origin of evolutionary novelties—structures that have no clear antecedents in ancestral animals and no clear related structures (homologues) in contemporary relatives. Although evolutionary theory provides a robust explanation for the appearance of minor variations in the size and shape of creatures and their component parts, it does not yet give as much guidance for understanding the emergence of entirely new structures, including digits, limbs, eyes and feathers.

Progress in solving the particularly puzzling origin of feathers has also been hampered by what now appear to be false leads, such as the assumption that the primitive feather evolved by elongation and division of the reptilian scale, and speculations that feathers evolved for a specific function, such as flight. A lack of primitive fossil feathers hindered progress as well. For many years the earliest bird fossil has been *Archaeopteryx lithographica,* which lived in the Late Jurassic period (about 148 million years ago). But *Archaeopteryx* offers no new insights on how feathers evolved, because its own feathers are nearly indistinguishable from those of today's birds.

Very recent contributions from several fields have put these traditional problems to rest. First, biologists have begun to find new evidence for the idea that developmental processes—the complex mechanisms by which an individual organism grows to its full size and form—can provide a window into the evolution of a species' anatomy. This idea has been reborn as the field of evolutionary developmental biology, or "evo-devo." It has given us a powerful tool for probing the origin of feathers. Second, paleontologists have unearthed a trove of feathered dinosaurs in China. These animals have a diversity of primitive feathers that are not as highly evolved as those of today's birds or even *Archaeopteryx*. They give us critical clues about the structure, function and evolution of modern birds' intricate appendages.

Together these advances have produced an highly detailed and revolutionary picture: feathers originated and diversified in carnivorous, bipedal theropod dinosaurs before the origin of birds or the origin of flight.

THE TOTALLY TUBULAR FEATHER

This surprising picture was pieced together thanks in large measure to a new appreciation of exactly what a feather is and how it develops in modern birds. Like hair, nails and scales, feathers are integumentary appendages—skin organs that form by controlled proliferation of cells in the epidermis, or outer skin layer, that produce the keratin proteins. A typical feather features a main shaft, called the rachis. Fused to the rachis are a series of branches, or barbs. In a fractal-like reflection of the branched rachis and barbs, the barbs themselves are also branched: a series of paired filaments called barbules are fused to the main shaft of the barb, the ramus. At the base of the feather, the rachis expands to form the hollow tubular calamus, or quill, which inserts into a follicle in the skin. A bird's feathers are replaced periodically during its life through molt—the growth of new feathers from the same follicles.

Variations in the shape and microscopic structure of the barbs, barbules and rachis create an astounding range of feathers. But despite this diversity, most feathers fall into two structural classes. A typical pennaceous feather has a prominent rachis and barbs that create a planar vane. The barbs in the vane are locked together by pairs of specialized barbules. The barbules that extend toward the tip of the feather have a series of tiny hooklets that interlock with grooves in the neighboring barbules. Pennaceous feathers cover the bodies of birds, and their tightly closed vanes create the aerodynamic surfaces of the wings and tail. In dramatic contrast to pennaceous feathers, a plumulaceous, or downy, feather has only a rudimentary rachis and a jumbled tuft of barbs with long barbules. The long, tangled barbules give these feathers their marvelous properties of lightweight thermal insulation and comfortable loft. Feathers can have a pennaceous vane and a plumulaceous base.

THE NATURE OF FEATHERS

FEATHERS DISPLAY AN AMAZING DIVERSITY and serve almost as wide a range of functions, from courtship to camouflage to flight. Variations in the shapes of a feather's components—the barbs, barbules and rachis—create this diversity. Most feathers, however, fall into two basic types. The pennaceous is the iconic feather of a quill pen or a bird's wing. The plumulaceous, or downy, feather has soft, tangled plumes that provide lightweight insulation.

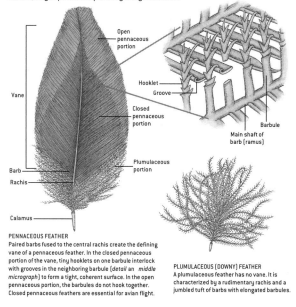

Open pennaceous portion

Hooklet

Groove

Vane

Closed pennaceous portion

Barbule

Main shaft of barb (ramus)

Plumulaceous portion

Barb

Rachis

Calamus

PENNACEOUS FEATHER
Paired barbs fused to the central rachis create the defining vane of a pennaceous feather. In the closed pennaceous portion of the vane, tiny hooklets on one barbule interlock with grooves in the neighboring barbule (*detail* an *middle micrograph*) to form a tight, coherent surface. In the open pennaceous portion, the barbules do not hook together. Closed pennaceous feathers are essential for avian flight.

PLUMULACEOUS (DOWNY) FEATHER
A plumulaceous feather has no vane. It is characterized by a rudimentary rachis and a jumbled tuft of barbs with elongated barbules.

In essence, all feathers are variations on a tube produced by proliferating epidermis with the nourishing dermal pulp in the center. And even though a feather is branched like a tree, it grows from its base like a hair. How do feathers accomplish this?

Feather growth begins with a thickening of the epidermis called the placode, which elongates into a tube—the feather germ. Proliferation of cells in a ring around the feather germ creates a cylindrical depression, the follicle, at its base. The growth of keratin cells, or keratinocytes, in the epidermis of the follicle—the follicle "collar"—forces older cells up and out, eventually creating the entire feather in an elaborate choreography that is one of the wonders of nature.

As part of that choreography, the follicle collar divides into a series of longitudinal ridges—barb ridges—that create the separate barbs. In a pennaceous feather, the barbs grow helically around the tubular feather germ and fuse on one side to form the rachis. Simultaneously, new barb ridges form on the other side of the tube. In a plumulaceous feather, barb ridges grow straight without any helical movement. In both types of feather, the barbules that extend from the barb ramus grow from a single layer of cells, called the barbule plate, on the periphery of the barb ridge.

EVO-DEVO COMES TO THE FEATHER

Together with various colleagues, we think the process of feather development can be mined to reveal the probable nature of the primitive structures that were the evolutionary precursors of feathers. Our developmental theory proposes that feathers evolved through a series of transitional stages, each marked by a developmental evolutionary novelty, a new mechanism of growth. Advances at one stage provided the basis for the next innovation.

In 1999 we proposed the following evolutionary scheme. Stage 1 was the tubular elongation of the placode from a feather germ and follicle.

HOW FEATHERS GROW

AS IN HAIR, NAILS AND SCALES, feathers grow by proliferation and differentiation of keratinocytes. These keratin-producing cells in the epidermis, or outer skin layer, achieve their purpose in life when they die, leaving behind a mass of deposited keratin. Keratins are filaments of proteins that polymerize to form solid structures. Feathers are made of beta-keratins, which are unique to reptiles and birds. The outer covering of the growing feather, called the sheath, is made of the softer alpha-keratin, which is found in all vertebrates and makes up our own skin and hair.

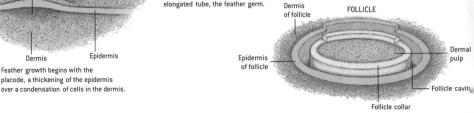

Feather germ

Condensation of cells

Placode

The placode then forms a unique elongated tube, the feather germ.

Proliferation of cells in a ring around the feather germ creates the follicle (*detail below*), the organ that generates the feather. At the base of the follicle, in the follicle collar, the continuing production of keratinoctyes forces older cells up and out, eventually forming the entire, tubular feather.

Dermis of follicle

FOLLICLE

Dermal pulp

Epidermis of follicle

Dermis   Epidermis

Feather growth begins with the placode, a thickening of the epidermis over a condensation of cells in the dermis.

Follicle cavity

Follicle collar

This yielded the first feather—an unbranched, hollow cylinder. Then, in stage 2, the follicle collar, a ring of epidermal tissue, differentiated (specialized): the inner layer became the longitudinal barb ridges, and the outer layer became a protective sheath. This stage produced a tuft of barbs fused to the hollow cylinder, or calamus.

The model has two alternatives for the next stage—either the origin of helical growth of barb ridges and formation of the rachis (stage 3a) or the origin of the barbules (stage 3b). The ambiguity about which came first arises because feather development does not indicate clearly which event occurred before the other. A stage 3a follicle would produce a feather with a rachis and a series of simple barbs. A stage 3b follicle would generate a tuft of barbs with branched barbules. Regardless of which stage came first, the evolution of both these features, stage 3a+b, would yield the first double-branched feathers, exhibiting a rachis, barbs and barbules. Because barbules were still undifferentiated at this stage, a feather would be

Sheath

Rachis ridge

Barb ridge

The outermost epidermal layer becomes the feather sheath, a temporary structure that protects the growing feather. Meanwhile the internal epidermal layer becomes partitioned into a series of compartments, called barb ridges, which subsequently grow to become the barbs of the feather.

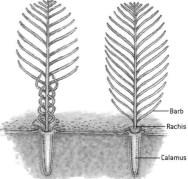

Barb

Rachis

Calamus

HELICAL GROWTH

Rachis ridge

Newly forming barb ridge

Follicle collar

Artery

In a pennaceous feather, the barb ridges grow helically around the collar until they fuse to form the rachis ridge. Subsequent barb ridges fuse to the rachis ridge. In a plumulaceous feather (*not shown*), barb ridges do not grow helically, and a simple rachis forms at the base of the feather.

As growth proceeds, the feather emerges from its superficial sheath. The feather then unfurls to obtain its planar shape. When the feather reaches its final size, the follicle collar forms the calamus, a simple tube at the base of the feather.

ANIMATIONS OF HOW FEATHERS GROW CAN BE VIEWED AT
http://fallon.anatomy.wisc.edu/feather.html

EVO-DEVO AND THE FEATHER

THE AUTHORS' THEORY of feather origin grew out of the realization that the mechanisms of development can help explain the evolution of novel features—a field dubbed evo-devo. The model proposes that the unique characteristics of feathers evolved through a series of evolutionary novelties in how they grow, each of which was essential for the appearance of the next stage. Thus, the theory bases its proposals on knowledge of the steps of feather development today rather than assumptions about what feathers might have been used for or about the groups of animals in which they might have evolved.

New fossil discoveries from Liaoning, China, provide the first insights into which theropod dinosaurs evolved the feathers of each hypothesized stage. Based on the similarities between the primitive feather predictions of the model and the shapes of the fossil skin appendages, the authors suggest that each stage evolved in a particular group of dinosaurs.

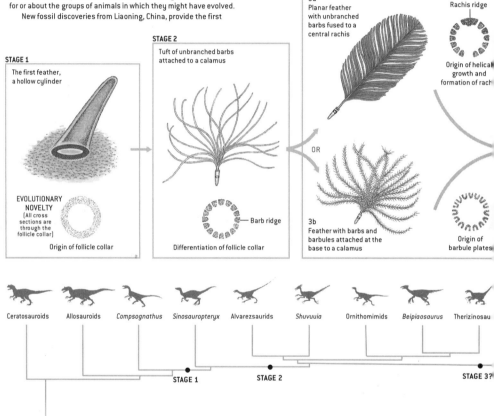

STAGE 1

The first feather, a hollow cylinder

EVOLUTIONARY NOVELTY
(All cross sections are through the follicle collar)

Origin of follicle collar

STAGE 2

Tuft of unbranched barbs attached to a calamus

Barb ridge

Differentiation of follicle collar

STAGE 3

3a
Planar feather with unbranched barbs fused to a central rachis

Rachis ridge

Origin of helical growth and formation of rach

OR

3b
Feather with barbs and barbules attached at the base to a calamus

Origin of barbule plates

Ceratosauroids   Allosauroids   Compsognathus   Sinosauropteryx   Alvarezsaurids   Shuvuuia   Ornithomimids   Beipiaosaurus   Therizinosau

STAGE 1   STAGE 2   STAGE 3?

open pennaceous—that is, its vane would not form a tight, coherent surface in which the barbules are locked together.

In stage 4 the capacity to grow differentiated barbules evolved. This advance enabled a stage 4 follicle to produce hooklets at the ends of barbules that could attach to the grooved barbules of the adjacent barbs

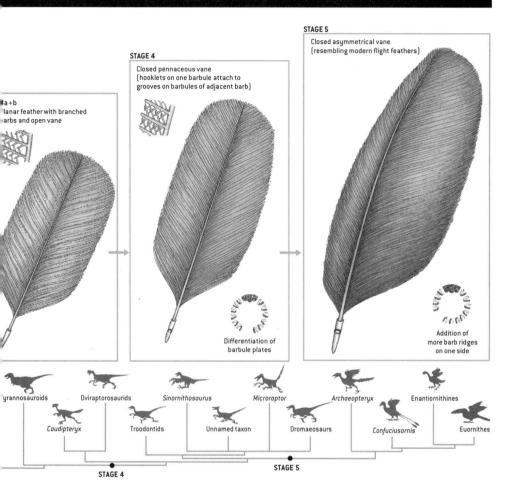

STAGE 5
Closed asymmetrical vane
(resembling modern flight feathers)

STAGE 4
Closed pennaceous vane
(hooklets on one barbule attach to
grooves on barbules of adjacent barb)

a + b
Planar feather with branched
barbs and open vane

Differentiation of
barbule plates

Addition of
more barb ridges
on one side

Tyrannosauroids   Oviraptorosaurids   *Sinornithosaurus*   *Microraptor*   *Archaeopteryx*   Enantiornithines

Caudipteryx   Troodontids   Unnamed taxon   Dromaeosaurs   *Confuciusornis*   Euornithes

STAGE 4

STAGE 5

and create a pennaceous feather with a closed vane. Only after stage 4 could additional feather variations evolve, including the many specializations we see at stage 5, such as the asymmetrical vane of a flight feather.

THE SUPPORTING CAST

Inspiration for the theory came from the hierarchical nature of feather development itself. The model hypothesizes, for example, that a simple tubular feather preceded the evolution of barbs because barbs are created by the differentiation of the tube into barb ridges. Likewise, a plumulaceous tuft of barbs evolved before the pennaceous feather with a rachis because the rachis is formed by the fusion of barb ridges. Similar logic underlies each of the hypothesized stages of the developmental model.

Support for the theory comes in part from the diversity of feathers among modern birds, which sport feathers representing every stage of the model. Obviously, these feathers are recent, evolutionarily derived simplifications that merely revert back to the stages that arise during evolution, because complex feather diversity (through stage 5) must have evolved before *Archaeopteryx*. These modern feathers demonstrate that all the hypothesized stages are within the developmental capacity of feather follicles. Thus, the developmental theory of feather evolution does not require any purely theoretical structures to explain the origin of all feather diversity.

Support also comes from exciting new molecular findings that have confirmed the first three stages of the evo-devo model. Recent technological advances allow us to peer inside cells and identify whether specific genes are expressed (turned on so that they can give rise to the products they encode). Several laboratories have combined these methods with experimental techniques that investigate the functions of the proteins made when their genes are expressed during feather development. Matthew Harris and John F. Fallon of the University of Wisconsin—Madison and one of us (Prum) have studied two important pattern formation genes—*sonic hedgehog (Shh)* and *bone morphogenetic protein 2 (Bmp2)*. These genes play a crucial role in the growth of vertebrate limbs, digits, and integumentary appendages such as hair, teeth and nails. We found that Shh and Bmp2 proteins work as a modular pair of signaling molecules that, like a general-purpose electronic component, is reused repeatedly throughout feather development. The Shh protein induces cell proliferation, and the Bmp2 protein regulates the extent of proliferation and fosters cell differentiation.

The expression of Shh and Bmp2 begins in the feather placode, where the pair of proteins is produced in a polarized anterior-posterior pattern. Next, Shh and Bmp2 are both expressed at the tip of the tubular feather germ during its initial elongation and, following that, in the epithelium that separates the forming barb ridges, establishing a pattern for the growth of the ridges. Then in pennaceous feathers, the Shh and Bmp2 signaling lays down a pattern for helical growth of barb ridges and rachis formation, whereas in plumulaceous feathers the Shh and Bmp2 signals create a simpler pattern of barb growth. Each stage in the development of a feather has a distinct pattern of Shh and Bmp2 signaling. Again and again the two proteins perform critical tasks as the feather unfolds to its final form.

These molecular data confirm that feather development is composed of a series of hierarchical stages in which subsequent events are mechanistically dependent on earlier stages. For example, the evolution of longitudinal stripes in Shh-Bmp2 expression is contingent on the prior development of an elongate tubular feather germ. Likewise, the variations in Shh-Bmp2 patterning during pennaceous feather growth are contingent on the prior establishment of the longitudinal stripes. Thus, the molecular data are beautifully consistent with the scenario that feathers evolved from an elongate hollow tube (stage 1), to a downy tuft of barbs (stage 2), to a pennaceous structure (stage 3a).

THE STARS OF THE DRAMA

New conceptual theories have spurred our thinking, and state-of-the-art laboratory techniques have enabled us to eavesdrop on the cell as it gives life and shape to a feather. But plain old-fashioned detective work in fossil-rich quarries in northern China has turned up the most spectacular evidence for the developmental theory. Chinese, American and Canadian paleontologists working in Liaoning Province have unearthed a startling trove of fossils in the early Cretaceous Yixian formation (124 to 128 million years old). Excellent conditions in the formation have preserved an array of ancient organisms, including the earliest placental mammal, the earliest flowering plant, an explosion of ancient birds [see "The Origin of Birds and Their Flight," by Kevin Padian and Luis M. Chiappe; *Scientific American,* February 1998], and a diversity of theropod dinosaur fossils with sharp integumentary details. Various dinosaur fossils clearly show fully modern feathers and a variety of primitive feather structures. The conclusions are inescapable: feathers originated and evolved their essentially modern structure in a lineage of terrestrial, bipedal, carnivorous dinosaurs before the appearance of birds or flight.

The first feathered dinosaur found there, in 1997, was a chicken-size coelurosaur (*Sinosauropteryx*); it had small tubular and perhaps branched structures emerging from its skin. Next the paleontologists discovered a turkey-size oviraptoran dinosaur (*Caudipteryx*) that had beautifully preserved modern-looking pennaceous feathers on the tip of its tail and forelimbs. Some skeptics have claimed that *Caudipteryx* was merely an early flightless bird, but many phylogenetic analyses place it among the oviraptoran theropod dinosaurs. Subsequent discoveries at Liaoning have revealed pennaceous feathers on specimens of dromaeosaurs, the theropods, which are hypothesized to be most closely related to birds but which clearly are not birds. In all, investigators found fossil feathers from more than a dozen nonavian theropod dinosaurs, among them the ostrich-size therizinosaur *Beipiaosaurus* and a variety of dromaeosaurs, including *Microraptor* and *Sinornithosaurus*.

The heterogeneity of the feathers found on these dinosaurs is striking and provides strong direct support for the developmental theory. The most primitive feathers known—those of *Sinosauropteryx*—are the simplest tubular structures and are remarkably like the predicted stage 1 of the developmental model. *Sinosauropteryx, Sinornithosaurus* and some other non-avian theropod specimens show open tufted structures that lack a rachis and are strikingly congruent with stage 2 of the model. There are also pennaceous feathers that obviously had differentiated barbules and coherent planar vanes, as in stage 4 of the model.

These fossils open a new chapter in the history of vertebrate skin. We now know that feathers first appeared in a group of theropod dinosaurs and diversified into essentially modern structural variety within other lineages of theropods before the origin of birds. Among the numerous feather-bearing dinosaurs, birds represent one particular group that evolved the ability to fly using the feathers of its specialized forelimbs and tail. *Caudipteryx, Protopteryx* and dromaeosaurs display a prominent "fan" of feathers at the tip of the tail, indicating that even some aspects of the plumage of modern birds evolved in theropods.

The consequence of these amazing fossil finds has been a simultaneous redefinition of what it means to be a bird and a reconsideration of the biology and life history of the theropod dinosaurs. Birds—the group that includes all species descended from the most recent common ancestor of *Archaeopteryx* and modern birds—used to be recognized as the flying, feathered vertebrates. Now we must acknowledge that birds are a group of the feathered theropod dinosaurs that evolved the capacity of powered flight. New fossil discoveries have continued to close the gap between birds and dinosaurs and ultimately make it more difficult even to define birds. Conversely, many of the most charismatic and culturally iconic dinosaurs, such as *Tyrannosaurus* and *Velociraptor,* are very likely to have had feathered skin but were not birds.

A FRESH LOOK

Thanks to the dividends provided by the recent findings, researchers can now reassess the various earlier hypotheses about the origin of feathers.

The new evidence from developmental biology is particularly damaging to the classical theory that feathers evolved from elongate scales. According to this scenario, scales became feathers by first elongating, then growing fringed edges, and finally producing hooked and grooved barbules. As we have seen, however, feathers are tubes; the two planar sides of the vane—in other words, the front and the back—are created by the inside and outside of the tube only after the feather unfolds from its cylindrical sheath. In contrast, the two planar sides of a scale develop from the top and bottom of the initial epidermal outgrowth that forms the scale.

The fresh evidence also puts to rest the popular and enduring theory that feathers evolved primarily or originally for flight. Only highly evolved feather shapes—namely, the asymmetrical feather with a closed vane, which did not occur until stage 5—could have been used for flight. Proposing that feathers evolved for flight now appears to be like hypothesizing that fingers evolved to play the piano. Rather feathers were "exapted" for their aerodynamic function only after the evolution of substantial developmental and structural complexity. That is, they evolved for some other purpose and were then exploited for a different use.

Numerous other proposed early functions of feathers remain plausible, including insulation, water repellency, courtship, camouflage and defense. Even with the wealth of new paleontological data, though, it seems unlikely that we will ever gain sufficient insight into the biology and natural history of the specific lineage in which feathers evolved to distinguish among these hypotheses. Instead our theory underscores that feathers evolved by a series of developmental innovations, each of which may have evolved for a different original function. We do know, however, that feathers emerged only after a tubular feather germ and follicle formed in the skin of some species. Hence, the first feather evolved because the first tubular appendage that grew out of the skin provided some kind of survival advantage.

Creationists and other evolutionary skeptics have long pointed to feathers as a favorite example of the insufficiency of evolutionary theory. There were no transitional forms between scales and feathers, they have argued. Further, they asked why natural selection for flight would first divide an elongate scale and then evolve an elaborate new mechanism to weave it back together. Now, in an ironic about-face, feathers offer a sterling example of how we can best study the origin of an evolutionary novelty: focus on understanding those features that are truly new and examine how they form during development in modern organisms. This new paradigm in evolutionary biology is certain to penetrate many more mysteries. Let our minds take wing.

FURTHER READING

Richard O. Prum. "Development and Evolutionary Origin of Feathers" in *Journal of Experimental Zoology (Molecular and Developmental Evolution),* Vol. 285, No. 4, pages 291–306; December 15, 1999.

Alan H. Brush. "Evolving a Protofeather and Feather Diversity" in *American Zoologist,* Vol. 40, No. 4, pages 631–639; 2000.

Matthew P. Harris, John F. Fallon and Richard O. Prum. "Rapid Communication: Shh-Bmp2 Signaling Module and the Evolutionary Origin and Diversification of Feathers" in *Journal of Experimental Zoology,* Vol. 294, No. 2, pages 160–176; August 15, 2002.

Richard O. Prum and Alan H. Brush. "The Evolutionary Origin and Diversification of Feathers" in *Quarterly Review of Biology,* Vol. 77, No. 3, pages 261–295; September 2002.

The Terror Birds of South America

LARRY G. MARSHALL

ORIGINALLY PUBLISHED IN FEBRUARY 1994

It is a summer day on the pampas of central Argentina some five million years ago. A herd of small, horselike mammals are grazing peacefully in the warm sun. None of the animals is aware of the vigilant creature standing 50 meters away in the tall grass. Most of the watcher's trim, feathered body is concealed by the vegetation. Its eyes, set on the sides of a disproportionately large head perched on a long and powerful neck, are fixed on the herd. The head moves from side to side in short, rapid jerks, permitting a fix on the prey without the aid of stereoscopic vision.

Soon the head drops to the level of the grass, and the creature moves forward a few meters, then raises its head again to renew the surveillance. At a distance of 30 meters, the animal is almost ready to attack. In preparation, it lowers its head to a large rock close to its feet, rubbing its deep beak there to sharpen the bladelike edges.

Now the terror bristles its feathers and springs. Propelled by its two long, muscular legs, it dashes toward the herd. Within seconds it is moving at 70 kilometers per hour. Its small wings, useless for flight, are extended to the sides in aid of balance and maneuverability.

The herd, stricken with fright, bolts in disarray as the predator bears down. The attacker fixes its attention on an old male lagging behind the fleeing animals and quickly gains on it. Although the old male runs desperately, the attacker is soon at its side. With a stunning sideswipe of its powerful left foot, it knocks the prey off balance, seizes it in its massive beak and, with swinging motions of its head, beats it on the ground until it is unconscious. Now the attacker can swallow the limp body whole—an easy feat, given the creature's meter-long head and half-meter gape. Content, the gorged predator returns to its round nest of twigs in the grass nearby and resumes the incubation of two eggs the size of basketballs.

Meet the terror birds, the most spectacular and formidable group of flightless, flesh-eating birds that ever lived. They are all extinct now,

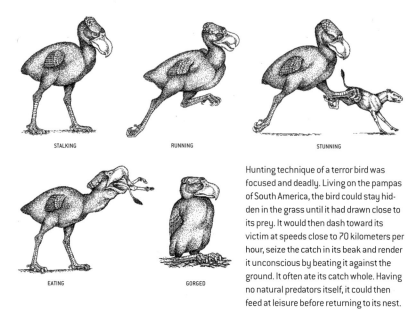

STALKING   RUNNING   STUNNING

EATING   GORGED

Hunting technique of a terror bird was focused and deadly. Living on the pampas of South America, the bird could stay hidden in the grass until it had drawn close to its prey. It would then dash toward its victim at speeds close to 70 kilometers per hour, seize the catch in its beak and render it unconscious by beating it against the ground. It often ate its catch whole. Having no natural predators itself, it could then feed at leisure before returning to its nest.

but they were once to the land what sharks are to the seas: engines of destruction and awesome eating machines. In their time, from 62 million years to about 2.5 million years ago, they became the dominant carnivores of South America. The story of their rise and decline is my subject here.

The terror birds are members of a group ornithologists call phorusrhacoids. The first phorusrhacoid to be described scientifically—in 1887 by the Argentine paleontologist Florentino Ameghino—was a fossil that he named *Phorusrhacos longissimus.* (*Longissimus* is the species, *Phorusrhacos* the genus. Taxonomists go on to classify living and extinct organisms in increasingly larger groups: family, order, class, phylum and kingdom.) The fossil came from the Santa Cruz Formation in Patagonia, the southernmost region of Argentina; the formation is about 17 million years old.

Ameghino and other researchers reconstructed the appearance of the birds from their fossil remains and their behavior from what creatures that might be living relatives do. The investigators initially interpreted the flesh-eating habits of the phorusrhacoids as an indication that they were related to modern eagles and hawks. Not all paleontologists agreed, and the issue was debated over the next 12 years. Charles W. Andrews of the British Museum resolved the controversy in 1899, concluding that among all living and extinct groups, the phorusrhacoids were most closely related to the South American seriema birds, which could also be regarded as the structural ancestors of the phorusrhacoids. Seriemas live today in the grasslands of northern Argentina, eastern Bolivia, Paraguay and central and eastern Brazil. Seriemas and phorusrhacoids are classified as members of the order Gruiformes, which includes cranes and rails and their kin.

There are two living seriema species, the red-legged seriema (*Cariama cristata*) and the black-legged, or Burmeister's, seriema (*Chunga burmeisteri*). These birds reach a height of 0.7 meter. They are light-bodied, long-legged and long-necked. Their wings are small relative to their body, and the birds resort to spurts of short-distance flight only when pressed. They are excellent runners, able to attain speeds in excess of 60 kilometers per hour. Seriemas build twig nests, four to six meters above the ground, in low trees. The young, usually two, mature in about two weeks, whereupon they leave the nest to live and hunt in the nearby grasslands. Like most carnivorous animals, seriemas are territorial. Their call has been described as eerie and piercing.

Like the phorusrhacoids, seriemas are carnivorous. They eat insects, reptiles, small mammals and other birds. Under favorable conditions, they will attack larger game. They seize their prey in their beaks and beat the animal on the ground until it is limp enough to be swallowed whole. This feeding strategy is also practiced today by the roadrunner (*Geococcyx californianus*) of the southwestern U.S. and the secretary bird (*Sagittarius serpentarius*) of Africa.

Seriemas are placed in the family Cariamidae, which now is restricted to South America. About 10 fossil species have been found there, the oldest being from the middle Paleocene epoch (some 62 million years ago) of Brazil. Relatives of this group are represented by two fossil families: the Bathornithidae, which appear in beds 40 to 20 million years old in North America, and the Idiornithidae, found in certain European rock formations 40 to 30 million years old. Some workers believe these families are so closely related that they should all be grouped in the family Cariamidae.

Most of the terror birds were considerably larger than their living relatives. The creatures ranged in height from one to three meters (just shy of 10 feet). The earliest known members are virtually as specialized as the latest, indicating that they originated before their first appearance in the fossil record.

About a dozen genera and 25 species of terror birds have been recognized. The relation among them is still not clear. They were classified in 1960 by Bryan Patterson of the Museum of Comparative Zoology at Harvard University and Jorge L. Kraglievich of the Municipal Museum of Natural and Traditional Sciences of Mar del Plata, Argentina. This classification ordered the terror birds in three families that, in comparison to families of mammals, include animals of medium, large and gigantic size. Other workers, basing their view on the period of greatest diversity among terror birds, achieved between five and three million years ago, recognize two families—gigantic and medium—as well as two subfamilies. Some researchers place all the fossils in one family.

In the three-family system the gigantic forms are members of the family Brontornithidae. Fossils of this family have been found in beds ranging in age from 27 to 17 million years. A heavy, ponderous build characterized the birds; the leg bones were fairly short, the beaks massive. This evidence suggests that the birds were cumbersome runners, slower afoot than the members of the other two families.

Next comes the family Phorusrhacidae. Its members ranged between two and three meters in height. Fossils have been found in rocks ranging in age from 27 million to three million years. The third family, Psilopteridae, comprised quite small members; most of them stood no more than one meter in height. Their known fossils range from 62 million to two million years in age. Within this family is the oldest known phorusrhacoid, *Paleopsilopterus,* found in Brazil. Members of these last two families were lightly built, swift runners. They were the ones that became the dominant running carnivores of their time, and they held that status for millions of years.

The fact that phorusrhacoids came in several sizes indicates that the adults were capable of preying on a wide variety of animals, from rodents to large herbivores. Although some of the adult herbivores were as big as some adult phorusrhacoids, the birds could easily have preyed on the young ones. Phorusrhacoids newly out of the nest would have had different food needs because they were smaller; they probably hunted rodents and other small vertebrates, much as their living seriema relatives still do.

During much of the age of mammals (the past 66 million years), phorusrhacoids thus occupied the role of fleet-footed carnivores in South America. They were able to assume this role by giving up the greatest virtue of being a bird—the power of flight. The door to dominance as carnivores opened to the phorusrhacoids when their predecessors in that role—the small, bipedal dinosaurs known as coelurosaurs—disappeared in the dinosaur extinction 66 million years ago. Paleobiologists call such a transition an evolutionary relay.

The body forms of the terror birds and the coelurosaurs were quite similar: trim, elongated bodies; long, powerful hind limbs; long necks;

large heads. Many coelurosaurs had reduced anterior limbs, indicating that the animals captured, killed and processed prey primarily with the hind limbs and mouth, as the phorusrhacoids did. Coelurosaurs apparently used their long tail as a balance while running; phorusrhacoids probably used their reduced wings for the same purpose. Different strategies and appendages were thus employed to serve the same functional purpose.

Terror birds and their relatives are also known outside South America. Their distribution is the key to the intriguing biogeographic history that accounts for the gradual ending of the terror birds' reign as South American carnivores.

In rocks from 55 million to 45 million years old in North America, Europe and Asia, large carnivorous birds are represented by the family Diatrymatidae—a family that, according to my Brazilian colleague Herculano M. F. Alvarenga, developed characteristics similar to those of the phorusrhacoids. Diatrymatidae family members attained heights of about two meters. Like the phorusrhacoids, they had massive skulls and large claws. Their legs, however, were relatively shorter and sturdier, suggesting that they were more methodical and cumbersome in their movements, much as the brontornithids were.

A reported phorusrhacoid, *Ameghinornis,* is known from the Phosphorites du Quercy rocks, 38 million to 35 million years old, in France. This animal was the size of a living seriema and was apparently capable of brief flight.

The Antarctic is also the scene of similar fossils. Two isolated footprints, 18 centimeters in length, are known in rocks about 55 million years old on the Fildes Peninsula of King George Island in West Antarctica. The three-toed bird was big, broad and elongated, either a ratite (a rhea or an ostrich or one of their relatives) or a phorusrhacoid.

The anterior part of a phorusrhacoid's beak was collected from rocks (40 million years old) of the La Meseta Formation on Seymour Island, which is on the south side of the Antarctic Peninsula. The proportions of the beak indicate that the bird was more than two meters tall.

Finally, a formidable phorusrhacoid named *Titanis walleri* is known from rocks aged 2.5 million to 1.5 million years in northern Florida. The estimated height of the bird is more than three meters. This record is the youngest yet found and represents the last of the known terror birds.

A scenario for this pattern of phorusrhacoid distribution can be constructed from the premises that these flightless birds required overland routes for dispersal and that the fossil record accurately reflects their occurrence in space and time.

Both biological and geologic evidence show that a continuously dry land bridge united North and South America about 62 million years ago. It ran by way of the Greater and Lesser Antilles, providing an opportunity for dispersal for various groups of terrestrial vertebrates. Among them were a seriema and a phorusrhacoid (probably a Psilopteridae) that dispersed north.

Forty-five million to 55 million years ago a land corridor between North America and Europe that included what is now Ellesmere Island provided another route by which the raptors could disperse. One group that appeared to have used the route was *Ameghinornis,* whose remains have been found in France. A note of caution here is in order: the supposition presumes that the phorusrhacoid group was present in North America. No fossils of that age have yet been found there.

From at least 45 million years ago, perhaps as much as 70 million, a body of land united southernmost South America and West Antarctica. The existence of a land connection at this time is supported by a group of marsupials, an armadillo and the southern beech in the same rock beds as the phorusrhacoid on Seymour Island. Together the land bridge and the cool, temperate climate of the time account for the presence of terror birds in West Antarctica 40 million years ago.

Eventually the land bridges uniting South America with North America and Antarctica disappeared. South America remained an island continent until the appearance of the Panamanian land bridge 2.5 million years ago. The bridge formed as a result of the continued tectonic uplift of the northern Andes, probably associated with a worldwide drop in sea level of as much as 50 meters resulting from the buildup of the polar ice caps. The final connection of the bridge was in the area of what has become southern Panama and northern Colombia.

A cooling of world climates at the time shrank tropical habitats and expanded the savannas. Grassland environments were established on the land bridge. After a time, a continuous corridor of savannas extended from Argentina to Florida. The reciprocal dispersal of terrestrial fauna made possible by these conditions is now known as the Great American Interchange. It represents the best-documented example in the fossil record of the intermingling of two long-separated continental biotas. Among the participants were the terror birds. One phorusrhacoid lineage survived beyond 2.5 million years ago in South America, and individuals dispersed north to give rise to *Titanis* in Florida.

Against this background, one can begin to see why a group of large, flightless birds rose to the top of the food pyramid in South America and why they finally lost that position. The answer lies in the historical development of the terrestrial fauna of South America. Recall that for most of the past 66 million years South America was, as Australia is today, an island continent. As a consequence of the groups that inhabited each continent 66 million years ago, the role of terrestrial mammal carnivores was filled in South America by marsupials and the role of large herbivores by placentals. This marsupial-placental combination was unique among continental faunas; both roles were filled by marsupials in Australia and by placentals in North America, Europe and Asia.

The group of South American marsupials that evolved to fill the place that placental dogs and cats eventually held on the northern continents is called borhyaenoid. Its doglike members are further grouped into three families. They ranged in size from that of a skunk to that of a bear. One specialized family, the thylacosmilids, had characteristics similar to those of the placental saber-toothed cats. It is particularly significant that all these animals were relatively short-legged and that none showed marked adaptation to running. These were the mammal occupants of the carnivore niche in South America.

Also in this niche were large terrestrial or semiterrestrial crocodiles of the family Sebecidae. They had deep skulls; their limbs were positioned more under the body than those of aquatic, flat-skulled crocodiles, and their laterally compressed teeth had serrated cutting edges, much like those of carnivorous dinosaurs. The other occupants of the carnivore niche were the terror birds. Thus, from about 66 million to about 2.5 million years ago, the role of terrestrial carnivore in South America was shared at various times, but not equally, by marsupial mammals, sebecid crocodiles and phorusrhacoid birds.

From about 27 million to 2.5 million years ago, the fossil record shows a protracted decrease in the size and diversity of the doglike borhyaenoids and a concurrent increase in the size and diversity of the phorusrhacoids. Consequently, by about five million years ago, phorusrhacoids had completely replaced the large carnivorous borhyaenoids on the savannas of South America. (The smaller ones, which were not competitive with the terror birds anyway, also became extinct before the Panamanian land bridge appeared.) This transition demonstrates another relay in the evolutionary history of the phorusrhacoids whereby they successfully replaced their marsupial counterparts, the borhyaenoids. Just why the phorusrhacoids

Downfall of terror birds was apparently caused by three examples of the many species of animals that crossed the Panamanian land bridge into South America. Greater intelligence, more speed and agility, or the ability to prey on terror bird eggs and hatchlings could explain how these migrants from North America displaced the terror birds.

were able to do so is unclear, but their superior running ability would certainly have been an advantage for capturing prey in the savanna environments that first came into prominence about 27 million years ago.

After the emergence of the Panamanian land bridge, placental dogs and cats of the families Canidae and Felidae dispersed into South America from North America. Because all the large marsupial carnivores of South America were by then long extinct, the only competition the dogs and cats had was the phorusrhacoids. It proved to be a losing battle for the birds.

Thus it was that the phorusrhacoids reached their peak in size and diversity just before the interchange, gradually declining thereafter because of the competition with the dogs and cats. Only one lineage survived beyond 2.5 million years in South America; it is the one that dispersed to Florida, where it is represented by *Titanis*. This was the only South American carnivorous animal to disperse northward. Its success there at coexisting with the advanced placental carnivores was brief. Why that was so is a major riddle. Perhaps the resident placental carnivores were too well established for the phorusrhacoids to find a permanent niche.

The fate of the phorusrhacoid relatives in North America and Europe between 55 million and 45 million years ago is also linked to the appearance of advanced placental carnivores. During that time on the northern continents, the large mammalian carnivores were the creodonts. This primitive group of placentals resembled the marsupial borhyaenoids in that they lacked special running abilities and had rather small brains.

The phorusrhacoid relatives on these continents disappeared with the appearance of advanced placental mammals beginning about 45 million years ago.

The terror birds thus flourished in the absence of advanced placental carnivores, which have repeatedly shown themselves to be better competitors. The marsupial borhyaenoids and placental creodonts were, in essence and in comparison with the terror birds, second rate.

Although plausible, this argument is speculative. One cannot identify with certainty a single factor that explains the extinction of any group of animals now found only as fossils. In the case of the terror birds, their disappearance on two occasions in time correlates directly with the appearance of advanced placental carnivores. Were the advanced placentals more intelligent than the terror birds and so better adapted to capturing the prey that the birds had had to themselves? Did the fact that they had four legs give them an advantage over the two-legged phorusrhacoids in speed or agility? Did the placentals eat the phorusrhacoids' eggs, which were readily accessible in ground nests because of the birds' large size? Did the placentals prey on the vulnerable hatchlings?

It is intriguing to think what might happen if all big carnivorous mammals were suddenly to vanish from South America. Would the seriemas again give rise to a group of giant flesh-eating birds that would rule the savannas as did the phorusrhacoids and their bygone allies?

FURTHER READING

Bryan Patterson and Jorge L. Kraglievich. "Systemáticas y Nomenclatura de las Aves Fororracoideas del Plioceno Argentino" in *Publicaciones del Museo Municipal de Ciencias Naturales y Tradicional de Mar del Plata*, Vol. 1, No. 1, pages 1–51; July 15, 1960.

Pierce Brodkorb. "A Giant Flightless Bird from the Pleistocene of Florida" in *The Auk*, Vol. 80, No. 2, pages 111–115; April 1963.

Larry G. Marshall. "The Terror Bird" in *Field Museum of Natural History Bulletin*, Vol. 49, No. 9, pages 6–15; October 1978.

Cecile Mourer-Chauviré. "Les Oiseaux Fossiles des Phosphorites du Quercy (Éocène Supérieur à Oligocène Supérieur): Implications Paléobiogéographiques" in *Geobios, Memoire Special*, No. 6, pages 413–426; 1982.

Judd A. Case, Michael O. Woodburne and Dan S. Chaney. "A Gigantic Phororhacoid(?) Bird from Antarctica" in *Journal of Paleontology*, Vol. 61, No. 6, pages 1280–1284; November 1987.

Larry G. Marshall. "Land Mammals and the Great American Interchange" in *American Scientist*, Vol. 76, No. 4, pages 380–388; July/August 1988.

Kent H. Redford and Pamela Shaw. "The Terror Bird Still Screams" in *International Wildlife*, Vol. 19, No. 3, pages 14–16; May/June 1989.

The Evolution of Life on the Earth

STEPHEN JAY GOULD

ORIGINALLY PUBLISHED IN OCTOBER 1994

Some creators announce their inventions with grand éclat. God proclaimed, "Fiat lux," and then flooded his new universe with brightness. Others bring forth great discoveries in a modest guise, as did Charles Darwin in defining his new mechanism of evolutionary causality in 1859: "I have called this principle, by which each slight variation, if useful, is preserved, by the term Natural Selection."

Natural selection is an immensely powerful yet beautifully simple theory that has held up remarkably well, under intense and unrelenting scrutiny and testing, for 135 years. In essence, natural selection locates the mechanism of evolutionary change in a "struggle" among organisms for reproductive success, leading to improved fit of populations to changing environments. (Struggle is often a metaphorical description and need not be viewed as overt combat, guns blazing. Tactics for reproductive success include a variety of non-martial activities such as earlier and more frequent mating or better cooperation with partners in raising offspring.) Natural selection is therefore a principle of local adaptation, not of general advance or progress.

Yet powerful though the principle may be, natural selection is not the only cause of evolutionary change (and may, in many cases, be overshadowed by other forces). This point needs emphasis because the standard misapplication of evolutionary theory assumes that biological explanation may be equated with devising accounts, often speculative and conjectural in practice, about the adaptive value of any given feature in its original environment (human aggression as good for hunting, music and religion as good for tribal cohesion, for example). Darwin himself strongly emphasized the multifactorial nature of evolutionary change and warned against too exclusive a reliance on natural selection, by placing the following statement in a maximally conspicuous place at the very end of his introduction: "I am convinced that Natural Selection has been the most important, but not the exclusive, means of modification."

234

Natural selection is not fully sufficient to explain evolutionary change for two major reasons. First, many other causes are powerful, particularly at levels of biological organization both above and below the traditional Darwinian focus on organisms and their struggles for reproductive success. At the lowest level of substitution in individual base pairs of DNA, change is often effectively neutral and therefore random. At higher levels, involving entire species or faunas, punctuated equilibrium can produce evolutionary trends by selection of species based on their rates of origin and extirpation, whereas mass extinctions wipe out substantial parts of biotas for reasons unrelated to adaptive struggles of constituent species in "normal" times between such events.

Second, and the focus of this article, no matter how adequate our general theory of evolutionary change, we also yearn to document and understand the actual pathway of life's history. Theory, of course, is relevant to explaining the pathway (nothing about the pathway can be inconsistent with good theory, and theory can predict certain general aspects of life's geologic pattern). But the actual pathway is strongly *underdetermined* by our general theory of life's evolution. This point needs some belaboring as a central yet widely misunderstood aspect of the world's complexity. Webs and chains of historical events are so intricate, so imbued with random and chaotic elements, so unrepeatable in encompassing such a multitude of unique (and uniquely interacting) objects, that standard models of simple prediction and replication do not apply.

History can be explained, with satisfying rigor if evidence be adequate, after a sequence of events unfolds, but it cannot be predicted with any precision beforehand. Pierre-Simon Laplace, echoing the growing and confident determinism of the late 18th century, once said that he could specify all future states if he could know the position and motion of all particles in the cosmos at any moment, but the nature of universal complexity shatters this chimerical dream. History includes too much chaos, or extremely sensitive dependence on minute and unmeasurable differences in initial conditions, leading to massively divergent outcomes based on tiny and unknowable disparities in starting points. And history includes too much contingency, or shaping of present results by long chains of unpredictable antecedent states, rather than immediate determination by timeless laws of nature.

Homo sapiens did not appear on the earth, just a geologic second ago, because evolutionary theory predicts such an outcome based on themes of progress and increasing neural complexity. Humans arose, rather, as a fortuitous and contingent outcome of thousands of linked events, any one of

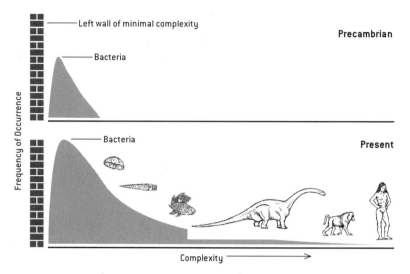

Progress does not rule (and is not even a primary thrust of) the evolutionary process. For reasons of chemistry and physics, life arises next to the "left wall" of its simplest conceivable and preservable complexity. This style of life (bacterial) has remained most common and most successful. A few creatures occasionally move to the right, thus extending the right tail in the distribution of complexity. Many always move to the left, but they are absorbed within space already occupied. Note that the bacterial mode has never changed in position, but just grown higher.

which could have occurred differently and sent history on an alternative pathway that would not have led to consciousness. To cite just four among a multitude: (1) If our inconspicuous and fragile lineage had not been among the few survivors of the initial radiation of multicellular animal life in the Cambrian explosion 530 million years ago, then no vertebrates would have inhabited the earth at all. (Only one member of our chordate phylum, the genus *Pikaia,* has been found among these earliest fossils. This small and simple swimming creature, showing its allegiance to us by possessing a notochord, or dorsal stiffening rod, is among the rarest fossils of the Burgess Shale, our best preserved Cambrian fauna.) (2) If a small and unpromising group of lobe-finned fishes had not evolved fin bones with a strong central axis capable of bearing weight on land, then vertebrates might never have become terrestrial. (3) If a large extraterrestrial body had not struck the earth 65 million years ago, then dinosaurs would still be dominant and mammals insignificant (the situation that had prevailed for 100 million years previously). (4) If a small lineage of primates had not evolved upright posture on the drying African savannas just two to four million years ago, then our ancestry might have ended in a line of apes that, like the chimpanzee and gorilla today, would have become ecologically marginal and probably doomed to extinction despite their remarkable behavioral complexity.

Therefore, to understand the events and generalities of life's pathway, we must go beyond principles of evolutionary theory to a paleontological examination of the contingent pattern of life's history on our planet—the single actualized version among millions of plausible alternatives that happened not to occur. Such a view of life's history is highly contrary both to conventional deterministic models of Western science and to the deepest social traditions and psychological hopes of Western culture for a history culminating in humans as life's highest expression and intended planetary steward.

Science can, and does, strive to grasp nature's factuality, but all science is socially embedded, and all scientists record prevailing "certainties," however hard they may be aiming for pure objectivity. Darwin himself, in the closing lines of *The Origin of Species*, expressed Victorian social preference more than nature's record in writing: "As natural selection works solely by and for the good of each being, all corporeal and mental endowments will tend to progress towards perfection."

Life's pathway certainly includes many features predictable from laws of nature, but these aspects are too broad and general to provide the "rightness" that we seek for validating evolution's particular results— roses, mushrooms, people and so forth. Organisms adapt to, and are constrained by, physical principles. It is, for example, scarcely surprising, given laws of gravity, that the largest vertebrates in the sea (whales) exceed the heaviest animals on land (elephants today, dinosaurs in the past), which, in turn, are far bulkier than the largest vertebrate that ever flew (extinct pterosaurs of the Mesozoic era).

Predictable ecological rules govern the structuring of communities by principles of energy flow and thermodynamics (more biomass in prey than in predators, for example). Evolutionary trends, once started, may have local predictability ("arms races," in which both predators and prey hone their defenses and weapons, for example—a pattern that Geerat J. Vermeij of the University of California at Davis has called "escalation" and documented in increasing strength of both crab claws and shells of their gastropod prey through time). But laws of nature do not tell us why we have crabs and snails at all, why insects rule the multicellular world and why vertebrates rather than persistent algal mats exist as the most complex forms of life on the earth.

Relative to the conventional view of life's history as an at least broadly predictable process of gradually advancing complexity through time, three features of the paleontological record stand out in opposition and shall therefore serve as organizing themes for the rest of this article: the constancy of modal complexity throughout life's history; the concentration of major events in short bursts interspersed with long periods of relative stability; and the role of external impositions, primarily mass extinctions, in disrupting patterns of "normal" times. These three features, combined with more general themes of chaos and contingency, require a new framework for conceptualizing and drawing life's history, and this article therefore closes with suggestions for a different iconography of evolution.

The primary paleontological fact about life's beginnings points to predictability for the onset and very little for the particular pathways thereafter. The earth is 4.6 billion years old, but the oldest rocks date to about 3.9 billion years because the earth's surface became molten early in its history, a result of bombardment by large amounts of cosmic debris during the solar system's coalescence, and of heat generated by radioactive decay of short-lived isotopes. These oldest rocks are too metamorphosed by subsequent heat and pressure to preserve fossils (though some scientists interpret the proportions of carbon isotopes in these rocks as signs of organic production). The oldest rocks sufficiently unaltered to retain cellular fossils—African and Australian sediments dated to 3.5 billion years old—do preserve prokaryotic cells (bacteria and cyanophytes) and stromatolites (mats of sediment trapped and bound by these cells in shallow marine waters). Thus, life on the earth evolved quickly and is as old as it could be. This fact alone seems to indicate an inevitability, or at least a predictability, for life's origin from the original chemical constituents of atmosphere and ocean.

No one can doubt that more complex creatures arose sequentially after this prokaryotic beginning—first eukaryotic cells, perhaps about two billion years ago, then multicellular animals about 600 million years ago, with a relay of highest complexity among animals passing from invertebrates, to marine vertebrates and, finally (if we wish, albeit parochially, to honor neural architecture as a primary criterion), to reptiles, mammals and humans. This is the conventional sequence represented in the old charts and texts as an "age of invertebrates," followed by an "age of fishes," "age of reptiles," "age of mammals," and "age of man" (to add the old gender bias to all the other prejudices implied by this sequence).

I do not deny the facts of the preceding paragraph but wish to argue that our conventional desire to view history as progressive, and to see humans as predictably dominant, has grossly distorted our interpretation

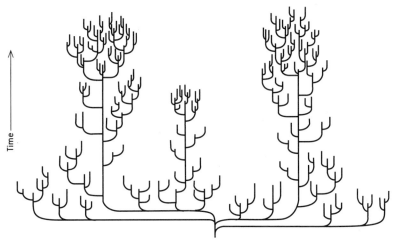

Anatomical Diversity

New iconography of life's tree shows that maximal diversity in anatomical forms (not in number of species) is reached very early in life's multicellular history. Later times feature extinction of most of these initial experiments and enormous success within surviving lines. This success is measured in the proliferation of species but not in the development of new anatomies. Today we have more species than ever before, although they are restricted to fewer basic anatomies.

of life's pathway by falsely placing in the center of things a relatively minor phenomenon that arises only as a side consequence of a physically constrained starting point. The most salient feature of life has been the stability of its bacterial mode from the beginning of the fossil record until today and, with little doubt, into all future time so long as the earth endures. This is truly the "age of bacteria"—as it was in the beginning, is now and ever shall be.

For reasons related to the chemistry of life's origin and the physics of self-organization, the first living things arose at the lower limit of life's conceivable, preservable complexity. Call this lower limit the "left wall" for an architecture of complexity. Since so little space exists between the left wall and life's initial bacterial mode in the fossil record, only one direction for future increment exists—toward greater complexity at the right. Thus, every once in a while, a more complex creature evolves and extends the range of life's diversity in the only available direction. In technical terms, the distribution of complexity becomes more strongly right skewed through these occasional additions.

But the additions are rare and episodic. They do not even constitute an evolutionary series but form a motley sequence of distantly related taxa, usually depicted as eukaryotic cell, jellyfish, trilobite, nautiloid, eurypterid (a large relative of horseshoe crabs), fish, an amphibian such as *Eryops,* a dinosaur, a mammal and a human being. This sequence cannot be construed as the major thrust or trend of life's history. Think rather of an occasional creature tumbling into the empty right region of complexity's space. Throughout this entire time, the bacterial mode has grown in height and remained constant in position. Bacteria represent the great success story of life's pathway. They occupy a wider domain of environments and span a broader range of biochemistries than any other group. They are adaptable, indestructible and astoundingly diverse. We cannot even imagine how anthropogenic intervention might threaten their extinction, although we worry about our impact on nearly every other form of life. The number of *Escherichia coli* cells in the gut of each human being exceeds the number of humans that has ever lived on this planet.

One might grant that complexification for life as a whole represents a pseudotrend based on constraint at the left wall but still hold that evolution within particular groups differentially favors complexity when the founding lineage begins far enough from the left wall to permit movement in both directions. Empirical tests of this interesting hypothesis are just beginning (as concern for the subject mounts among paleontologists), and we do not yet have enough cases to advance a generality. But the first two studies—by Daniel W. McShea of the University of Michigan on mammalian vertebrae and by George F. Boyajian of the University of Pennsylvania on ammonite suture lines—show no evolutionary tendencies to favor increased complexity.

Moreover, when we consider that for each mode of life involving greater complexity, there probably exists an equally advantageous style based on greater simplicity of form (as often found in parasites, for example), then preferential evolution toward complexity seems unlikely a priori. Our impression that life evolves toward greater complexity is probably only a bias inspired by parochial focus on ourselves, and consequent overattention to complexifying creatures, while we ignore just as many lineages adapting equally well by becoming simpler in form. The morphologically degenerate parasite, safe within its host, has just as much prospect for evolutionary success as its gorgeously elaborate relative coping with the slings and arrows of outrageous fortune in a tough external world.

Even if complexity is only a drift away from a constraining left wall, we might view trends in this direction as more predictable and characteristic of life's pathway as a whole if increments of complexity accrued in a persistent and gradually accumulating manner through time. But nothing about life's history is more peculiar with respect to this common (and false) expectation than the actual pattern of extended stability and rapid episodic movement, as revealed by the fossil record.

Life remained almost exclusively unicellular for the first five sixths of its history—from the first recorded fossils at 3.5 billion years to the first well-documented multicellular animals less than 600 million years ago. (Some simple multicellular algae evolved more than a billion years ago, but these organisms belong to the plant kingdom and have no genealogical connection with animals.) This long period of unicellular life does include, to be sure, the vitally important transition from simple prokaryotic cells without organelles to eukaryotic cells with nuclei, mitochondria and other complexities of intracellular architecture—but no recorded attainment of multicellular animal organization for a full three billion years. If complexity is such a good thing, and multicellularity represents its initial phase in our usual view, then life certainly took its time in making this crucial step. Such delays speak strongly against general progress as the major theme of life's history, even if they can be plausibly explained by lack of sufficient atmospheric oxygen for most of Precambrian time or by failure of unicellular life to achieve some structural threshold acting as a prerequisite to multicellularity.

More curiously, all major stages in organizing animal life's multicellular architecture then occurred in a short period beginning less than 600 million years ago and ending by about 530 million years ago—and the steps within this sequence are also discontinuous and episodic, not gradually accumulative. The first fauna, called Ediacaran to honor the Australian locality of its initial discovery but now known from rocks on all continents, consists of highly flattened fronds, sheets and circlets composed of numerous slender segments quilted together. The nature of the Ediacaran fauna is now a subject of intense discussion. These creatures do not seem to be simple precursors of later forms. They may constitute a separate and failed experiment in animal life, or they may represent a full range of diploblastic (two-layered) organization, of which the modern phylum Cnidaria (corals, jellyfishes and their allies) remains as a small and much altered remnant.

In any case, they apparently died out well before the Cambrian biota evolved. The Cambrian then began with an assemblage of bits and pieces, frustratingly difficult to interpret, called the "small shelly fauna." The subsequent main pulse, starting about 530 million years ago, constitutes the famous Cambrian explosion, during which all but one modern phylum of animal life made a first appearance in the fossil record. (Geologists

Great diversity quickly evolved at the dawn of multicellular animal life during the Cambrian period (530 million years ago). The creatures shown here are all found in the Middle Cambrian Burgess Shale fauna of Canada. They include some familiar forms (sponges, brachiopods) that have survived. But many creatures (such as the giant *Anomalocaris*, at the lower right, largest of all the Cambrian animals) did not live for long and are so anatomically peculiar (relative to survivors) that we cannot classify them among known phyla.

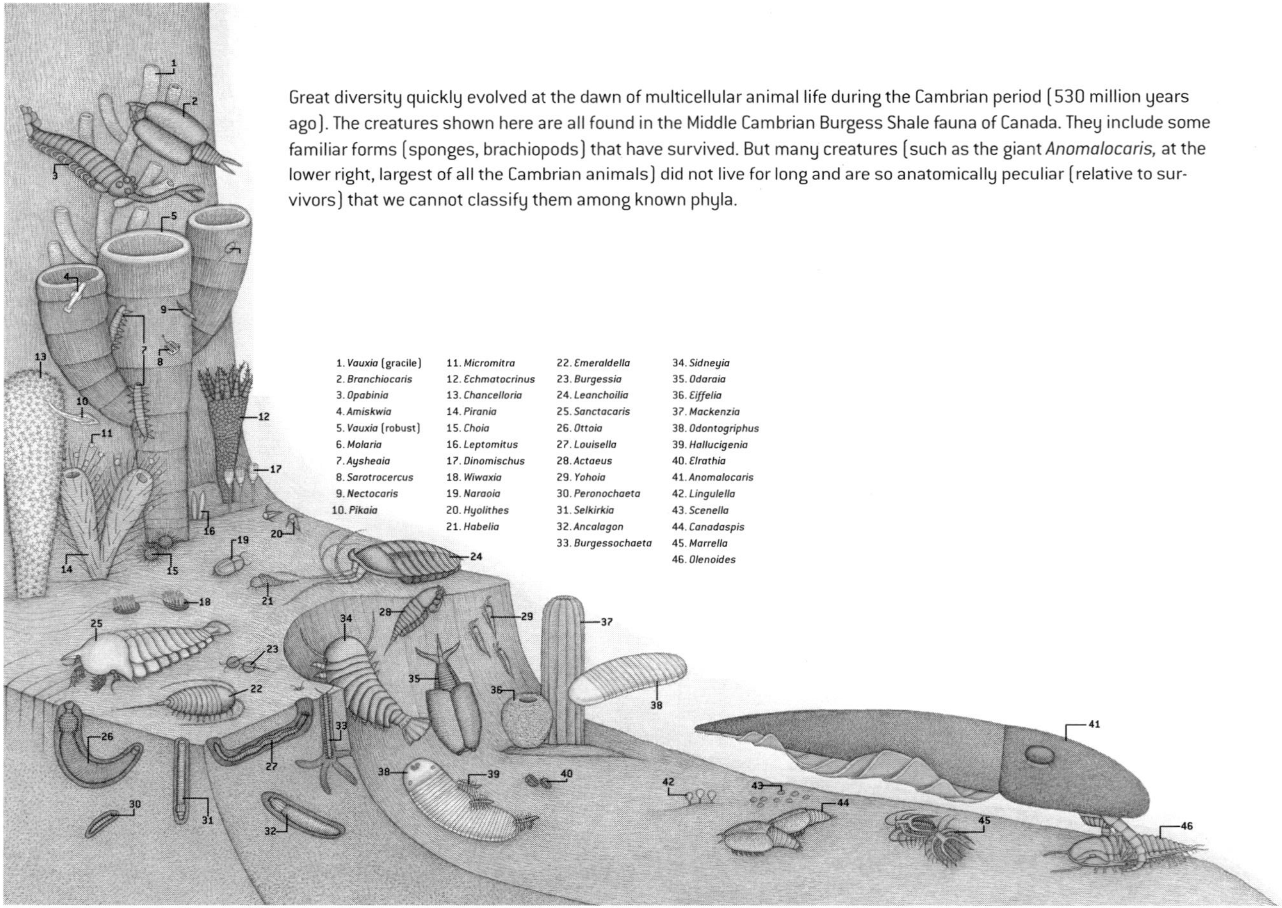

1. *Vauxia* (gracile)
2. *Branchiocaris*
3. *Opabinia*
4. *Amiskwia*
5. *Vauxia* (robust)
6. *Molaria*
7. *Aysheaia*
8. *Sarotrocercus*
9. *Nectocaris*
10. *Pikaia*
11. *Micromitra*
12. *Echmatocrinus*
13. *Chancelloria*
14. *Pirania*
15. *Choia*
16. *Leptomitus*
17. *Dinomischus*
18. *Wiwaxia*
19. *Naraoia*
20. *Hyolithes*
21. *Habelia*
22. *Emeraldella*
23. *Burgessia*
24. *Leanchoilia*
25. *Sanctacaris*
26. *Ottoia*
27. *Louisella*
28. *Actaeus*
29. *Yohoia*
30. *Peronochaeta*
31. *Selkirkia*
32. *Ancalagon*
33. *Burgessochaeta*
34. *Sidneyia*
35. *Odaraia*
36. *Elfelia*
37. *Mackenzia*
38. *Odontogriphus*
39. *Hallucigenia*
40. *Elrathia*
41. *Anomalocaris*
42. *Lingulella*
43. *Scenella*
44. *Canadaspis*
45. *Marrella*
46. *Olenoides*

had previously allowed up to 40 million years for this event, but an elegant study, published in 1993, clearly restricts this period of phyletic flowering to a mere five million years.) The Bryozoa, a group of sessile and colonial marine organisms, do not arise until the beginning of the subsequent, Ordovician period, but this apparent delay may be an artifact of failure to discover Cambrian representatives.

Although interesting and portentous events have occurred since, from the flowering of dinosaurs to the origin of human consciousness, we do not exaggerate greatly in stating that the subsequent history of animal life amounts to little more than variations on anatomical themes established during the Cambrian explosion within five million years. Three billion years of unicellularity, followed by five million years of intense creativity and then capped by more than 500 million years of variation on set anatomical themes can scarcely be read as a predictable, inexorable or continuous trend toward progress or increasing complexity.

We do not know why the Cambrian explosion could establish all major anatomical designs so quickly. An "external" explanation based on ecology seems attractive: the Cambrian explosion represents an initial filling of the "ecological barrel" of niches for multicellular organisms, and any experiment found a space. The barrel has never emptied since; even the great mass extinctions left a few species in each principal role, and their occupation of ecological space forecloses opportunity for fundamental novelties. But an "internal" explanation based on genetics and development also seems necessary as a complement: the earliest multicellular animals may have maintained a flexibility for genetic change and embryological transformation that became greatly reduced as organisms "locked in" to a set of stable and successful designs.

In any case, this initial period of both internal and external flexibility yielded a range of invertebrate anatomies that may have exceeded (in just a few million years of production) the full scope of animal form in all the earth's environments today (after more than 500 million years of additional time for further expansion). Scientists are divided on this question. Some claim that the anatomical range of this initial explosion exceeded that of modern life, as many early experiments died out and no new phyla have ever arisen. But scientists most strongly opposed to this view allow that Cambrian diversity at least equaled the modern range—so even the most cautious opinion holds that 500 million subsequent years of opportunity have not expanded the Cambrian range, achieved in just five million years. The Cambrian explosion was the most remarkable and puzzling event in the history of life.

Moreover, we do not know why most of the early experiments died, while a few survived to become our modern phyla. It is tempting to say that the victors won by virtue of greater anatomical complexity, better ecological fit or some other predictable feature of conventional Darwinian struggle. But no recognized traits unite the victors, and the radical alternative must be entertained that each early experiment received little more than the equivalent of a ticket in the largest lottery ever played out on our planet—and that each surviving lineage, including our own phylum of vertebrates, inhabits the earth today more by the luck of the draw than by any predictable struggle for existence. The history of multicellular animal life may be more a story of great reduction in initial possibilities, with stabilization of lucky survivors, than a conventional tale of steady ecological expansion and morphological progress in complexity.

Finally, this pattern of long stasis, with change concentrated in rapid episodes that establish new equilibria, may be quite general at several scales of time and magnitude, forming a kind of fractal pattern in self-similarity. According to the punctuated equilibrium model of speciation, trends within lineages occur by accumulated episodes of geologically instantaneous speciation, rather than by gradual change within continuous populations (like climbing a staircase rather than rolling a ball up an inclined plane).

Even if evolutionary theory implied a potential internal direction for life's pathway (although previous facts and arguments in this article cast doubt on such a claim), the occasional imposition of a rapid and substantial, perhaps even truly catastrophic, change in environment would have intervened to stymie the pattern. These environmental changes trigger mass extinction of a high percentage of the earth's species and may so derail any internal direction and so reset the pathway that the net pattern of life's history looks more capricious and concentrated in episodes than steady and directional. Mass extinctions have been recognized since the dawn of paleontology; the major divisions of the geologic time scale were established at boundaries marked by such events. But until the revival of interest that began in the late 1970s, most paleontologists treated mass extinctions only as intensifications of ordinary events, leading (at most) to a speeding up of tendencies that pervaded normal times. In this gradualistic theory of mass extinction, these events really took a few million years to unfold (with the appearance of suddenness interpreted as an artifact of an imperfect fossil record), and they only made the ordinary occur faster (more intense Darwinian competition in tough times, for example, leading to even more efficient replacement of less adapted by superior forms).

The reinterpretation of mass extinctions as central to life's pathway and radically different in effect began with the presentation of data by Luis and Walter Alvarez in 1979, indicating that the impact of a large extraterrestrial object (they suggested an asteroid seven to 10 kilometers in diameter) set off the last great extinction at the Cretaceous-Tertiary boundary 65 million years ago. Although the Alvarez hypothesis initially received very skeptical treatment from scientists (a proper approach to highly unconventional explanations), the case now seems virtually proved by discovery of the "smoking gun," a crater of appropriate size and age located off the Yucatán peninsula in Mexico.

This reawakening of interest also inspired paleontologists to tabulate the data of mass extinction more rigorously. Work by David M. Raup, J. J. Sepkoski, Jr., and David Jablonski of the University of Chicago has established that multicellular animal life experienced five major (end of Ordovician, late Devonian, end of Permian, end of Triassic and end of Cretaceous) and many minor mass extinctions during its 530-million-year history. We have no clear evidence that any but the last of these events was triggered by catastrophic impact, but such careful study leads to the general conclusion that mass extinctions were more frequent, more rapid, more extensive in magnitude and more different in effect than paleontologists had previously realized. These four properties encompass the radical implications of mass extinction for understanding life's pathway as more contingent and chancy than predictable and directional.

Mass extinctions are not random in their impact on life. Some lineages succumb and others survive as sensible outcomes based on presence or absence of evolved features. But especially if the triggering cause of extinction be sudden and catastrophic, the reasons for life or death may be random with respect to the original value of key features when first evolved in Darwinian struggles of normal times. This "different rules" model of mass extinction imparts a quirky and unpredictable character to life's pathway based on the evident claim that lineages cannot anticipate future contingencies of such magnitude and different operation.

To cite two examples from the impact-triggered Cretaceous-Tertiary extinction 65 million years ago: First, an important study published in 1986 noted that diatoms survived the extinction far better than other single-celled plankton (primarily coccoliths and radiolaria). This study found that many diatoms had evolved a strategy of dormancy by encystment, perhaps to survive through seasonal periods of unfavorable conditions (months of darkness in polar species as otherwise fatal to these photosynthesizing cells; sporadic availability of silica needed to construct their skeletons). Other planktonic cells had not evolved any mechanisms for dormancy. If the terminal Cretaceous impact produced a dust cloud that blocked light for several months or longer (one popular idea for a "killing scenario" in the extinction), then diatoms may have survived as a fortuitous result of dormancy mechanisms evolved for the entirely different function of weathering seasonal droughts in ordinary times. Diatoms are not superior to radiolaria or other plankton that succumbed in far greater numbers; they were simply fortunate to possess a favorable feature, evolved for other reasons, that fostered passage through the impact and its sequelae.

Second, we all know that dinosaurs perished in the end Cretaceous event and that mammals therefore rule the vertebrate world today. Most people assume that mammals prevailed in these tough times for some reason of general superiority over dinosaurs. But such a conclusion seems most unlikely. Mammals and dinosaurs had coexisted for 100 million years, and mammals had remained rat-sized or smaller, making no evolutionary "move" to oust dinosaurs. No good argument for mammalian prevalence by general superiority has ever been advanced, and fortuity seems far more likely. As one plausible argument, mammals may have survived partly as a result of their small size (with much larger, and therefore extinction-resistant, populations as a consequence, and less ecological specialization with more places to hide, so to speak). Small size may not have been a positive mammalian adaptation at all, but more a sign of inability ever to penetrate the dominant domain of dinosaurs. Yet this "negative" feature of normal times may be the key reason for mammalian survival and a prerequisite to my writing and your reading this article today.

Sigmund Freud often remarked that great revolutions in the history of science have but one common, and ironic, feature: they knock human arrogance off one pedestal after another of our previous conviction about our own self-importance. In Freud's three examples, Copernicus moved our home from center to periphery; Darwin then relegated us to "descent from an animal world"; and, finally (in one of the least modest statements of intellectual history), Freud himself discovered the unconscious and exploded the myth of a fully rational mind.

In this wise and crucial sense, the Darwinian revolution remains woefully incomplete because, even though thinking humanity accepts the fact of evolution, most of us are still unwilling to abandon the comforting view that evolution means (or at least embodies a central principle of) progress defined to render the appearance of something like human consciousness either virtually inevitable or at least predictable. The pedestal is not smashed until we abandon progress or complexification as a central principle and come to entertain the strong possibility that *H. sapiens* is but a tiny, late-arising twig on life's enormously arborescent bush—a small bud that would almost surely not appear a second time if we could replant the bush from seed and let it grow again.

Primates are visual animals, and the pictures we draw betray our deepest convictions and display our current conceptual limitations. Artists have always painted the history of fossil life as a sequence from invertebrates, to fishes, to early terrestrial amphibians and reptiles, to dinosaurs, to mammals and, finally, to humans. There are no exceptions; all sequences painted since the inception of this genre in the 1850s follow the convention.

Yet we never stop to recognize the almost absurd biases coded into this universal mode. No scene ever shows another invertebrate after fishes evolved, but invertebrates did not go away or stop evolving! After terrestrial reptiles emerge, no subsequent scene ever shows a fish (later oceanic tableaux depict only such returning reptiles as ichthyosaurs and plesiosaurs). But fishes did not stop evolving after one small lineage managed to invade the land. In fact, the major event in the evolution of fishes, the origin and rise to dominance of the teleosts, or modern bony fishes, occurred during the time of the dinosaurs and is therefore never shown at all in any of these sequences—even though teleosts include more than half of all species of vertebrates. Why should humans appear at the end of all sequences? Our order of primates is ancient among mammals, and many other successful lineages arose later than we did.

We will not smash Freud's pedestal and complete Darwin's revolution until we find, grasp and accept another way of drawing life's history. J. B. S. Haldane proclaimed nature "queerer than we can suppose," but these limits may only be socially imposed conceptual locks rather then inherent restrictions of our neurology. New icons might break the locks. Trees—or rather copiously and luxuriantly branching bushes—rather than ladders and sequences hold the key to this conceptual transition.

We must learn to depict the full range of variation, not just our parochial perception of the tiny right tail of most complex creatures. We must recognize that this tree may have contained a maximal number of branches near the beginning of multicellular life and that subsequent history is for the most part a process of elimination and lucky survivorship of a few, rather than continuous flowering, progress and expansion of a growing multitude. We must understand that little twigs are contingent nubbins, not predictable goals of the massive bush beneath. We must remember the greatest of all Biblical statements about wisdom: "She is a tree of life to them that lay hold upon her; and happy is every one that retaineth her."

FURTHER READING

Henry B. Whittington. *The Burgess Shale.* Yale University Press, 1985.
Steven M. Stanley. *Extinction: A Scientific American Book.* W. H. Freeman and Company, 1987.
S. J. Gould. *Wonderful Life: The Burgess Shale and the Nature of History,* W. W. Norton, 1989.
Stephen Jay Gould, ed. *The Book of Life.* W. W. Norton, 1993.

HUMAN EVOLUTION

An Ancestor to Call Our Own

KATE WONG

ORIGINALLY PUBLISHED IN JANUARY 2003

Poitiers, France—Michel Brunet removes the cracked, brown skull from its padlocked, foam-lined metal carrying case and carefully places it on the desk in front of me. It is about the size of a coconut, with a slight snout and a thick brow visoring its stony sockets. To my inexpert eye, the face is at once foreign and inscrutably familiar. To Brunet, a paleontologist at the University of Poitiers, it is the visage of the lost relative he has sought for 26 years. "He is the oldest one," the veteran fossil hunter murmurs, "the oldest hominid."

Brunet and his team set the field of paleoanthropology abuzz when they unveiled their find last July. Unearthed from sandstorm-scoured deposits in northern Chad's Djurab Desert, the astonishingly complete cranium— dubbed *Sahelanthropus tchadensis* (and nicknamed Toumaï, which means "hope of life" in the local Goran language)—dates to nearly seven million years ago. It may thus represent the earliest human forebear on record, one who Brunet says "could touch with his finger" the point at which our lineage and the one leading to our closest living relative, the chimpanzee, diverged.

Less than a century ago simian human precursors from Africa existed only in the minds of an enlightened few. Charles Darwin predicted in 1871 that the earliest ancestors of humans would be found in Africa, where our chimpanzee and gorilla cousins live today. But evidence to support that idea didn't come until more than 50 years later, when anatomist Raymond Dart of the University of the Witwatersrand described a fossil skull from Taung, South Africa, as belonging to an extinct human he called *Australopithecus africanus,* the "southern ape from Africa." His claim met variously with frosty skepticism and outright rejection—the remains were those of a juvenile gorilla, critics countered. The discovery of another South African specimen, now recognized as *A. robustus,* eventually vindicated Dart, but it wasn't until the 1950s that the notion of ancient, apelike human ancestors from Africa gained widespread acceptance.

In the decades that followed, pioneering efforts in East Africa headed by members of the Leakey family, among others, turned up additional fossils. By the late 1970s the australopithecine cast of characters had grown to include A. *boisei,* A. *aethiopicus* and A. *afarensis* (Lucy and her kind, who lived between 2.9 million and 3.6 million years ago during the Pliocene epoch and gave rise to our own genus, *Homo*). Each was adapted to its own environmental niche, but all were bipedal creatures with thick jaws, large molars and small canines—radically different from the generalized, quadrupedal Miocene apes known from farther back on the family tree. To probe human origins beyond A. *afarensis,* however, was to fall into a gaping hole in the fossil record between 3.6 million and 12 million years ago. Who, researchers wondered, were Lucy's forebears?

Despite widespread searching, diagnostic fossils of the right age to answer that question eluded workers for nearly two decades. Their luck finally began to change around the mid-1990s, when a team led by Meave Leakey of the National Museums of Kenya announced its discovery of A. *anamensis,* a four-million-year-old species that, with its slightly more archaic characteristics, made a reasonable ancestor for Lucy [see "Early Hominid Fossils from Africa," by Meave Leakey and Alan Walker; *Scientific American,* June 1997]. At around the same time, Tim D. White of the University of California at Berkeley and his colleagues described a collection of 4.4-million-year-old fossils from Ethiopia representing an even more primitive hominid, now known as *Ardipithecus ramidus ramidus.* Those findings gave scholars a tantalizing glimpse into Lucy's past. But estimates from some molecular biologists of when the chimp-human split occurred suggested that even older hominids lay waiting to be discovered.

Those predictions have recently been borne out. Over the past few years, researchers have made a string of stunning discoveries—Brunet's among them—that may go a long way toward bridging the remaining gap between humans and their African ape ancestors. These fossils, which range from roughly five million to seven million years old, are upending long-held ideas about when and where our lineage arose and what the last common ancestor of humans and chimpanzees looked like. Not surprisingly, they have also sparked vigorous debate. Indeed, experts are deeply divided over where on the family tree the new species belong and even what constitutes a hominid in the first place.

STANDING TALL

The first hominid clue to come from beyond the 4.4-million-year mark was announced in the spring of 2001. Paleontologists Martin Pickford and

Brigitte Senut of the National Museum of Natural History in Paris found in Kenya's Tugen Hills the six-million-year-old remains of a creature they called *Orrorin tugenensis*. To date the researchers have amassed 19 specimens, including bits of jaw, isolated teeth, finger and arm bones, and some partial upper leg bones, or femurs. According to Pickford and Senut, *Orrorin* exhibits several characteristics that clearly align it with the hominid family—notably those suggesting that, like all later members of our group, it walked on two legs. "The femur is remarkably humanlike," Pickford observes. It has a long femoral neck, which would have placed the shaft at an angle relative to the lower leg (thereby stabilizing the hip), and a groove on the back of that femoral neck, where a muscle known as the obturator externus pressed against the bone during upright walking. In other respects, *Orrorin* was a primitive animal: its canine teeth are large and pointed relative to human canines, and its arm and finger bones retain adaptations for climbing. But the femur characteristics signify to Pickford and Senut that when it was on the ground, *Orrorin* walked like a man.

In fact, they argue, *Orrorin* appears to have had a more humanlike gait than the much younger Lucy did. Breaking with paleoanthropological dogma, the team posits that *Orrorin* gave rise to *Homo* via the proposed genus *Praeanthropus* (which comprises a subset of the fossils currently assigned to *A. afarensis* and *A. anamensis*), leaving Lucy and her kin on an evolutionary sideline. *Ardipithecus,* they believe, was a chimpanzee ancestor.

Not everyone is persuaded by the femur argument. C. Owen Lovejoy of Kent State University counters that published computed tomography scans through *Orrorin*'s femoral neck—which Pickford and Senut say reveal humanlike bone structure—actually show a chimplike distribution of cortical bone, an important indicator of the strain placed on that part of the femur during locomotion. Cross sections of *A. afarensis*'s femoral neck, in contrast, look entirely human, he states. Lovejoy suspects that *Orrorin* was frequently—but not habitually—bipedal and spent a significant amount of time in the trees. That wouldn't exclude it from hominid status, because full-blown bipedalism almost certainly didn't emerge in one fell swoop. Rather *Orrorin* may have simply not yet evolved the full complement of traits required for habitual bipedalism. Viewed that way, *Orrorin* could still be on the ancestral line, albeit further removed from *Homo* than Pickford and Senut would have it.

Better evidence of early routine bipedalism, in Lovejoy's view, surfaced a few months after the *Orrorin* report, when Berkeley graduate student Yohannes Haile-Selassie announced the discovery of slightly younger fossils from Ethiopia's Middle Awash region. Those 5.2-million- to 5.8-million-year-old remains, which have been classified as a subspecies of *Ardipithecus ramidus, A. r. kadabba,* include a complete foot phalanx, or toe bone, bearing a telltale trait. The bone's joint is angled in precisely the way one would expect if *A. r. kadabba* "toed off" as humans do when walking, reports Lovejoy, who has studied the fossil.

Other workers are less impressed by the toe morphology. "To me, it looks for all the world like a chimpanzee foot phalanx," comments David Begun of the University of Toronto, noting from photographs that it is longer, slimmer and more curved than a biped's toe bone should be. Clarification may come when White and his collaborators publish findings on an as yet undescribed partial skeleton of *Ardipithecus,* which White says they hope to do within the next year or two.

Differing anatomical interpretations notwithstanding, if either *Orrorin* or *A. r. kadabba* were a biped, that would not only push the origin of our strange mode of locomotion back by nearly 1.5 million years, it would also lay to rest a popular idea about the conditions under which our striding gait evolved. Received wisdom holds that our ancestors became bipedal on the African savanna, where upright walking may have kept the blistering sun off their backs, given them access to previously out-of-reach foods, or afforded them a better view above the tall grass. But paleoecological analyses indicate that *Orrorin* and *Ardipithecus* dwelled in forested habitats, alongside monkeys and other typically woodland creatures. In fact, Giday Wolde-Gabriel of Los Alamos National Laboratory and his colleagues, who studied the soil chemistry and animal remains at the *A. r. kadabba* site, have noted that early hominids may not have ventured beyond these relatively wet and wooded settings until after 4.4 million years ago.

If so, climate change may not have played as important a role in driving our ancestors from four legs to two as has been thought. For his part, Lovejoy observes that a number of the savanna-based hypotheses focusing on posture were not especially well conceived to begin with. "If your eyes were in your toes, you could stand on your hands all day and look over tall grass, but you'd never evolve into a hand-walker," he jokes. In other words, selection for upright posture alone would not, in his view, have led to bipedal locomotion. The most plausible explanation for the emergence of bipedalism, Lovejoy says, is that it freed the hands and allowed males to collect extra food with which to woo mates. In this model, which he developed in the 1980s, females who chose good providers could devote more energy to child rearing, thereby maximizing their reproductive success.

THE OLDEST ANCESTOR?

The paleoanthropological community was still digesting the implications of the *Orrorin* and *A. r. kadabba* discoveries when Brunet's fossil find from Chad came to light. With *Sahelanthropus* have come new answers— and new questions. Unlike *Orrorin* and *A. r. kadabba,* the *Sahelanthropus* material does not include any postcranial bones, making it impossible at this point to know whether the animal was bipedal, the traditional hallmark of humanness. But Brunet argues that a suite of features in the teeth and skull, which he believes belongs to a male, judging from the massive brow ridge, clearly links this creature to all later hominids. Characteristics of *Sahelanthropus*'s canines are especially important in his assessment. In all modern and fossil apes, and therefore presumably in the last common ancestor of chimps and humans, the large upper canines are honed against the first lower premolars, producing a sharp edge along the back of the canines. This so-called honing canine-premolar complex is pronounced in males, who use their canines to compete with one another for females. Humans lost these fighting teeth, evolving smaller, more incisorlike canines that occlude tip to tip, an arrangement that creates a distinctive wear pattern over time. In their size, shape and wear, the *Sahelanthropus* canines are modified in the human direction, Brunet asserts.

At the same time, *Sahelanthropus* exhibits a number of apelike traits, such as its small braincase and widely spaced eye sockets. This mosaic of primitive and advanced features, Brunet says, suggests a close relationship to the last common ancestor. Thus, he proposes that *Sahelanthropus* is the earliest member of the human lineage and the ancestor of all later hominids, including *Orrorin* and *Ardipithecus.* If Brunet is correct, humanity may have arisen more than a million years earlier than a number of molecular studies had estimated. More important, it may have originated in a different locale than has been posited. According to one model of human origins, put forth in the 1980s by Yves Coppens of the College of France, East Africa was the birthplace of humankind. Coppens, noting that the oldest human fossils came from East Africa, proposed that the continent's Rift Valley—a gash that runs from north to south—split a single ancestral ape species into two populations. The one in the east gave rise to humans; the one in the west spawned today's apes [see "East Side Story: The Origin of Humankind," by Yves Coppens; *Scientific American,* May 1994]. Scholars have recognized for some time that the apparent geographic separation might instead be an artifact of the scant fossil record. The discovery of a seven-million-year-old hominid in Chad, some 2,500 kilometers west of the Rift Valley, would deal the theory a fatal blow.

Most surprising of all may be what *Sahelanthropus* reveals about the last common ancestor of humans and chimpanzees. Paleoanthropologists have typically imagined that that creature resembled a chimp in having, among other things, a strongly projecting lower face, thinly enameled molars and large canines. Yet *Sahelanthropus,* for all its generally apelike traits, has only a moderately prognathic face, relatively thick enamel, small canines and a brow ridge larger than that of any living ape. "If *Sahelanthropus* shows us anything, it shows us that the last common ancestor was not a chimpanzee," Berkeley's White remarks. "But why should we have expected otherwise?" Chimpanzees have had just as much time to evolve as humans have had, he points out, and they have become highly specialized, fruit-eating apes.

Brunet's characterization of the Chadian remains as those of a human ancestor has not gone unchallenged, however. "Why *Sahelanthropus* is necessarily a hominid is not particularly clear," comments Carol V. Ward of the University of Missouri. She and others are skeptical that the canines are as humanlike as Brunet claims. Along similar lines, in a letter published last October in the journal *Nature,* in which Brunet's team initially reported its findings, University of Michigan paleoanthropologist Milford H. Wolpoff, along with *Orrorin* discoverers Pickford and Senut, countered that *Sahelanthropus* was an ape rather than a hominid. The massive brow and certain features on the base and rear of *Sahelanthropus*'s skull, they observed, call to mind the anatomy of a quadrupedal ape with a difficult-to-chew diet, whereas the small canine suggests that it was a female of such a species, not a male human ancestor. Lacking proof that *Sahelanthropus* was bipedal, so their reasoning goes, Brunet doesn't have a leg to stand on. (Pickford and Senut further argue that the animal was specifically a gorilla ancestor.) In a barbed response, Brunet likened his detractors to those Dart encountered in 1925, retorting that *Sahelanthropus*'s apelike traits are simply primitive holdovers from its own ape predecessor and therefore uninformative with regard to its relationship to humans.

The conflicting views partly reflect the fact that researchers disagree over what makes the human lineage unique. "We have trouble defining hominids," acknowledges Roberto Macchiarelli, also at the University of Poitiers. Traditionally paleoanthropologists have regarded bipedalism as the characteristic that first set human ancestors apart from other apes. But subtler changes—the metamorphosis of the canine, for instance—may have preceded that shift.

To understand how animals are related to one another, evolutionary biologists employ a method called cladistics, in which organisms are grouped according to shared, newly evolved traits. In short, creatures that have these derived characteristics in common are deemed more closely related to one another than they are to those that exhibit only primitive traits inherited from a more distant common ancestor. The first occurrence in the fossil record of a shared, newly acquired trait serves as a baseline indicator of the biological division of an ancestral species into two daughter species—in this case, the point at which chimps and humans diverged from their common ancestor—and that trait is considered the defining characteristic of the group.

Thus, cladistically "what a hominid is from the point of view of skeletal morphology is summarized by those characters preserved in the skeleton that are present in populations that directly succeeded the genetic splitting event between chimps and humans," explains William H. Kimbel of Arizona State University. With only an impoverished fossil record to work from, paleontologists can't know for certain what those traits were. But the two leading candidates for the title of seminal hominid characteristic, Kimbel says, are bipedalism and the transformation of the canine. The problem researchers now face in trying to suss out what the initial changes were and which, if any, of the new putative hominids sits at the base of the human clade is that so far *Orrorin, A. r. kadabba* and *Sahelanthropus* are represented by mostly different bony elements, making comparisons among them difficult.

HOW MANY HOMINIDS?

Meanwhile the arrival of three new taxa to the table has intensified debate over just how diverse early hominids were. Experts concur that between three million and 1.5 million years ago, multiple hominid species existed alongside one another at least occasionally. Now some scholars argue that this rash of discoveries demonstrates that human evolution was a complex affair from the outset. Toronto's Begun—who believes that the Miocene ape ancestors of modern African apes and humans spent their evolutionarily formative years in Europe and western Asia before reentering Africa—observes that *Sahelanthropus* bears exactly the kind of motley features that one would expect to see in an animal that was part of an adaptive radiation of apes moving into a new milieu. "It would not surprise me if there were 10 or 15 genera of things that are more closely related to *Homo* than to chimps," he says. Likewise, in a commentary that accompanied the report by Brunet and his team in *Nature,* Bernard Wood of George Washington University wondered whether *Sahelanthropus* might hail from the African ape equivalent of Canada's famed Burgess Shale, which has yielded myriad invertebrate fossils from the Cambrian period, when the major modern animal groups exploded into existence. Viewed that way, the human evolutionary tree would look more like an unkempt bush, with some, if not all, of the new discoveries occupying terminal twigs instead of coveted spots on the meandering line that led to humans.

Other workers caution against inferring the existence of multiple, coeval hominids on the basis of what has yet been found. "That's *X-Files* paleontology," White quips. He and Brunet both note that between seven million and four million years ago, only one hominid species is known to have existed at any given time. "Where's the bush?" Brunet demands. Even at humanity's peak diversity, two million years ago, White says, there were only three taxa sharing the landscape. "That ain't the Cambrian explosion," he remarks dryly. Rather, White suggests, there is no evidence that the base of the family tree is anything other than a trunk. He thinks that the new finds might all represent snapshots of the *Ardipithecus* lineage through time, with *Sahelanthropus* being the earliest hominid and with *Orrorin* and *A. r. kadabba* representing its lineal descendants. (In this configuration, *Sahelanthropus* and *Orrorin* would become species of *Ardipithecus*.)

Investigators agree that more fossils are needed to elucidate how *Orrorin, A. r. kadabba* and *Sahelanthropus* are related to one another and to ourselves, but obtaining a higher-resolution picture of the roots of humankind won't be easy. "We're going to have a lot of trouble diagnosing the very earliest members of our clade the closer we get to that last common ancestor," Missouri's Ward predicts. Nevertheless, "it's really important to sort out what the starting point was," she observes. "Why the human lineage began is the question we're trying to answer, and these new finds in some ways may hold the key to answering that question—or getting closer than we've ever gotten before."

It may be that future paleoanthropologists will reach a point at which identifying an even earlier hominid will be well nigh impossible. But it's unlikely that this will keep them from trying. Indeed, it would seem that the search for the first hominids is just heating up. "The *Sahelanthropus* cranium is a messenger [indicating] that in central Africa there is a desert full of fossils of the right age to answer key questions about the genesis of our clade," White reflects. For his part, Brunet, who for more than a quarter of a century has doggedly pursued his vision through political unrest, sweltering heat and the blinding sting of an unrelenting desert wind, says that ongoing work in Chad will keep his team busy for years to come. "This is the beginning of the story," he promises, "just the beginning." As I sit in Brunet's office contemplating the seven-million-year-old skull of *Sahelanthropus,* the fossil hunter's quest doesn't seem quite so unimaginable. Many of us spend the better part of a lifetime searching for ourselves.

FURTHER READING

Yohannes Haile-Selassie. "Late Miocene Hominids from the Middle Awash, Ethiopia" in *Nature,* Vol. 412, pages 178–181; July 12, 2001.

Ian Tattersall and Jeffrey H. Schwartz. *Extinct Humans.* Westview Press, 2001.

Martin Pickford, Brigitte Senut, Dominique Gommercy and Jacques Treil. "Bipedalism in *Orrorin tugenensis* Revealed by Its Femora" in *Comptes Rendus: Palevol,* Vol. 1, No. 1, pages 1–13; 2002.

Michel Brunet, Franck Guy, David Pilbeam, Hassane Taisso Mackaye et al. "A New Hominid from the Upper Miocene of Chad, Central Africa" in *Nature,* Vol. 418, pages 145–151; July 11, 2002.

Walter C. Hartwig, ed. *The Primate Fossil Record.* Cambridge University Press, 2002.

Early Hominid Fossils from Africa

MEAVE LEAKEY AND ALAN WALKER

ORIGINALLY PUBLISHED IN JUNE 1992; UPDATED IN 2003

The year was 1965. Bryan Patterson, a paleoanthropologist from Harvard University, unearthed a fragment of a fossil arm bone at a site called Kanapoi in northern Kenya. He and his colleagues knew it would be hard to make a great deal of anatomical or evolutionary sense out of a small piece of elbow joint. Nevertheless, they did recognize some features reminiscent of a species of early hominid (a hominid is any upright-walking primate) known as *Australopithecus,* first discovered 40 years earlier in South Africa by Raymond Dart of the University of the Witwatersrand. In most details, however, Patterson and his team considered the fragment of arm bone to be more like those of modern humans than the one other *Australopithecus* humerus known at the time.

And yet the age of the Kanapoi fossil proved somewhat surprising. Although the techniques for dating the rocks where the fossil was uncovered were still fairly rudimentary, the group working in Kenya was able to show that the bone was probably older than the various *Australopithecus* specimens that had previously been found. Despite this unusual result, however, the significance of Patterson's discovery was not to be confirmed for another 30 years. In the interim, researchers identified the remains of so many important early hominids that the humerus from Kanapoi was rather forgotten.

Yet Patterson's fossil would eventually help establish the existence of a new species of *Australopithecus*—the oldest yet to be identified—and push back the origins of upright walking to more than four million years ago. But to see how this happened, we need to trace the steps that paleoanthropologists have taken in constructing an outline for the story of hominid evolution.

AN EVOLVING STORY

Scientists classify the immediate ancestors of the genus *Homo* (which includes our own species, *Homo sapiens*) in the genus *Australopithecus.* For

260

several decades it was believed that these ancient hominids first inhabited the earth at least three and a half million years ago. The specimens found in South Africa by Dart and others indicated that there were at least two types of *Australopithecus*—*A. africanus* and *A. robustus.* The leg bones of both species suggested that they had the striding, bipedal locomotion that is a hallmark of humans among living mammals. (The upright posture of these creatures was vividly confirmed in 1978 at the Laetoli site in Tanzania, where a team led by archaeologist Mary Leakey discovered a spectacular series of footprints made 3.6 million years ago by three *Australopithecus* individuals as they walked across wet volcanic ash.) Both *A. africanus* and *A. robustus* were relatively small-brained and had canine teeth that differed from those of modern apes in that they hardly projected past the rest of the tooth row. The younger of the two species, *A. robustus,* had bizarre adaptations for chewing—huge molar and premolar teeth combined with bony crests on the skull where powerful chewing muscles would have been attached.

Paleoanthropologists identified more species of *Australopithecus* over the next several decades. In 1959 Mary Leakey unearthed a skull from yet another East African species closely related to *robustus.* Skulls of these species uncovered during the past 45 years in the north-eastern part of Africa, in Ethiopia and Kenya, differed considerably from those found in South Africa; as a result, researchers think that two separate *robustus*-like species—a northern one and a southern one—existed.

In 1978 Donald C. Johanson, now at the Institute of Human Origins at Arizona State University, along with his colleagues, identified still another species of *Australopithecus.* Johanson and his team had been studying a small number of hominid bones and teeth discovered at Laetoli, as well as a large and very important collection of specimens from the Hadar region of Ethiopia (including the famous "Lucy" skeleton). The group named the new species *afarensis.* Radiometric dating revealed that the species had lived between 3.6 and 2.9 million years ago, making it the oldest *Australopithecus* known at the time.

This early species is probably the best studied of all the *Australopithecus* recognized so far, and it is certainly the one that has generated the most controversy over the past 30 years. The debates have ranged over many issues: whether the *afarensis* fossils were truly distinct from the *africanus* fossils from South Africa; whether there was one or several species at Hadar; whether the Tanzanian and Ethiopian fossils were of the same species; and whether the fossils had been dated correctly.

But the most divisive debate concerns the issue of how extensively the bipedal *afarensis* climbed in trees. Fossils of *afarensis* include various bone and joint structures typical of tree climbers. Some scientists argue that

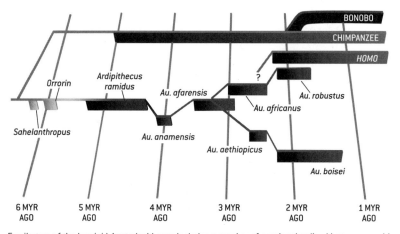

Family tree of the hominid *Australopithecus* includes a number of species that lived between roughly 4 million and 1.25 million years (Myr) ago. Just over 2 Myr ago a new genus, *Homo* (which includes our own species, *H. sapiens*), evolved from one of the species of *Australopithecus*.

such characteristics indicate that these hominids must have spent at least some time in the trees. But others view these features as simply evolutionary baggage, left over from arboreal ancestors. Underlying this discussion is the question of where *Australopithecus* lived — in forests or on the open savanna.

By the beginning of the 1990s, researchers knew a fair amount about the various species of *Australopithecus* and how each had adapted to its environmental niche. A description of any one of the species would mention that the creatures were bipedal and that they had ape-size brains and large, thickly enameled teeth in strong jaws, with nonprojecting canines. Males were typically larger than females, and individuals grew and matured rapidly. But the origins of *Australopithecus* were only hinted at, because the gap between the earliest well-known species in the group (*afarensis,* from about 3.6 million years ago) and the postulated time of the last common ancestor of chimpanzees and humans (about six million years ago, according to molecular evidence) was still very great. Fossil hunters had unearthed only a few older fragments of bone, tooth and jaw from the intervening 1.5 million years to indicate the anatomy and course of evolution of the earliest hominids.

FILLING THE GAP

Discoveries in Kenya over the past several years have filled in some of the missing interval between 3.5 million and 5 million years ago. Beginning in

1982, expeditions run by the National Museums of Kenya to the Lake Turkana basin in northern Kenya began finding hominid fossils nearly four million years old. But because these fossils were mainly isolated teeth—no jawbones or skulls were preserved—very little could be said about them except that they resembled the remains of *afarensis* from Laetoli. But our excavations at an unusual site, just inland from Allia Bay on the east side of Lake Turkana, yielded more complete fossils.

The site at Allia Bay is a bone bed, where millions of fragments of weathered tooth and bone from a wide variety of animals, including hominids, spill out of the hillside. Exposed at the top of the hill lies a layer of hardened volcanic ash called the Moiti Tuff, which has been dated radiometrically to just over 3.9 million years old. The fossil fragments lie several meters below the tuff, indicating that the remains are older than the tuff. We do not yet understand fully why so many fossils are concentrated in this spot, but we can be certain that they were deposited by the precursor of the present-day Omo River.

Today the Omo drains the Ethiopian highlands located to the north, emptying into Lake Turkana, which has no outlet. But this has not always been so. Our colleagues Frank Brown of the University of Utah and Craig Feibel of Rutgers University have shown that the ancient Omo River dominated the Turkana area for much of the Pliocene (roughly 5.3 to 1.8 million years ago) and the early Pleistocene (1.8 to 0.7 million years ago). Only infrequently was a lake present in the area at all. Instead, for most of the past four million years, an extensive river system flowed across the broad floodplain, proceeding to the Indian Ocean without dumping its sediments into a lake.

The Allia Bay fossils are located in one of the channels of this ancient river system. Most of the fossils collected from Allia Bay are rolled and weathered bones and teeth of aquatic animals—fish, crocodiles, hippopotamuses and the like—that were damaged during transport down the river from some distance away. But some of the fossils are much better preserved; these come from the animals that lived on or near the riverbanks. Among these creatures are several different species of leaf-eating monkeys, related to modern colobus monkeys, as well as antelopes whose living relatives favor closely wooded areas. Reasonably well preserved hominid fossils can also be found here, suggesting that, at least occasionally, early hominids inhabited a riparian habitat.

Where do these *Australopithecus* fossils fit in the evolutionary history of hominids? The jaws and teeth from Allia Bay, as well as a nearly complete radius (the outside bone of the forearm) from the nearby sediments of

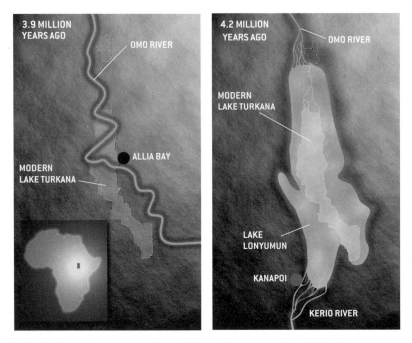

Turkana basin was home to *anamensis* roughly four million years ago. Around 3.9 million years ago a river sprawled across the basin (*left*). The fossil site Allia Bay sat within the strip of forest that lined this river. Some 4.2 million years ago a large lake filled the basin (*right*); a second site, Kanapoi, was located on a river delta that fed into the lake.

Sibilot just to the north, show an interesting mixture of characteristics. Some of the traits are primitive ones—that is, they are ancestral features thought to be present before the split occurred between the chimpanzee and human lineages. Yet these bones also share characteristics seen in later hominids and are therefore said to have more advanced features. As our team continues to unearth more bones and teeth at Allia Bay, these new fossils add to our knowledge of the wide range of traits present in early hominids.

Across Lake Turkana, some 145 kilometers (about 90 miles) south of Allia Bay, lies the site of Kanapoi, where our story began. One of us (Leakey) has mounted expeditions from the National Museums of Kenya to explore the sediments located southwest of Lake Turkana and to document the faunas present during the earliest stages of the basin's history. Kanapoi, virtually unexplored since Patterson's day, has proved to be one of the most rewarding sites in the Turkana region.

A series of deep erosion gullies, known as badlands, has exposed the sediments at Kanapoi. Fossil hunting is difficult here, though, because of a carapace of lava pebbles and gravel that makes it hard to spot small bones and teeth. Studies of the layers of sediment, also carried out by Feibel, reveal that the fossils here have been preserved by deposits from a river ancestral to the present-day Kerio River, which once flowed into the Turkana basin and emptied into an ancient lake that we call Lonyumun. This lake reached its maximum size about 4.1 million years ago and thereafter shrank as it filled with sediments.

Excavations at Kanapoi have primarily yielded the remains of carnivore meals, so the fossils are rather fragmentary. But workers at the site have also recovered two nearly complete lower jaws, one complete upper jaw and lower face, the upper and lower thirds of a tibia, bits of skull and several sets of isolated teeth. After careful study of the fossils from both Allia Bay and Kanapoi—including Patterson's fragment of an arm bone—we felt that in details of anatomy, these specimens were different enough from previously known hominids to warrant designating a new species. So in 1995, in collaboration with both Feibel and Ian McDougall of the Australian National University, we named this new species *Australopithecus anamensis,* drawing on the Turkana word for "lake" (*anam*) to refer to both the present and ancient lakes.

To establish the age of these fossils, we relied on the extensive efforts of Brown, Feibel and McDougall, who have been investigating the paleogeographic history of the entire lake basin. If their study of the basin's development is correct, the *anamensis* fossils should be between 4.2 and 3.9 million years old. McDougall has determined the age of the so-called Kanapoi Tuff—the layer of volcanic ash that covers most of the fossils at this site—to be just over four million years old. Now that he has successfully ascertained the age of the tuff, we are confident in both the age of the fossils and Brown's and Feibel's understanding of the history of the lake basin.

A major question in paleoanthropology today is how the anatomical mosaic of the early hominids evolved. By comparing the nearly contemporaneous Allia Bay and Kanapoi collections of *anamensis,* we can piece together a fairly accurate picture of certain aspects of the species, even though we have not yet uncovered a complete skull.

The jaws of *anamensis* are primitive—the sides sit close together and parallel to each other (as in modern apes), rather than widening at the back of the mouth (as in later hominids, including humans). In its lower jaw, *anamensis* is also chimplike in terms of the shape of the region where the left and right sides of the jaw meet (technically known as the mandibular symphysis).

Teeth from *anamensis,* however, appear more advanced. The enamel is relatively thick, as it is in all other species of *Australopithecus;* in contrast, the tooth enamel of African great apes is much thinner. The thickened enamel suggests *anamensis* had already adapted to a changed diet—possibly much harder food—even though its jaws and some skull features were still very apelike. We also know that *anamensis* had only a tiny external ear canal. In this regard, it is more like chimpanzees and unlike all later hominids, including humans, which have large external ear canals. (The size of the external canal is unrelated to the size of the fleshy ear.)

The most informative bone of all the ones we have uncovered from this new hominid is the nearly complete tibia—the larger of the two bones in the lower leg. The tibia is revealing because of its important role in weight bearing: the tibia of a biped is distinctly different from the tibia of an animal that walks on all four legs. In size and practically all details of the knee and ankle joints, the tibia found at Kanapoi closely resembles the one from the fully bipedal *afarensis* found at Hadar, even though the latter specimen is almost a million years younger.

Fossils of other animals collected at Kanapoi point to a somewhat different paleoecological scenario from the setting across the lake at Allia Bay. The channels of the river that laid down the sediments at Kanapoi were probably lined with narrow stretches of forest that grew close to the riverbanks in otherwise open country. Researchers have recovered the remains of the same spiral-horned antelope found at Allia Bay that very likely lived in dense thickets. But open-country antelopes and hartebeest appear to have lived at Kanapoi as well, suggesting that more open savanna prevailed away from the rivers. These results offer equivocal evidence regarding the preferred habitat of *anamensis:* we know that bushland was present at both sites that have yielded fossils of the species, but there are clear signs of more diverse habitats at Kanapoi.

AN EVEN OLDER HOMINID?

At about the same time that we were finding new hominids at Allia Bay and Kanapoi, a team led by our colleague Tim D. White of the University of California at Berkeley discovered fossil hominids in Ethiopia that are even older than *anamensis*. In 1992 and 1993 White led an expedition to the Middle Awash area of Ethiopia, where his team uncovered hominid fossils at a site known as Aramis. The group's finds include isolated teeth, a piece of a baby's mandible (the lower jaw), fragments from an adult's skull and some arm bones, all of which have been dated to around 4.4 million years ago. In 1994, together with his colleagues Berhane Asfaw of the Paleoanthropology Laboratory in Addis Ababa and Gen Suwa of the

University of Tokyo, White gave these fossils a new name: *Australopithecus ramidus.* In 1995 the group renamed the fossils, moving them to a new genus, *Ardipithecus.* Earlier fossils of this genus have now been found dating back to 5.8 million years ago. Other fossils buried near the hominids, such as seeds and the bones of forest monkeys and antelopes, strongly imply that these hominids, too, lived in a closed-canopy woodland.

This new species represents the most primitive hominid known—a link between the African apes and *Australopithecus.* Many of the *Ardipithecus ramidus* fossils display similarities to the anatomy of the modern African great apes, such as thin dental enamel and strongly built arm bones. In other features, though—such as the opening at the base of the skull, technically known as the foramen magnum, through which the spinal cord connects to the brain—the fossils resemble later hominids.

Describing early hominids as either primitive or more advanced is a complex issue. Scientists now have almost decisive molecular evidence that humans and chimpanzees once had a common ancestor and that this lineage had previously split from gorillas. This is why we often use the two living species of chimpanzee (*Pan troglodytes* and *P. paniscus*) to illustrate ancestral traits. But we must remember that since their last common ancestor with humans, chimpanzees have had exactly the same amount of time to evolve as humans have. Determining which features were present in the last common ancestor of humans and chimpanzees is not easy.

But *Ardipithecus,* with its numerous chimplike features, appears to have taken the human fossil record back close to the time of the chimp-human split. More recently, White and his group have found parts of a single *Ardipithecus* skeleton in the Middle Awash region. As White and his team extract these exciting new fossils from the enclosing stone, reconstruct them and prepare them for study, the paleoanthropological community eagerly anticipates the publication of the group's analysis of these astonishing finds.

But even pending White's results, new fossil discoveries are offering other surprises. A team led by Michel Brunet of the University of Poitiers has found fragments of *Australopithecus* fossils in Chad. Surprisingly, these fossils were recovered far from either eastern or southern Africa, the only areas where *Australopithecus* had appeared. The Chad sites lie 2,500 kilometers west of the western part of the Rift Valley, thus extending the range of *Australopithecus* well into the center of Africa.

These fossils debunk a hypothesis about human evolution postulated by Dutch primatologist Adriaan Kortlandt and expounded in *Scientific American* by Yves Coppens of the College of France [see "East Side Story: The

Origin of Humankind," May 1994]. This idea was that the formation of Africa's Rift Valley subdivided a single ancient species, isolating the ancestors of hominids on the east side from the ancestors of modern apes on the west side.

Brunet's latest discovery, an important cranium older than six million years, is also from Chad and shows that early hominids were probably present across much of the continent. This cranium, which the team called *Sahelanthropus tchadensis,* together with fragmentary jaws and limb bones from about six million years ago in Kenya, are even older than the *Ardipithecus* fossils. The significance of these exciting discoveries is now the center of an active debate.

The fossils of *anamensis* that we have identified should also provide some answers in the long-standing debate over whether early *Australopithecus* species lived in wooded areas or on the open savanna. The outcome of this discussion has important implications: for many years, paleoanthropologists have accepted that upright-walking behavior originated on the savanna, where it most likely provided benefits such as keeping the hot sun off the back or freeing hands for carrying food. Yet our evidence suggests that the earliest bipedal hominid known to date lived at least part of the time in wooded areas. The discoveries of the past several years represent a remarkable spurt in the sometimes painfully slow process of uncovering human evolutionary past. But clearly there is still much more to learn.

FURTHER READING

Tim D. White, Gen Suwa and Berhane Asfaw. *"Australopithecus ramidus,* a New Species of Early Hominid from Aramis, Ethiopia" in *Nature,* Vol. 371, pages 306–312; September 22, 1994.

Meave G. Leakey, Craig S. Feibel, Ian McDougall and Alan Walker. "New Four-Million-Year-Old Hominid Species from Kanapoi and Allia Bay, Kenya" in *Nature,* Vol. 376, pages 565–571; August 17, 1995.

Donald C. Johanson and Blake Edgar. *From Lucy to Language.* Simon & Schuster, 1996.

C. V. Ward, M. G. Leakey and A. Walker. "The Earliest Known *Australopithecus, A. anamensis"* in *Journal of Human Evolution,* Vol. 41, pages 255–368; 2001.

Planet of the Apes

DAVID R. BEGUN

ORIGINALLY PUBLISHED IN AUGUST 2003

" It is therefore probable that Africa was formerly inhabited by extinct apes closely allied to the gorilla and chimpanzee; as these two species are now man's closest allies, it is somewhat more probable that our early progenitors lived on the African continent than elsewhere."

So mused Charles Darwin in his 1871 work, *The Descent of Man*. Although no African fossil apes or humans were known at the time, remains recovered since then have largely confirmed his sage prediction about human origins. There is, however, considerably more complexity to the story than even Darwin could have imagined. Current fossil and genetic analyses indicate that the last common ancestor of humans and our closest living relative, the chimpanzee, surely arose in Africa, around six million to eight million years ago. But from where did this creature's own forebears come? Paleoanthropologists have long presumed that they, too, had African roots. Mounting fossil evidence suggests that this received wisdom is flawed.

Today's apes are few in number and in kind. But between 22 million and 5.5 million years ago, a time known as the Miocene epoch, apes ruled the primate world. Up to 100 ape species ranged throughout the Old World, from France to China in Eurasia and from Kenya to Namibia in Africa. Out of this dazzling diversity, the comparatively limited number of apes and humans arose. Yet fossils of great apes—the large-bodied group represented today by chimpanzees, gorillas and orangutans (gibbons and siamangs make up the so-called lesser apes)—have turned up only in western and central Europe, Greece, Turkey, South Asia and China. It is thus becoming clear that, by Darwin's logic, Eurasia is more likely than Africa to have been the birthplace of the family that encompasses great apes and humans, the hominids. (The term "hominid" has traditionally been reserved for humans and protohumans, but scientists are increasingly placing our great ape kin in the definition as well and using another word, "hominin," to refer to the human subset. The word "hominoid" encompasses all apes—including gibbons and siamangs—and humans.)

Perhaps it should not come as a surprise that the apes that gave rise to hominids may have evolved in Eurasia instead of Africa: the combined effects of migration, climate change, tectonic activity and ecological shifts on a scale unsurpassed since the Miocene made this region a hotbed of hominoid evolutionary experimentation. The result was a panoply of apes, two lineages of which would eventually find themselves well positioned to colonize Southeast Asia and Africa and ultimately to spawn modern great apes and humans.

Paleoanthropology has come a long way since Georges Cuvier, the French natural historian and founder of vertebrate paleontology, wrote in 1812 that "*l'homme fossile n'existe pas*" ("fossil man does not exist"). He included all fossil primates in his declaration. Although that statement seems unreasonable today, evidence that primates lived alongside animals then known to be extinct—mastodons, giant ground sloths and primitive ungulates, or hoofed mammals, for example—was quite poor. Ironically, Cuvier himself described what scholars would later identify as the first fossil primate ever named, *Adapis parisiensis* Cuvier 1822, a lemur from the chalk mines of Paris that he mistook for an ungulate. It wasn't until 1837, shortly after Cuvier's death, that his disciple Édouard Lartet described the first fossil higher primate recognized as such. Now known as *Pliopithecus,* this jaw from southeastern France, and other specimens like it, finally convinced scholars that such creatures had once inhabited the primeval forests of Europe. Nearly 20 years later Lartet unveiled the first fossil great ape, *Dryopithecus,* from the French Pyrénées.

In the remaining years of the 19th century and well into the 20th, paleontologists recovered many more fragments of ape jaws and teeth, along with a few limb bones, in Spain, France, Germany, Austria, Slovakia, Hungary, Georgia and Turkey. By the 1920s, however, attention had shifted from Europe to South Asia (India and Pakistan) and Africa (mainly Kenya), as a result of spectacular finds in those regions, and the apes of Eurasia were all but forgotten. But fossil discoveries of the past two decades have rekindled intense interest in Eurasian fossil apes, in large part because paleontologists have at last recovered specimens complete enough to address what these animals looked like and how they are related to living apes and humans.

THE FIRST APES

To date, researchers have identified as many as 40 genera of Miocene fossil apes from localities across the Old World—eight times the number that

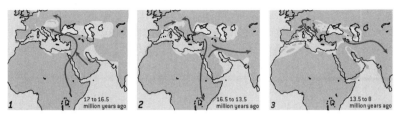

APES ON THE MOVE: Africa was the cradle of apekind, having spawned the first apes more than 20 million years ago. But it was not long before these animals colonized the rest of the Old World. Changes in sea level alternately connected Africa to and isolated it from Eurasia and thus played a critical role in ape evolution. A land bridge joining East Africa to Eurasia between 17 million and 16.5 million years ago enabled early Miocene apes to invade Eurasia (*1*). Over the next few million years, they spread to western Europe and the Far East, and great apes evolved; some primitive apes returned to Africa (*2*). Isolated from Africa by elevated sea levels, the early Eurasian great apes radiated into a number of forms (*3*). Drastic climate changes at the end of the Late Miocene wiped out most of the Eurasian great apes. The two lineages that survived—those represented by *Sivapithecus* and *Dryopithecus*—did so by moving into Southeast Asia and the African tropics (*4*).

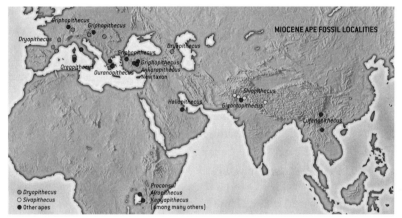

MIOCENE APE FOSSIL LOCALITIES

survive today. Such diversity seems to have characterized the ape family from the outset: almost as soon as apes appear in the fossil record, there are lots of them. So far 14 genera are known to have inhabited Africa during the early Miocene alone, between 22 million and 17 million years ago. And considering the extremely imperfect nature of the fossil record, chances are that this figure significantly underrepresents the number of apes that actually existed at that time.

Like living apes, these creatures varied considerably in size. The smallest weighed in at a mere three kilograms, hardly more than a small housecat; the largest tipped the scales at a gorillalike heft of 80 kilograms. They were even more diverse than their modern counterparts in terms of what they ate, with some specializing in leaves and others in fruits and nuts, although the majority subsisted on ripe fruits, as most apes do today. The biggest difference between those first apes and extant ones lay in their posture and means of getting around. Whereas modern apes exhibit a rich repertoire of locomotory modes—from the highly acrobatic brachiation employed by the arboreal gibbon to the gorilla's terrestrial knuckle walking—early Miocene apes were obliged to travel along tree branches on all fours.

To understand why the first apes were restricted in this way, consider the body plan of the early Miocene ape. The best-known ape from this period is *Proconsul*, exceptionally complete fossils of which have come from sites on Kenya's Rusinga Island [see "The Hunt for Proconsul," by Alan Walker and Mark Teaford; *Scientific American*, January 1989]. Specialists currently recognize four species of *Proconsul*, which ranged in size from about 10 kilograms to possibly as much as 80 kilograms. *Proconsul* gives us a good idea of the anatomy and locomotion of an early ape. Like all extant apes, this one lacked a tail. And it had more mobile hips, shoulders, wrists, ankles, hands and feet than those of monkeys, presaging the fundamental adaptations that today's apes and humans have for flexibility in these joints. In modern apes, this augmented mobility enables their unique pattern of movement, swinging from branch to branch. In humans, these capabilities have been exapted, or borrowed, in an evolutionary sense, for enhanced manipulation in the upper limb—something that allowed our ancestors to start making tools, among other things.

At the same time, however, *Proconsul* and its cohorts retained a number of primitive, monkeylike characteristics in the backbone, pelvis and forelimbs, leaving them, like their monkey forebears, better suited to traveling along the tops of tree branches than hanging and swinging from limb to limb. (Intriguingly, one enigmatic early Miocene genus from Uganda, *Morotopithecus,* may have been more suspensory, but the evidence is inconclusive.) Only when early apes shed more of this evolutionary baggage could they begin to adopt the forms of locomotion favored by contemporary apes.

PASSAGE TO EURASIA

Most of the early Miocene apes went extinct. But one of them—perhaps *Afropithecus* from Kenya—was ancestral to the species that first made its way over to Eurasia some 16.5 million years ago. At around that time global sea levels dropped, exposing a land bridge between Africa and Eurasia. A mammalian exodus ensued. Among the creatures that migrated out of their African homeland were elephants, rodents, ungulates such as pigs and antelopes, a few exotic animals such as aardvarks, and primates.

The apes that journeyed to Eurasia from Africa appear to have passed through Saudi Arabia, where the remains of *Heliopithecus,* an ape similar to *Afropithecus,* have been found. Both *Afropithecus* and *Heliopithecus* (which some workers regard as members of the same genus) had a thick covering of enamel on their teeth—good for processing hard foods, such as nuts, and tough foods protected by durable husks. This dental innovation may have played a key role in helping their descendants establish a foothold in the forests of Eurasia by enabling them to exploit food resources not available to *Proconsul* and most earlier apes. By the time the seas rose to swallow the bridge linking Africa to Eurasia half a million years later, apes had ensconced themselves in this new land.

The movement of organisms into new environments drives speciation, and the arrival of apes in Eurasia was no exception. Indeed, within a geologic blink of an eye, these primates adapted to the novel ecological conditions and diversified into a plethora of forms—at least eight known in just 1.5 million years. This flurry of evolutionary activity laid the groundwork for the emergence of great apes and humans. But only recently have researchers begun to realize just how important Eurasia was in this regard. Paleontologists traditionally thought that apes more sophisticated in their food-processing abilities than *Afropithecus* and *Heliopithecus* reached Eurasia about 15 million years ago, around the time they first appear in Africa. This fit with the notion that they arose in Africa and then dispersed northward. New fossil evidence, however, indicates that advanced apes (those with massive jaws and large, grinding teeth) were actually in Eurasia far earlier than that. In 2001 and 2003 my colleagues and I described a more modern-looking ape, *Griphopithecus,* from 16.5-million-year-old sites in Germany and Turkey, pushing the Eurasian ape record back by more than a million years.

The apparent absence of such newer models in Africa between 17 million and 15 million years ago suggests that, contrary to the long-held view of this region as the wellspring of all ape forms, some hominoids began evolving modern cranial and dental features in Eurasia and returned to Africa changed into more advanced species only after the sea receded again. (A few genera—such as *Kenyapithecus* from Fort Ternan, Kenya—may have gone on to develop some postcranial adaptations to life on the ground, but for the most part, these animals still looked like their early Miocene predecessors from the neck down.)

RISE OF THE GREAT APES

By the end of the middle Miocene, roughly 13 million years ago, we have evidence for great apes in Eurasia, notably Lartet's fossil great ape, *Dryopithecus,* in Europe and *Sivapithecus* in Asia. Like living great apes, these animals had long, strongly built jaws that housed large incisors, bladelike (as opposed to tusklike) canines, and long molars and premolars with relatively simple chewing surfaces—a feeding apparatus well suited to a diet of soft, ripe fruits. They also possessed shortened snouts, reflecting the reduced importance of olfaction in favor of vision. Histological studies of the teeth of *Dryopithecus* and *Sivapithecus* suggest that these creatures grew fairly slowly, as living great apes do, and that they probably had life histories similar to those of the great apes—maturing at a leisurely rate, living long lives, bearing one large offspring at a time, and so forth. Other evidence hints that were they around today, these early great apes might have even matched wits with modern ones: fossil braincases of *Dryopithecus* indicate that it was as large-brained as a chimpanzee of comparable proportions. We lack direct clues to brain size in *Sivapithecus,* but given that life history correlates strongly with brain size, it is likely that this ape was similarly brainy.

Examinations of the limb skeletons of these two apes have revealed additional great ape–like characteristics. Most important, both *Dryopithecus* and *Sivapithecus* display adaptations to suspensory locomotion, especially in the elbow joint, which was fully extendable and stable throughout the full range of motion. Among primates, this morphology is unique to apes, and it figures prominently in their ability to hang and swing below branches. It also gives humans the ability to throw with great speed and accuracy. For its part, *Dryopithecus* exhibits numerous other adaptations to suspension, both in the limb bones and in the hands and feet, which had powerful grasping capabilities. Together these features strongly suggest that *Dryopithecus* negotiated the forest canopy in much the way that living great apes do. Exactly how *Sivapithecus* got around is less clear. Some characteristics of this animal's limbs are indicative of suspension, whereas others imply that it had more quadrupedal habits. In all likelihood, *Sivapithecus* employed a mode of locomotion for which no modern analogue exists—the product of its own unique ecological circumstances.

The *Sivapithecus* lineage thrived in Asia, producing offshoots in Turkey, Pakistan, India, Nepal, China and Southeast Asia. Most phylogenetic analyses concur that it is from *Sivapithecus* that the living orangutan, *Pongo*

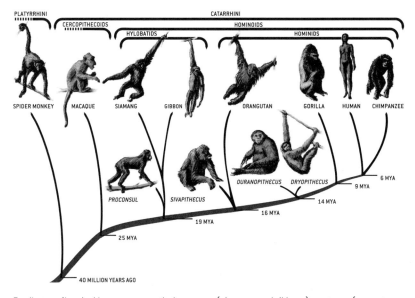

PLATYRRHINI   CATARRHINI
        CERCOPITHECOIDS   HOMINOIDS
                    HYLOBATIDS   HOMINIDS

SPIDER MONKEY   MACAQUE   SIAMANG   GIBBON   ORANGUTAN   GORILLA   HUMAN   CHIMPANZEE

OURANOPITHECUS DRYOPITHECUS   6 MYA
                                                                       9 MYA
PROCONSUL   SIVAPITHECUS   14 MYA
                                              16 MYA
                                19 MYA
              25 MYA

40 MILLION YEARS AGO

Family tree of hominoids encompasses the lesser apes (siamangs and gibbons), great apes (orangutans, gorillas and chimpanzees), and humans. Most Miocene apes were evolutionary dead ends. But researchers have identified a handful of them as candidate ancestors of living apes and humans. *Proconsul*, a primitive Miocene ape, is thought to have been the last common ancestor of the living hominoids; *Sivapithecus*, an early great ape, is widely regarded as an orangutan forebear; and either *Dryopithecus* or *Ouranopithecus* may have given rise to African apes and humans.

pygmaeus, is descended. Today this ape, which dwells in the rain forests of Borneo and Sumatra, is the sole survivor of that successful group.

In the west the radiation of great apes was similarly grand. From the earliest species of *Dryopithecus, D. fontani*, the one found by Lartet, several other species emerged over about three million years. More specialized descendants of this lineage followed suit. Within two million years four new species of *Dryopithecus* would evolve and span the region from northwestern Spain to the Republic of Georgia. But where *Dryopithecus* belongs on the hominoid family tree has proved controversial. Some studies link *Dryopithecus* to Asian apes; others position it as the ancestor of all living great apes. My own phylogenetic analysis of these animals—the most comprehensive in terms of the number of morphological characteristics considered—indicates that *Dryopithecus* is most closely related to an ape known as *Ouranopithecus* from Greece and that one of these two European genera was the likely ancestor of African apes and humans.

A *Dryopithecus* skull from Rudabánya, Hungary, that my colleagues and I discovered in 1999 bolsters that argument. Nicknamed "Gabi" after its discoverer, Hungarian geologist Gabor Hernyák, it is the first specimen to preserve a key piece of anatomy: the connection between the face and the braincase. Gabi shows that the cranium of *Dryopithecus,* like that of African apes and early fossil humans, had a long and low braincase, a flatter nasal region and an enlarged lower face. Perhaps most significant, it reveals that also like African apes and early humans, *Dryopithecus* was klinorhynch, meaning that viewed in profile its face tilts downward. Orangutans, in contrast—as well as *Proconsul,* gibbons and siamangs—have faces that tilt upward, a condition known as airorhinchy. That fundamental aspect of *Dryopithecus*'s cranial architecture speaks strongly to a close evolutionary relationship between this animal and the African apes and humans lineage. Additional support for that link comes from the observation that the *Dryopithecus* skull resembles that of an infant or juvenile chimpanzee—a common feature of ancestral morphology. It follows, then, that the unique aspects of adult cranial form in chimpanzees, gorillas and fossil humans evolved as modifications to the ground plan represented by *Dryopithecus* and living African ape youngsters.

One more Miocene ape deserves special mention. The best-known Eurasian fossil ape, in terms of the percentage of the skeleton recovered, is seven-million-year-old *Oreopithecus* from Italy. First described in 1872 by renowned French paleontologist Paul Gervais, *Oreopithecus* was more specialized for dining on leaves than was any other Old World fossil monkey or ape. It survived very late into the Miocene in the dense and isolated forests of the islands of Tuscany, which would eventually be joined to one another and the rest of Europe by the retreat of the sea to form the backbone of the Italian peninsula. Large-bodied and small-brained, this creature is so unusual looking that it is not clear whether it is a primitive form that predates the divergence of gibbons and great apes or an early great ape or a close relative of *Dryopithecus.* Meike Köhler and Salvador Moyà-Solà of the Miquel Crusafont Institute of Paleontology in Barcelona have proposed that *Oreopithecus* walked bipedally along tree limbs and had a humanlike hand capable of a precision grip. Most paleoanthropologists, however, believe that it was instead a highly suspensory animal. Whatever *Oreopithecus* turns out to be, it is a striking reminder of how very diverse and successful at adapting to new surroundings the Eurasian apes were.

So what happened to the myriad species that did not evolve into the living great apes and humans, and why did the ancestors of extant species persevere? Clues have come from paleoclimatological studies. Throughout the middle Miocene, the great apes flourished in Eurasia, thanks to its then lush subtropical forest cover and consistently warm temperatures. These conditions assured a nearly continuous supply of ripe fruits and an easily traversed arboreal habitat with several tree stories. Climate changes in the late Miocene brought an end to this easy living. The combined effects of Alpine, Himalayan and East African mountain building, shifting ocean currents, and the early stages of polar ice cap formation precipitated the birth of the modern Asian monsoon cycle, the desiccation of East Africa and the development of a temperate climate in Europe. Most of the Eurasian great apes went extinct as a result of this environmental overhaul. The two lineages that did persevere—those represented by *Sivapithecus* and *Dryopithecus*—did so by moving south of the Tropic of Cancer, into Southeast Asia from China and into the African tropics from Europe, both groups tracking the ecological settings to which they had adapted in Eurasia.

The biogeographical model outlined above provides an important perspective on a long-standing question in paleoanthropology concerning how and why humans came to walk on two legs. To address that issue, we need to know from what form of locomotion bipedalism evolved. Lacking unambiguous fossil evidence of the earliest biped and its ancestor, we cannot say with certainty what that ancestral condition was, but researchers generally fall into one of two theoretical camps: those who think two-legged walking arose from arboreal climbing and suspension and those who think it grew out of a terrestrial form of locomotion, perhaps knuckle walking.

YOUR GREAT, GREAT GRAND APE

The Eurasian forebear of African apes and humans moved south in response to a drying and cooling of its environs that led to the replacement of forests with woodlands and grasslands. I believe that adaptations to life on the ground—knuckle walking in particular—were critical in enabling this lineage to withstand that loss of arboreal habitat and make it to Africa. Once there, some apes returned to the forests, others settled into varied woodland environments, and one ape—the one from which humans descended—eventually invaded open territory by committing to life on the ground.

Flexibility in adaptation is the consistent message in ape and human evolution. Early Miocene apes left Africa because of a new adaptation in their jaws and teeth that allowed them to exploit a diversity of ecological settings. Eurasian great apes evolved an array of skeletal adaptations that permitted them to live in varied environments as well as large brains to grapple with complex social and ecological challenges. These modifications made it possible for a few of them to survive the dramatic climate changes that took place at the end of the Miocene and return to Africa, around nine million years ago. Thus, the lineage that produced African apes and humans was preadapted to coping with the problems of a radically changing environment. It is therefore not surprising that one of these species eventually evolved very large brains and sophisticated forms of technology.

As an undergraduate more than 20 years ago, I began to look at fossil apes out of the conviction that to understand why humans evolved we have to know when, where, how and from what we arose. Scientists commonly look to living apes for anatomical and behavioral insights into the earliest humans. There is much to be gained from this approach. But living great apes have also evolved since their origins. The study of fossil great apes gives us both a unique view of the ancestors of living great apes and humans and a starting point for understanding the processes and circumstances that led to the emergence of this group. For example, having established the connection between European great apes and living African apes and humans, we can now reconstruct the last common ancestor of chimps and humans: it was a knuckle-walking, fruit-eating, forest-living chimplike primate that used tools, hunted animals, and lived in highly complex and dynamic social groups, as do living chimps and humans.

TANGLED BRANCHES

We still have much to learn. Many fossil apes are represented only by jaws and teeth, leaving us with little or no idea about their posture and locomotion, brain size or body mass. Moreover, paleontologists have yet to recover any remains of ancient African great apes. Indeed, there is a substantial geographic and temporal gap in the fossil record between representatives of the early members of the African hominid lineage in Europe (*Dryopithecus* and *Ouranopithecus*) and the earliest African fossil hominids.

Moving up the family tree (or, more accurately, family bush), we find more confusion in that the earliest putative members of the human family are not obviously human. For instance, the recently discovered *Sahelanthropus tchadensis,* a six-million- to seven-million-year-old find from Chad, is humanlike in having small canine teeth and perhaps a more centrally located foramen magnum (the hole at the base of the skull through which the spinal cord exits), which could indicate that the animal was bipedal. Yet *Sahelanthropus* also exhibits a number of chimplike characteristics, including a small brain, projecting face, sloped forehead and large neck muscles.

Another creature, *Orrorin tugenensis,* fossils of which come from a Kenyan site dating to six million years ago, exhibits a comparable mosaic of chimp and human traits, as does 5.8-million-year-old *Ardipithecus ramidus kadabba* from Ethiopia. Each of these taxa has been described by its discoverers as a human ancestor [see "An Ancestor to Call Our Own," by Kate Wong, in this volume]. But in truth, we do not yet know enough about any of these creatures to say whether they are protohumans, African ape ancestors or dead-end apes. The earliest unambiguously human fossil, in my view, is 4.4-million-year-old *Ardipithecus ramidus ramidus,* also from Ethiopia.

The idea that the ancestors of great apes and humans evolved in Eurasia is controversial, but not because there is inadequate evidence to support it. Skepticism comes from the legacy of Darwin, whose prediction noted at the beginning of this article is commonly interpreted to mean that humans and African apes must have evolved solely in Africa. Doubts also come from fans of the aphorism "absence of evidence is not evidence of absence." To wit, just because we have not found fossil great apes in Africa does not mean that they are not there. This is true. But there are many fossil sites in Africa dated to between 14 million and seven million years ago — some of which have yielded abundant remains of forest-dwelling animals — and not one contains great ape fossils. Although it is possible that Eurasian great apes, which bear strong resemblances to living great apes, evolved in parallel with as yet undiscovered African ancestors, this seems unlikely.

It would be helpful if we had a more complete fossil record from which to piece together the evolutionary history of our extended family. Ongoing fieldwork promises to fill some of the gaps in our knowledge. But until then, we must hypothesize based on what we know. The view expressed here is testable, as required of all scientific hypotheses, through the discovery of more fossils in new places.

FURTHER READING

David R. Begun, Carol V. Ward and Michael D. Rose, eds. *Function, Phylogeny and Fossils: Miocene Hominoid Evolution and Adaptations.* Plenum Press, 1997.

The Primate Fossil Record. Edited by Walter Carl Hartwig. Cambridge University Press, 2002.

László Kordos and David R. Begun. "Rudabánya: A Late Miocene Subtropical Swamp Deposit with Evidence of the Origin of the African Apes and Humans" in *Evolutionary Anthropology,* Vol. 11, Issue 1, pages 45–57; 2002.

Once We Were Not Alone

IAN TATTERSALL

ORIGINALLY PUBLISHED IN JANUARY 2000

Homo sapiens has had the earth to itself for the past 25,000 years or so, free and clear of competition from other members of the hominid family. This period has evidently been long enough for us to have developed a profound feeling that being alone in the world is an entirely natural and appropriate state of affairs.

So natural and appropriate, indeed, that during the 1950s and 1960s a school of thought emerged that, in essence, claimed that only one species of hominid could have existed at a time because there was simply no ecological space on the planet for more than one culture-bearing species. The "single-species hypothesis" was never very convincing—even in terms of the rather sparse hominid fossil record of 35 years ago. But the implicit scenario of the slow, single-minded transformation of the bent and benighted ancestral hominid into the graceful and gifted modern *H. sapiens* proved powerfully seductive—as fables of frogs becoming princes always are.

So seductive that it was only in the late 1970s, following the discovery of incontrovertible fossil evidence that hominid species coexisted some 1.8 million years ago in what is now northern Kenya, that the single-species hypothesis was abandoned. Yet even then, paleoanthropologists continued to cleave to a rather minimalist interpretation of the fossil record. Their tendency was to downplay the number of species and to group together distinctively different fossils under single, uninformative epithets such as "archaic *Homo sapiens.*" As a result, they tended to lose sight of the fact that many kinds of hominids had regularly contrived to coexist.

Although the minimalist tendency persists, recent discoveries and fossil reappraisals make clear that the biological history of hominids resembles that of most other successful animal families. It is marked by diversity rather than by linear progression. Despite this rich history—during which hominid species developed and lived together and competed and rose and fell—*H. sapiens* ultimately emerged as the sole hominid. The reasons for this are generally unknowable, but different interactions between the last coexisting hominids—*H. sapiens* and *H. neanderthalensis*—in two distinct geographical regions offer some intriguing insights.

A SUITE OF SPECIES

From the beginning, almost from the very moment the earliest hominid biped—the first "australopith"—made its initial hesitant steps away from the forest depths, we have evidence for hominid diversity. The oldest-known potential hominid is *Ardipithecus ramidus,* represented by some fragmentary fossils from the 4.4-million-year-old site of Aramis in Ethiopia. Only slightly younger is the better-known *Australopithecus anamensis,* from sites in northern Kenya that are about 4.2 million years old.

Ardipithecus, though claimed on indirect evidence to have been an upright walker, is quite apelike in many respects. In contrast, *A. anamensis* looks reassuringly similar to the 3.8- to 3.0-million-year-old *Australopithecus afarensis,* a small-brained, big-faced bipedal species to which the famous "Lucy" belonged. Many remnants of *A. afarensis* have been found in various eastern African sites, but some researchers have suggested that the mass of fossils described as *A. afarensis* may contain more than one species, and it is only a matter of time until the subject is raised again. In any event, *A. afarensis* was not alone in Africa. A distinctive jaw, from an australopith named *A. bahrelghazali,* was recently found in Chad. It is probably between 3.5 and 3.0 million years old and is thus roughly coeval with Lucy.

In southern Africa, scientists have just reported evidence of another primitive bipedal hominid species. As yet unnamed and undescribed, this distinctive form is 3.3 million years old. At about 3 million years ago, the same region begins to yield fossils of *A. africanus,* the first australopith to be discovered (in 1924). This species may have persisted until not much more than 2 million years ago. A recently named 2.5-million-year-old species from Ethiopia, *Australopithecus garhi,* is claimed to fall in an intermediate position between *A. afarensis,* on the one hand, and a larger group that includes more recent australopiths and *Homo,* on the other. Almost exactly the same age is the first representative of the "robust" group of australopiths, *Paranthropus aethiopicus.* This early form is best known from the 2.5-million-year-old "Black Skull" of northern Kenya, and in the period between about 2 and 1.4 million years ago the robusts were represented all over eastern Africa by the familiar *P. boisei.* In South Africa, during the period around 1.6 million years ago, the robusts included the distinctive *P. robustus* and possibly also a closely related second species, *P. crassidens.*

PARANTHROPUS BOISEI had massive jaws, equipped with huge grinding teeth for a presumed vegetarian diet. Its skull is accordingly strongly built, but it is not known if in body size it was significantly larger than the "gracile" australopiths.

HOMO RUDOLFENSIS was a relatively large-brained hominid, typified by the famous KNM-ER 1470 cranium. Its skull was distinct from the apparently smaller-brained H. habilis, but its body proportions are effectively unknown.

HOMO HABILIS ("handy man") was so named because it was thought to be the maker of the 1.8-million-year-old stone tools discovered at Olduvai Gorge in Tanzania. This hominid fashioned sharp flakes by banging one rock cobble against another.

HOMO ERGASTER, sometimes called "African H. erectus," had a high, rounded cranium and a skeleton broadly similar to that of modern humans. Although H. ergaster clearly ate meat, its chewing teeth are relatively small. The best specimen of this hominid is that of an adolescent from about 1.6 million years ago known as Turkana boy.

Sharing a single landscape, four kinds of hominids lived about 1.8 million years ago in what is now part of northern Kenya. Although paleoanthropologists have no idea how—or if—these different species interacted, they do know that *Paranthropus boisei, Homo rudolfensis, H. habilis* and *H. ergaster* foraged in the same area around Lake Turkana.

I apologize for inflicting this long list of names on you, but in fact it actually underestimates the number of australopith species that existed. What is more, we don't know how long each of these creatures lasted. Nevertheless, even if average species longevity was only a few hundred thousand years, it is clear that from the very beginning the continent of Africa was at least periodically—and most likely continually—host to multiple kinds of hominids.

The appearance of the genus *Homo* did nothing to perturb this pattern. The 2.5- to 1.8-million-year-old fossils from eastern and southern Africa that announce the earliest appearance of *Homo* are an oddly assorted lot and probably a lot more diverse than their conventional assignment to the two species *H. habilis* and *H. rudolfensis* indicates. Still, at Kenya's East Turkana, in the period between 1.9 and 1.8 million years ago, these two species were joined not only by the ubiquitous *P. boisei* but by *H. ergaster*,

Tuc d'Audoubert Cave in France was entered sometime between perhaps 11,000 and 13,000 years ago by *H. sapiens,* also called Cro Magnons, who sculpted small clay bison in a recess almost a mile underground. Hominids of modern body form most likely emerged in Africa at around 150,000 years ago and coexisted with other hominids for a time before emerging as the only species of our family. Until about 30,000 years ago, they overlapped with *H. neanderthalensis* (*left*) in Europe and in the Levant, and they may have been contemporaneous with the *H. erectus* (*right*) then living in Java.

the first hominid of essentially modern body form. Here, then, is evidence for four hominid species sharing not just the same continent but the same landscape.

The first exodus of hominids from Africa, presumably in the form of *H. ergaster* or a close relative, opened a vast prospect for further diversification. One could wish for a better record of this movement, and particularly of its dating, but there are indications that hominids of some kind had reached China and Java by about 1.8 million years ago. A lower jaw that may be about the same age from Dmanisi in ex-Soviet Georgia is distinctively different from anything else yet found [see "Out of Africa Again . . . and Again?," by Ian Tattersall, this volume]. By the million-year mark *H. erectus* was established in both Java and China, and it is possible that a more robust hominid species was present in Java as well. At the other end of the Eurasian continent, the oldest-known European hominid fragments—from about 800,000 years ago—are highly distinctive and have been dubbed *H. antecessor* by their Spanish discoverers.

About 600,000 years ago, in Africa, we begin to pick up evidence for *H. heidelbergensis,* a species also seen at sites in Europe—and possibly China—between 500,000 to 200,000 years ago. As we learn more about *H. heidelbergensis,* we are likely to find that more than one species is actually represented in this group of fossils. In Europe, *H. heidelbergensis* or a relative gave rise to an endemic group of hominids whose best-known representative was *H. neanderthalensis,* a European and western Asian species that flourished between about 200,000 and 30,000 years ago. The sparse record from Africa suggests that at this time independent developments were taking place there, too—including the emergence of *H. sapiens.* And in Java, possible *H. erectus* fossils from Ngandong have just been dated to around 40,000 years ago, implying that this area had its own indigenous hominid evolutionary history for perhaps millions of years as well.

The picture of hominid evolution just sketched is a far cry from the *"Australopithecus africanus* begat *Homo erectus* begat *Homo sapiens"* scenario that prevailed 40 years ago—and it is, of course, based to a great extent on fossils that have been discovered since that time. Yet the dead hand of linear thinking still lies heavily on paleoanthropology, and even today many of my colleagues would argue that this scenario overestimates diversity. There are various ways of simplifying the picture, most of them involving

Speculative family tree shows the variety of hominid species that have populated the planet—some known only by a fragment of skull or jaw. As the tree suggests, the emergence of *H. sapiens* has not been a single, linear transformation of one species into another but rather a meandering, multifaceted evolution.

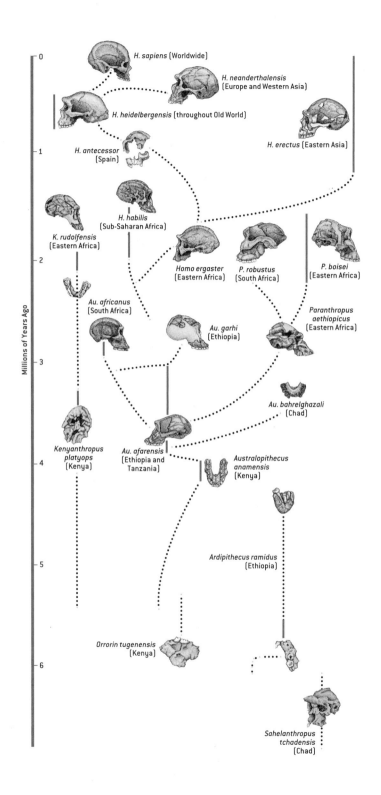

Millions of Years Ago

H. sapiens (Worldwide)

H. neanderthalensis
(Europe and Western Asia)

H. heidelbergensis (throughout Old World)

H. antecessor
(Spain)

H. erectus (Eastern Asia)

H. habilis
(Sub-Saharan Africa)

K. rudolfensis
(Eastern Africa)

Homo ergaster
(Eastern Africa)

P. robustus
(South Africa)

P. boisei
(Eastern Africa)

Au. africanus
(South Africa)

Au. garhi
(Ethiopia)

*Paranthropus
aethiopicus*
(Eastern Africa)

Au. bahrelghazali
(Chad)

*Kenyanthropus
platyops*
(Kenya)

Au. afarensis
(Ethiopia and
Tanzania)

*Australopithecus
anamensis*
(Kenya)

Ardipithecus ramidus
(Ethiopia)

Orrorin tugenensis
(Kenya)

*Sahelanthropus
tchadensis*
(Chad)

the cop-out of stuffing all variants of *Homo* of the last half a million or even two million years into the species *H. sapiens.*

My own view, in contrast, is that the 20 or so hominid species invoked (if not named) above represent a minimum estimate. Not only is the human fossil record as we know it full of largely unacknowledged morphological indications of diversity, but it would be rash to claim that every hominid species that ever existed is represented in one fossil collection or another. And even if only the latter is true, it is still clear that the story of human evolution has not been one of a lone hero's linear struggle.

Instead it has been the story of nature's tinkering: of repeated evolutionary experiments. Our biological history has been one of sporadic events rather than gradual accretions. Over the past five million years, new hominid species have regularly emerged, competed, coexisted, colonized new environments and succeeded—or failed. We have only the dimmest of perceptions of how this dramatic history of innovation and interaction unfolded, but it is already evident that our species, far from being the pinnacle of the hominid evolutionary tree, is simply one more of its many terminal twigs.

THE ROOTS OF OUR SOLITUDE

Although this is all true, *H. sapiens* embodies something that is undeniably unusual and is neatly captured by the fact that we are alone in the world today. Whatever that something is, it is related to how we interact with the external world: it is behavioral, which means that we have to look to our archaeological record to find evidence of it. This record begins some 2.5 million years ago with the production of the first recognizable stone tools: simple sharp flakes chipped from parent "cores." We don't know exactly who the inventor was, but chances are that he or she was something we might call an australopith.

This innovation represented a major cognitive leap and had profound long-term consequences for hominids. It also inaugurated a pattern of highly intermittent technological change. It was a full million years before the next significant technological innovation came along: the creation about 1.5 million years ago, probably by *H. ergaster,* of the hand axe. These symmetrical implements, shaped from large stone cores, were the first to conform to a "mental template" that existed in the toolmaker's mind. This template remained essentially unchanged for another million years or more, until the invention of "prepared-core" tools by *H. heidelbergensis* or a relative. Here a stone core was elaborately shaped in such a way that a single blow would detach what was an effectively finished implement.

Among the most accomplished practitioners of prepared-core technology were the large-brained, big-faced and low-skulled Neanderthals, who occupied Europe and western Asia until about 30,000 years ago. Because they left an excellent record of themselves and were abruptly replaced by modern humans who did the same, the Neanderthals furnish us with a particularly instructive yardstick by which to judge our own uniqueness. The stoneworking skills of the Neanderthals were impressive, if somewhat stereotyped, but they rarely if ever made tools from other preservable materials. And many archaeologists question the sophistication of their hunting skills.

Further, despite misleading early accounts of bizarre Neanderthal "bear cults" and other rituals, no substantial evidence has been found for symbolic behaviors among these hominids or for the production of symbolic objects—certainly not before contact had been made with modern humans. Even the occasional Neanderthal practice of burying the dead may have been simply a way of discouraging hyena incursions into their living spaces, or have a similar mundane explanation, for Neanderthal burials lack the "grave goods" that would attest to ritual and belief in an afterlife. The Neanderthals, in other words, though admirable in many ways and for a long time successful in the difficult circumstances of the late Ice Ages, lacked the spark of creativity that, in the end, distinguished *H. sapiens.*

Although the source of *H. sapiens* as a physical entity is obscure, most evidence points to an African origin perhaps between 150,000 and 200,000 years ago. Modern behavior patterns did not emerge until much later. The best evidence comes from Israel and environs, where Neanderthals lived about 200,000 years ago or perhaps even earlier. By about 100,000 years ago, they had been joined by anatomically modern *H. sapiens,* and the remarkable thing is that the tools and sites the two hominid species left behind are essentially identical. As far as can be told, these two hominids behaved in similar ways despite their anatomical differences. And as long as they did so, they somehow contrived to share the Levantine environment.

The situation in Europe could hardly be more different. The earliest *H. sapiens* sites there date from only about 40,000 years ago, and just 10,000 or so years later the formerly ubiquitous Neanderthals were gone. Significantly, the *H. sapiens* who invaded Europe brought with them abundant evidence of a fully formed and unprecedented modern sensibility. Not only did they possess a new "Upper Paleolithic" stoneworking technology based on the production of multiple long, thin blades from cylindrical cores, but they made tools from bone and antler, with an exquisite sensitivity to the properties of these materials.

Even more significant, they brought with them art, in the form of carvings, engravings and spectacular cave paintings; they kept records on bone and stone plaques; they made music on wind instruments; they crafted elaborate personal adornments; they afforded some of their dead elaborate burials with grave goods (hinting at social stratification in addition to belief in an afterlife, for not all burials were equally fancy); their living sites were highly organized, with evidence of sophisticated hunting and fishing. The pattern of intermittent technological innovation was gone, replaced by constant refinement. Clearly, these people were *us*.

In all these ways, early Upper Paleolithic people contrasted dramatically with the Neanderthals. Some Neanderthals in Europe seem to have picked up new ways of doing things from the arriving *H. sapiens,* but we have no direct clues as to the nature of the interaction between the two species. In light of the Neanderthals' rapid disappearance, though, and of the appalling subsequent record of *H. sapiens,* we can reasonably surmise that such interactions were rarely happy for the former. Certainly the repeated pattern at archaeological sites is one of short-term replacement, and there is no convincing biological evidence of any intermixing in Europe.

In the Levant, the coexistence ceased—after about 60,000 years or so— at right about the time that Upper Paleolithic–like tools began to appear. About 40,000 years ago the Neanderthals of the Levant yielded to a presumably culturally rich *H. sapiens,* just as their European counterparts had.

The key to the difference between the European and the Levantine scenarios lies, most probably, in the emergence of modern cognition—which, it is reasonable to assume, is equivalent to the advent of symbolic thought. Business had continued more or less as usual right through the appearance of modern bone structure, and only later, with the acquisition of fully modern behavior patterns, did *H. sapiens* become completely intolerant of competition from its nearest—and, evidently, not its dearest.

To understand how this change in sensibility occurred, we have to recall certain things about the evolutionary process. First, as in this case, all innovations must necessarily arise *within* preexisting species—for where else can they do so? And second, many novelties arise as "exaptations," features acquired in one context before (often long before) being co-opted in a different one. For example, hominids possessed essentially modern vocal tracts for hundreds of thousands of years before the behavioral record gives us any reason to believe that they employed the articulate speech that the peculiar form of this tract permits. Finally, we need to bear in mind the phenomenon of emergence whereby a chance coincidence gives rise to something totally unexpected. The classic example here is water, whose properties are unpredicted by those of hydrogen and oxygen atoms alone.

If we combine these various observations we can see that, profound as the consequences of achieving symbolic thought may have been, the process whereby it came about was unexceptional. We have no idea at present how the modern human brain converts a mass of electrical and chemical discharges into what we experience as consciousness. We do know, however, that somehow our lineage passed to symbolic thought from some nonsymbolic precursor state. The only plausible possibility is that with the arrival of anatomically modern *H. sapiens,* existing exaptations were fortuitously linked by some relatively minor genetic innovation to create an unprecedented potential.

Yet even in principle this cannot be the full story, because anatomically modern humans behaved archaically for a long time before adopting modern behaviors. That discrepancy may be the result of the late appearance of some key hardwired innovation not reflected in the skeleton, which is all that fossilizes. But this seems unlikely, because it would have necessitated a wholesale Old World–wide replacement of hominid populations in a very short time, something for which there is no evidence.

It is much more likely that the modern human capacity was born at— or close to—the origin of *H. sapiens,* as an ability that lay fallow until it was activated by a cultural stimulus of some kind. If sufficiently advantageous, this behavioral novelty could then have spread rapidly by cultural contact among populations that already had the potential to acquire it. No population replacement would have been necessary.

It is impossible to be sure what this innovation might have been, but the best current bet is that it was the invention of language. For language is not simply the medium by which we express our ideas and experiences to each other. Rather it is fundamental to the thought process itself. It involves categorizing and naming objects and sensations in the outer and inner worlds and making associations between resulting mental symbols. It is, in effect, impossible for us to conceive of thought (as we are familiar with it) in the absence of language, and it is the ability to form mental symbols that is the fount of our creativity, for only once we create such symbols can we recombine them and ask such questions as "What if . . .?"

We do not know exactly how language might have emerged in one local population of *H. sapiens,* although linguists have speculated widely. But we do know that a creature armed with symbolic skills is a formidable competitor—and not necessarily an entirely rational one, as the rest of the living world, including *H. neanderthalensis,* has discovered to its cost.

FURTHER READING

Randall White. *Dark Caves, Bright Visions: Life in Ice Age Europe.* W. W. Norton/American Museum of Natural History, 1986.

Derek Bickerton. *Language and Species.* University of Chicago Press, 1990.

Ian Tattersall. *The Fossil Trail: How We Know What We Think We Know about Human Evolution.* Oxford University Press, 1995.

Christopher Stringer and Robin McKie. *African Exodus: The Origins of Modern Humanity.* Henry Holt, 1997.

William Howells. *Getting Here: The Story of Human Evolution.* Updated edition. Compass Press, 1997.

Ian Tattersall. *The Last Neanderthal: The Rise, Success and Mysterious Extinction of Our Closest Human Relatives.* Macmillan, 1995. (Second edition by Westview Press due December 1999.)

Nina G. Jablonski and Leslie C. Aiello, eds. *The Origin and Diversification of Language.* University of California Press, 1998.

Out of Africa Again ... and Again?

IAN TATTERSALL

ORIGINALLY PUBLISHED IN APRIL 1997

I t all used to seem so simple. The human lineage evolved in Africa. Only at a relatively late date did early humans finally migrate from the continent of their birth, in the guise of the long-known species *Homo erectus,* whose first representatives had arrived in eastern Asia by around one million years ago. All later kinds of humans were the descendants of this species, and almost everyone agreed that all should be classified in our own species, *H. sapiens.* To acknowledge that some of these descendants were strikingly different from ourselves, they were referred to as "archaic *H. sapiens,*" but members of our own species they were nonetheless considered to be.

Such beguiling simplicity was, alas, too good to last, and over the past few years it has become evident that the later stages of human evolution have been a great deal more eventful than conventional wisdom for so long had it. This is true for the earlier stages, too, although there is still no reason to believe that humankind's birthplace was elsewhere than in Africa. Indeed, for well over the first half of the documented existence of the hominid family (which includes all upright-walking primates), there is no record at all outside that continent. But recent evidence does seem to indicate that it was not necessarily *H. erectus* who migrated from Africa—and that these peregrinations began earlier than we had thought.

A CONFUSED EARLY HISTORY

Recent discoveries in Kenya of fossils attributed to the new species *Australopithecus anamensis* have now pushed back the record of upright-walking hominids to about 4.2 to 3.9 million years (Myr) ago. More dubious finds in Ethiopia, dubbed *Ardipithecus ramidus,* may extend this to 4.4 Myr ago or so. The *A. anamensis* fossils bear a strong resemblance to the later and far better known species *Australopithecus afarensis,* found at sites in Ethiopia and Tanzania in the 3.9- to 3.0-Myr range and most famously represented by the "Lucy" skeleton from Hadar, Ethiopia.

Lucy and her kind were upright walkers, as the structures of their pelvises and knee joints particularly attest, but they retained many ancestral features, notably in their limb proportions and in their hands and feet, that would have made them fairly adept tree climbers. Together with their ape-size brains and large, protruding faces, these characteristics have led many to call such creatures "bipedal chimpanzees." This is probably a fairly accurate characterization, especially given the increasing evidence that early hominids favored quite heavily wooded habitats. Their preferred way of life was evidently a successful one, for although these primates were less adept arborealists than the living apes and less efficient bipeds than later hominids, their basic "eat your cake and have it" adaptation endured for well over two million years, even as species of this general kind came and went in the fossil record.

It is not even clear to what extent lifestyles changed with the invention of stone tools, which inaugurate our archaeological record at about 2.5 Myr ago. No human fossils are associated with the first stone tools known, from sites in Kenya and Ethiopia. Instead there is a motley assortment of hominid fossils from the period following about 2 Myr ago, mostly associated with the stone tools and butchered mammal bones found at Tanzania's Olduvai Gorge and in Kenya's East Turkana region. By one reckoning, at least some of the first stone toolmakers in these areas were hardly bigger or more advanced in their body skeletons than the tiny Lucy; by another, the first tools may have been made by taller, somewhat larger-brained hominids with more modern body structures. Exactly how many species of early hominids there were, which of them made the tools, and how they walked remains one of the major conundrums of human evolution.

Physically, at least, the picture becomes clearer following about 1.9 Myr ago, when the first good evidence occurs in northern Kenya of a species that is recognizably like ourselves. Best exemplified by the astonishingly complete 1.6-Myr-old skeleton known as the Turkana Boy, discovered in 1984, these humans possessed an essentially modern body structure, indicative of modern gait, combined with moderately large-faced skulls that contained brains double the size of those of apes (though not much above half the modern human average). The Boy himself had died as an adolescent, but it is estimated that had he lived to maturity he would have attained a height of six feet, and his limbs were long and slender, like those of people who live today in hot, arid African climates, although this common adaptation does not, of course, indicate any special relationship. Here at last we have early hominids who were clearly at home on the open savanna.

A long-standing paleoanthropological tradition seeks to minimize the number of species in the human fossil record and to trace a linear, progressive pattern of descent among those few that are recognized. In keeping with this practice, the Boy and his relatives were originally assigned to the species *H. erectus*. This species was first described from a skullcap and thighbone found in Java a century ago. Fossils later found in China—notably the now lost 500,000-year-old (500 Kyr old) "Peking Man"—and elsewhere in Java were soon added to the species, and eventually *H. erectus* came to embrace a wide variety of hominid fossils, including a massive brain-case from Olduvai Gorge known as OH9. The latter has been redated to about 1.4 Myr, although it was originally thought to have been a lot younger. All these fossil forms possessed brains of moderate size (about 900 to 1,200 milliliters in volume, compared with an average of around 1,400 milliliters for modern humans and about 400 milliliters for apes), housed in long, low skull vaults with sharp angles at the back and heavy brow ridges in front. The few limb bones known were robust but essentially like our own.

Whether *H. erectus* had ever occupied Europe was vigorously debated, the alternative being to view all early human fossils from that region (the earliest of them being no more than about 500 Kyr old) as representatives of archaic *H. sapiens*. Given that the Javan fossils were conventionally dated in the range of 1 Myr to 700 Kyr and younger and that the earliest Chinese fossils were reckoned to be no more than 1 Myr old, the conclusion appeared clear: *H. erectus* (as exemplified by OH9 and also by the earlier Turkana Boy and associated fossils) had evolved in Africa and had exited that continent not much more than 1 Myr ago, rapidly spreading to eastern Asia and spawning all subsequent developments in human evolution, including those in Europe.

Yet on closer examination the specimens from Kenya turned out to be distinctively different in braincase construction from those of classic eastern Asian *H. erectus*. In particular, certain anatomical features that appear specialized in the eastern Asian *H. erectus* look ancestral in the African fossils of comparable age. Many researchers began to realize that we are dealing with two kinds of early human here, and the earlier Kenyan form is now increasingly placed in its own species, *H. ergaster*. This species makes a plausible ancestor for all subsequent humans, whereas the cranial specializations of *H. erectus* suggest that this species, for so long regarded as the standard-issue hominid of the 1- to 0.5-Myr period, was in fact a local (and, as I shall explain below, ultimately terminal) eastern Asian development.

AN EASTERN ASIAN CUL-DE-SAC

The plot thickened in early 1994, when Carl C. Swisher of the Berkeley Geochronology Center and his colleagues applied the newish argon/argon dating method to volcanic rock samples taken from two hominid sites in Java. The results were 1.81 and 1.66 Myr: far older than anyone had really expected, although the earlier date did confirm one made many years before. Unfortunately, the fossils from these two sites are rather undiagnostic as to species: the first is a braincase of an infant (juveniles never show all the adult characteristics on which species are defined), and the second is a horrendously crushed and distorted cranium that has never been satisfactorily reconstructed. Both specimens have been regarded by most as *H. erectus,* but more for reasons of convenience than anything else. Over the decades, sporadic debate has continued regarding whether the Javan record contains one or more species of early hominid. Further, major doubt has recently been cast on whether the samples that yielded the older date were actually obtained from the same spot as the infant specimen. Still, these dates do fit with other evidence pointing to the probability that hominids of some kind were around in eastern Asia much earlier than anyone had thought.

Independent corroboration of this scenario comes, for instance, from the site of Dmanisi in the former Soviet republic of Georgia, where in 1991 a hominid lower jaw was found that its describers allocated to *H. erectus.* Three different methods suggested that this jaw was as old as 1.8 Myr; although not everyone has been happy with this dating, taken together the Georgian and new Javan dates imply an unexpectedly early hominid exodus from Africa. And the most parsimonious reading of the admittedly imperfect record suggests that these pioneering emigrants must have been *H. ergaster* or something very much like it.

A very early hominid departure from Africa has the advantage of explaining an apparent anomaly in the archaeological record. The stone tools found in sediments coeval with the earliest *H. ergaster* (just under 2 Myr ago) are effectively identical with those made by the first stone toolmakers many hundreds of thousands of years before. These crude tools consisted principally of sharp flakes struck with a stone "hammer" from small cobbles. Effective cutting tools though these may have been (experimental archaeologists have shown that even elephants can be quite efficiently butchered using them), they were not made to a standard form but were apparently produced simply to obtain a sharp cutting edge. Following about 1.4 Myr ago, however, standardized stone tools began to be made in Africa, typified by the hand axes and cleavers of the Acheulean industry (first identified in the mid-19th century from St. Acheul in France). These were larger implements, carefully shaped on both sides to a teardrop form. Oddly, stone tool industries in eastern Asia lacked such utensils, which led many to wonder why the first human immigrants to the region had not brought this technology with them, if their ancestors had already wielded it for half a million years. The new dates suggest, however, that the first emigrants had left Africa before the invention of the Acheulean technology, in which case there is no reason why we should expect to find this technology in eastern Asia. Interestingly, a few years ago the archaeologist Robin W. Dennell caused quite a stir by reporting very crude stone tools from Riwat in Pakistan that are older than 1.6 Myr. Their great age is now looking decreasingly anomalous.

Of course, every discovery raises new questions, and in this case the problem is to explain what it was that enabled human populations to expand beyond Africa for the first time. Most scholars had felt it was technological advances that allowed the penetration of the cooler continental areas to the north. If, however, the first emigrants left Africa equipped with only the crudest of stone-working technologies, we have to look to something other than technological prowess for the magic ingredient. And because the first human diaspora apparently followed hard on the heels of the acquisition of more or less modern body form, it seems reasonable to conclude that the typically human wanderlust emerged in concert with the emancipation of hominids from the forest edges that had been their preferred habitat. Of course, the fact that the Turkana Boy and his kin were adapted in their body proportions to hot, dry environments does nothing to explain why *H. ergaster* was able to spread rapidly into the cooler temperate zones beyond the Mediterranean; evidently the new body form that made possible remarkable endurance in open habitats was in itself enough to make the difference.

The failure of the Acheulean ever to diffuse as far as eastern Asia reinforces the notion, consistent with the cranial specializations of *H. erectus,* that this part of the world was a kind of paleo-anthropological cul-de-sac. In this region ancient human populations largely followed their own course, independent of what was going on elsewhere in the world. Further datings tend to confirm this view. Thus, Swisher and his colleagues have very recently reported dates for the Ngandong *H. erectus* site in Java that center on only about 40 Kyr ago. These dates, though very carefully obtained, have aroused considerable skepticism; but, if accurate, they have considerable implications for the overall pattern of human evolution. For they are so recent as to suggest that the long-lived *H. erectus* might even have suffered a fate similar to that experienced by the Neanderthals in Europe: extinction at the hands of late-arriving *H. sapiens*. Here we find reinforcement of the gradually emerging picture of human evolution as one of repeated experimentation, with regionally differentiated species, in this case on opposite sides of the Eurasian continent, being ultimately replaced by other hominid lineages that had evolved elsewhere.

At the other end of the scale, an international group led by Huang Wanpo of Beijing's Academia Sinica last year reported a remarkably ancient date for Longgupo Cave in China's Sichuan Province. This site had previously yielded an incisor tooth and a tiny lower jaw fragment with two teeth that were initially attributed to *H. erectus,* plus a few very crude stone artifacts. Huang and his colleagues concluded that the fossils and tools might be as many as 1.9 Myr old, and their reanalysis of the fossils suggested to them a closer resemblance to earliest African *Homo* species than to *H. erectus.*

This latter claim has not gone unexamined. As my colleague Jeffrey H. Schwartz of the University of Pittsburgh and I pointed out, for instance, the teeth in the jaw fragment resemble African *Homo* in primitive features rather than in the specialized ones that indicate a special relationship. What is more, they bear a striking resemblance to the teeth of an orangutan-related hominoid known from a much later site in Vietnam. And although the incisor appears hominid, it is fairly generic, and there is nothing about it that aligns it with any particular human species. Future fossil finds from Longgupo will, with luck, clarify the situation; meanwhile the incisor and stone tools are clear evidence of the presence of humans in China at what may be a very early date indeed. These ancient eastern Asians were the descendants of the first emigrants from Africa, and, whatever the hominids of Longgupo eventually turn out to have been, it is a good bet that Huang and his colleagues are right in guessing that they represent a precursor form to *H. erectus* rather than that species itself.

All this makes sense, but one anomaly remains. If *H. erectus* was an indigenous eastern Asian development, then we have to consider whether we have correctly identified the Olduvai OH9 braincase as belonging to this species. If we have, then *H. erectus* evolved in eastern Asia at quite an early date (remember, OH9 is now thought to be almost 1.4 Myr old), and one branch of the species migrated back to Olduvai in Africa. But if these new Asian dates are accurate, it seems more probable that as we come to know more about OH9 and its kind we will find that they belonged to a different species of hominid altogether.

The opposite end of the Eurasian continent was, as I have hinted, also isolated from the human evolutionary mainstream. As we saw, humans seem to have arrived in Europe fairly late. In this region, the first convincing archaeological sites, with rather crude tools, show up at about 800 Kyr ago or thereabouts (although in the Levant, within hailing distance of Africa, the site of 'Ubeidiya has yielded Acheulean tools dated to around 1.4 Myr ago, just about as early as any found to the south). The problem has been that there has been no sign of the toolmakers themselves.

This gap has now began to be filled by finds made by Eudald Carbonell of the University of Tarragona in Spain and his co-workers at the Gran Dolina cave site in the Atapuerca Hills of northern Spain. In 1994 excavations there produced numerous rather simple stone tools, plus a number of human fossil fragments, the most complete of which is a partial upper face of an immature individual. All came from a level that was dated to more than 780 Kyr ago. No traces of Acheulean technology were found among the tools, and the investigators noted various primitive traits in the fossils, which they provisionally attributed to *H. heidelbergensis*. This is the species into which specimens formerly classified as archaic *H. sapiens* are increasingly being placed. Carbonell and his colleagues see their fossils as the starting point of an indigenous European lineage that gradually evolved into the Neanderthals. These latter, large-brained hominids are known only from Europe and western Asia, where they flourished in the period between about 200 Kyr and 30 Kyr ago, when they were extinguished in some way by invading *H. sapiens*.

This is not the only possibility, however. With only a preliminary description of the very fragmentary Gran Dolina fossils available, it is hard to be sure, but it seems at least equally possible that they are the remains of hominids who made an initial foray out of Africa into Europe but failed to establish themselves there over the long term. Representatives of *H. heidelbergensis* are known in Africa as well, as long ago as 600 Kyr, and this species quite likely recolonized Europe later on. There it would have given rise to the Neanderthals, whereas a less specialized African population founded the lineage that ultimately produced *H. sapiens*.

At another site, only a kilometer from Gran Dolina, Juan-Luis Arsuaga of Universidad Complutense de Madrid and his colleagues have discovered a huge cache of exquisitely preserved human fossils, about 300 Kyr old. These are said to anticipate the Neanderthals in certain respects, but they are not fully Neanderthal by any means. And although they emphasize that the Neanderthals (and possibly other related species) were an indigenous European development, these fossils from Sima de los Huesos ("Pit of the Bones") do not establish an unequivocal backward connection to their Gran Dolina neighbors.

BORN IN AFRICA

Every longtime reader of *Scientific American* will be familiar with the competing models of "regional continuity" and "single African origin" for the emergence of our own species, *H. sapiens* [see "The Multiregional Evolution of Humans," by Alan G. Thorne and Milford H. Wolpoff, and "The Recent African Genesis of Humans," by Allan C. Wilson and Rebecca L. Cann; April 1992]. The first of these models holds that the highly archaic *H. erectus* (including *H. ergaster*) is nothing more than an ancient variant of *H. sapiens* and that for the past two million years the history of our lineage has been one of a braided stream of evolving populations of this species in all areas of the Old World, each adapting to local conditions, yet all consistently linked by gene exchange. The variation we see today among the major geographic populations of humans is, by this reckoning, simply the latest permutation of this lengthy process.

The other notion, which happens to coincide much better with what we know of evolutionary processes in general, proposes that all modern human populations are descended from a single ancestral population that emerged in one place at some time between about 150 Kyr and 100 Kyr ago. The fossil evidence, thin as it is, suggests that this place of origin was somewhere in Africa (although the neighboring Levant is an alternative possibility); proponents of this scenario point to the support afforded by comparative molecular studies for the notion that all living humans are descended from an African population.

In view of what I have already said about the peripheral roles played in human evolution by early populations both in eastern Asia and Europe, it should come as no surprise that between these two possibilities my strong preference is for a single and comparatively recent origin for *H. sapiens,* very likely in Africa—the continent that, from the very beginning, has been the engine of mainstream innovation in human evolution. The rise of modern humans is a recent drama that played out against a long and complex backdrop of evolutionary diversification among the hominids, but the fossil record shows that from the earliest times Africa was consistently the center from which new lineages of hominids sprang. Clearly, interesting evolutionary developments occurred in both Europe and eastern Asia,

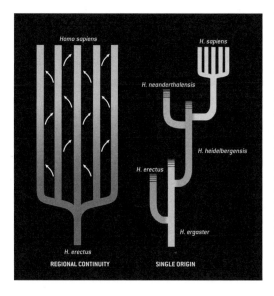

The leading theories of the origins of modern humans are contrasted in these diagrams. According to the notion of "regional continuity," all modern human populations trace their beginnings to *H. erectus,* but each regional population evolved along its own distinctive lines, exchanging enough genes with its neighbors (*arrows represent gene exchange*) to remain part of the same species; all eventually became *H. sapiens.* The "single origin" theory holds that *H. sapiens* descended from a single ancestral population that emerged in one place, probably Africa.

but they involved populations that were not only derived from but also eventually supplanted by emigrants from Africa. In Africa our lineage was born, and ever since its hominids were first emancipated from the forest edges, that continent has pumped out successive waves of emigrants to all parts of the Old World. What we see in the human fossil record as it stands today is without doubt a shadowy reflection at best of what must have been a very complex sequence of events.

Most important, the new dates from eastern Asia show that human population mobility dates right back to the very origins of effectively modern bodily form. Those from Europe demonstrate that although distinctive regional variants evolved there, the history of occupation of that region may itself not have been at all a simple one. As ever, though, new evidence of the remote human past has served principally to underline the complexity of events in our evolution. We can only hope that an improving fossil record will flesh out the details of what was evidently a richly intricate process of hominid speciation and population movement over the past two million years.

FURTHER READING

J.-L. Arsuaga et al. "Three New Human Skulls from the Sima de los Huesos Middle Pleistocene Site in Sierra de Atapuerca, Spain" in *Nature,* Vol. 362, No. 6420, pages 534–537; April 8, 1993.

C. C. Swisher III et al. "Age of the Earliest Known Hominids in Java, Indonesia" in *Science,* Vol. 263, No. 5150, pages 1118–1121; February 25, 1994.

L. Gabunia and A. Vekua. "A Plio-Pleistocene Hominid from Dmanisi, East Georgia, Caucasus" in *Nature,* Vol. 373, No. 6514, pages 509–512; February 9, 1995.

E. Carbonell et al. "Lower Pleistocene Hominids and Artifacts from Atapuerca-TD6 (Spain)" in *Science*, Vol. 269, No. 5225, pages 826–830; August 11, 1995.

W. Huang et al. "Early *Homo* and Associated Artefacts from Asia" in *Nature,* Vol. 378, No. 6554, pages 275–278; November 16, 1995.

J. H. Schwartz and I. Tattersall. "Whose Teeth" in *Nature,* Vol. 381, No. 6579, pages 201–202; May 16, 1996.

C. C. Swisher III et al. "Latest *Homo erectus* of Java: Potential Contemporaneity with *Homo sapiens* in Southeast Asia" in *Science*, Vol. 274, No. 5294, pages 1870–1874; December 13, 1996.

Who Were the Neandertals?

KATE WONG

ORIGINALLY PUBLISHED IN APRIL 2000

I t was such a neat and tidy story. No match for the anatomically modern humans who swept in with a sophisticated culture and technology, the Neandertals—a separate species—were quickly driven to extinction by the invading moderns. But neat and tidy stories about the past have a way of unraveling, and the saga of the Neandertals, it appears, is no exception. For more than 200,000 years, these large-brained hominids occupied Europe and western Asia, battling the bitter cold of glacial maximums and the daily perils of prehistoric life. Today they no longer exist. Beyond these two facts, however, researchers fiercely debate who the Neandertals were, how they lived and exactly what happened to them.

The steadfast effort to resolve these elusive issues stems from a larger dispute over how modern humans evolved. Some researchers posit that our species arose recently (around 200,000 years ago) in Africa and subsequently replaced archaic hominids around the world, whereas others propose that these ancient populations contributed to the early modern human gene pool. As the best known of these archaic groups, Neandertals are critical to the origins controversy. Yet this is more than an academic argument over certain events of our primeval past, for in probing Neandertal biology and behavior, researchers must wrestle with the very notion of what it means to be fully human and determine what, if anything, makes us moderns unique. Indeed, spurred by recent discoveries, paleoanthropologists and archaeologists are increasingly asking, How much like us were they?

Comparisons of Neandertals and modern humans first captured the attention of researchers when a partial Neandertal skeleton turned up in Germany's Neander Valley in 1856. Those remains—a heavily built skull with the signature arched browridge and massive limb bones—were clearly different, and Neandertals were assigned to their own species, *Homo neanderthalensis* (although even then there was disagreement: several German scientists argued that these were the remains of a crippled

Cossack horseman). But it was the French discovery of the famous "Old Man" of La Chapelle-aux-Saints some 50 years later that led to the characterization of Neandertals as primitive protohumans. Reconstructions showed them as stooped, lumbering, apelike brutes, in stark contrast to upright, graceful *Homo sapiens*. The Neandertal, it seemed, represented the ultimate "other," a dim-witted ogre lurking behind the evolutionary threshold of humanity.

Decades later reevaluation of the La Chapelle individual revealed that certain anatomical features had been misinterpreted. In fact, Neandertal posture and movement would have been the same as ours. Since then, paleoanthropologists have struggled to determine whether the morphological features that do characterize Neandertals as a group—such as the robustness of their skeletons, their short limbs and barrel chests, prominent browridges and low, sloping foreheads, protruding midfaces and chinless jaws—warrant designating them as a separate species. Researchers agree that some of these characteristics represent environmental adaptations. The Neandertals' stocky body proportions, for example, would have allowed them to retain heat more effectively in the extremely cold weather brought on by glacial cycles. But other traits, such as the form of the Neandertal browridge, lack any clear functional significance and seem to reflect the genetic drift typical of isolated populations.

For those scholars who subscribe to the replacement model of modern human origins, the distinctive Neandertal morphology clearly resulted from following an evolutionary trajectory separate from that of moderns. But for years, another faction of researchers has challenged this interpretation, arguing that many of the features that characterize Neandertals are also seen in the early modern Europeans that followed them. "They clearly have a suite of features that are, overall, different, but it's a frequency difference, not an absolute difference," contends David W. Frayer, a paleoanthropologist at the University of Kansas. "Virtually everything you can find in Neandertals you can find elsewhere."

He points to one of the earliest-known modern Europeans, a fossil from a site in southwestern Germany called Vogelherd, which combines the skull shape of moderns with features that are typically Neandertal, such as the distinct space between the last molar and the ascending part of the lower jaw known as a retromolar gap, and the form of the mandibular foramen—a nerve canal in the lower jaw. Additional evidence, according to Frayer and Milford H. Wolpoff of the University of Michigan, comes from a group of early moderns discovered in Moravia (Czech Republic) at a site called Mladeč. The Mladeč people, they say, exhibit characteristics on their skulls that other scientists have described as uniquely Neandertal traits.

Although such evidence was once used to argue that Neandertals could have independently evolved into modern Europeans, this view has shifted somewhat. "It's quite clear that people entered Europe as well, so the people that are there later in time are a mix of Neandertals and those populations coming into Europe," says Wolpoff, who believes the two groups differed only as much as living Europeans and aboriginal Australians do. Evidence for mixing also appears in later Neandertal fossils, according to Fred H. Smith, a paleoanthropologist at Northern Illinois University. Neandertal remains from Vindija cave in northwestern Croatia reflect "the assimilation of some early modern features," he says, referring to their more modern-shaped browridges and the slight presence of a chin on their mandibles.

Those who view Neandertals as a separate species, however, maintain that the Vindija fossils are too fragmentary to be diagnostic and that any similarities that do exist can be attributed to convergent evolution. These researchers likewise dismiss the mixing argument for the early moderns from Mladeč. "When I look at the morphology of these people, I see robustness, I don't see Neandertal," counters Christopher B. Stringer of the Natural History Museum in London.

Another reason to doubt these claims for interbreeding, some scientists say, is that they contradict the conclusions reached by Svante Pääbo, then at the University of Munich, and his colleagues, who in July 1997 announced that they had retrieved and analyzed mitochondrial DNA (mtDNA) from a Neandertal fossil. The cover of the journal *Cell,* which contained their report, said it all: "Neandertals Were Not Our Ancestors." From the short stretch of mtDNA they sequenced, the researchers determined that the difference between the Neandertal mtDNA and living moderns' mtDNA was considerably greater than the differences found among living human populations. But though it seemed on the surface that the species question had been answered, undercurrents of doubt have persisted.

New fossil evidence from western Europe has intensified interest in whether Neandertals and moderns mixed. In January 1999 researchers announced the discovery in central Portugal's Lapedo Valley of a largely complete skeleton from a four-year-old child buried 24,500 years ago in the Gravettian style known from other early modern Europeans. According to Erik Trinkaus of Washington University, Cidália Duarte of the Portuguese Institute of Archaeology in Lisbon and their colleagues, the specimen, known as Lagar Velho 1, bears a combination of Neandertal and modern human traits that could only have resulted from extensive inter-breeding between the two populations.

If the mixed ancestry interpretation for Lagar Velho 1 holds up after further scrutiny, the notion of Neandertals as a variant of our species will gain new strength. Advocates of the replacement model do allow for iso-lated instances of interbreeding between moderns and the archaic spe-cies, because some other closely related mammal species interbreed on oc-casion. But unlike central and eastern European specimens that are said to show a combination of features, the Portuguese child dates to a time when Neandertals are no longer thought to have existed. For Neandertal features to have persisted thousands of years after those people disap-peared, Trinkaus and Duarte say, coexisting populations of Neandertals and moderns must have mixed significantly.

Their interpretation has not gone unchallenged. In a commentary ac-companying the team's report in the *Proceedings of the National Academy of Sciences USA* last June, paleoanthropologists Ian Tattersall of the American Museum of Natural History in New York City and Jeffrey H. Schwartz of the University of Pittsburgh argued that Lagar Velho 1 is instead most likely "a chunky Gravettian child." The robust body proportions that Trinkaus and his colleagues view as evidence for Neandertal ancestry, Stringer says, might instead reflect adaptation to Portugal's then cold climate. But this interpretation is problematic, according to Jean-Jacques Hublin of France's CNRS, who points out that although some cold-adapted moderns exhibit such proportions, none are known from that period in Europe. Rather Hublin is troubled that Lagar Velho 1 represents a child, noting that "we do not know anything about the variation in children of a given age in this range of time."

SURVIVAL SKILLS

Taxonomic issues aside, much research has focused on Neandertal behav-ior, which remained largely misunderstood until relatively recently. Ne-andertals were often portrayed as incapable of hunting or planning ahead, recalls archaeologist John J. Shea of the State University of New York at Stony Brook. "We've got reconstructions of Neandertals as people who couldn't survive a single winter, let alone a quarter of a million years in the worst environments in which humans ever lived," he observes. Analysis of animal remains from the Croatian site of Krapina, however, in-dicates that Neandertals were skilled hunters capable of killing even large animals such as rhinoceroses, according to University of Cambridge archaeologist Preston T. Miracle. And Shea's studies suggest that some Neandertals employed sophisticated stone-tipped spears to conquer their quarry—a finding supported last year when researchers reported the discovery in Syria of a Neandertal-made stone point lodged in a neckbone of a prehistoric wild ass. Moreover, additional research conducted by Shea and investigations carried out by University of Arizona archaeologists Mary C. Stiner and Steven L. Kuhn have shown that Neandertal subsistence strategies varied widely with the environment and the changing seasons.

Such demonstrations refute the notion that Neandertals perished because they could not adapt. But it may be that moderns were better at it. One popular theory posits that modern humans held some cognitive advantage over Neandertals, perhaps a capacity for the most human trait of all: symbolic thought, including language. Explanations such as this one arose from observations that after 40,000 years ago, whereas Neandertal culture remained relatively static, that of modern Europeans boasted a bevy of new features, many of them symbolic. It appeared that only moderns performed elaborate burials, expressed themselves through body ornaments, figurines and cave paintings, and crafted complex bone and antler tools—an industry broadly referred to as Upper Paleolithic. Neandertal assemblages, in contrast, contained only Middle Paleolithic stone tools made in the Mousterian style.

Yet hints that Neandertals thought symbolically had popped up. Neandertal burials, for example, are well known across Europe, and several, it has been argued, contain grave goods. (Other researchers maintain that for Neandertals, interment merely constituted a way of concealing the decomposing body, which might have attracted unwelcome predators. They view the purported grave goods as miscellaneous objects that happened to be swept into the grave.) Evidence for art, in the form of isolated pierced teeth and engraved bone fragments, and red and yellow ocher, has been reported from a few sites, too, but given their relative rarity, researchers tend to assign alternative explanations to these items.

The possibility that Neandertals might have engaged in modern practices was taken more seriously in 1980, when researchers reported a Neandertal from the Saint-Césaire rock-shelter in Charente-Maritime, France, found in association with stone tools manufactured according to a cultural tradition known as the Châtelperronian, which was assumed to have been the handiwork of moderns. Then, in 1996, Hublin and his colleagues made an announcement that catapulted the Châtelperronian into

Day in the life of Neandertals at the Grotte du Renne in France is imagined here. The Châtelperronian stratigraphic levels have yielded a trove of pendants and advanced bone and stone tools. Such items, along with evidence of huts and hearths, were once linked to modern humans alone, but the Grotte du Renne remains suggest that some Neandertals were similarly industrious.

the archaeological limelight. Excavations that began in the late 1940s at a site called the Grotte du Renne at Arcysur-Cure near Auxerre, France, had yielded numerous blades, body ornaments and bone tools and revealed evidence of huts and hearths—all hallmarks of the Upper Paleolithic. The scant human remains found amid the artifacts were impossible to identify initially, but using computed tomography to examine the hidden inner-ear region preserved inside an otherwise uninformative skull fragment, Hublin's team identified the specimen as Neandertal.

In response, a number of scientists suggested that Neandertals had acquired the modern-looking items either by stealing them, collecting artifacts discarded by moderns or perhaps trading for them. But this view has come under fire, most recently from archaeologists Francesco d'Errico of the University of Bordeaux and João Zilhão of the Portuguese Institute of Archaeology, who argue that the Châtelperronian artifacts at the Grotte du Renne and elsewhere, though superficially similar to those from the Aurignacian, reflect an older, different method of manufacture.

Most researchers are now convinced that Neandertals manufactured the Châtelperronian tools and ornaments, but what prompted this change after hundreds of thousands of years is unclear. Cast in this light, "it's more economical to see that as a result of imitation or acculturation from modern humans than to assume that Neandertals invented it for themselves," reasons Cambridge archaeologist Paul A. Mellars. "It would be an extraordinary coincidence if they invented all these things shortly before the modern humans doing the same things arrived." Furthermore, Mellars disagrees with d'Errico and Zilhão's proposed order of events. "The dating evidence proves to me that [Neandertals] only started to do these things after the modern humans had arrived in western Europe or at least in northern Spain," he asserts. (Unfortunately, because scientists have been unable to date these sites with sufficient precision, researchers can interpret the data differently.)

From his own work on the Grotte du Renne body ornaments, New York University archaeologist Randall White argues that these artifacts reflect manufacturing methods known—albeit at lower frequencies—from Aurignacian ornaments. Given the complicated stratigraphy of the Grotte du Renne site, the modern-looking items might have come from overlying Aurignacian levels. But more important, according to White, the Châtelperronian does not exist outside of France, Belgium, Italy and northern Spain. Once you look at the Upper Paleolithic from a pan-European perspective, he says, "the Châtelperronian becomes post-Aurignacian by a long shot."

Still, post-Aurignacian does not necessarily mean after contact with moderns. The earliest Aurignacian sites do not include any human remains. Researchers have assumed that they belonged to moderns because moderns are known from younger Aurignacian sites. But "who the Aurignacians were biologically between 40,000 and 35,000 years ago remains very much an unanswered question," White notes.

He adds that if you look at the Near East around 90,000 years ago, anatomically modern humans and Neandertals were both making Mousterian stone tools, which, though arguably less elaborate than Aurignacian tools, actually require a considerable amount of know-how. "I cannot imagine that Neandertals were producing these kinds of technologically complex tools and passing that on from generation to generation without talking about it," White declares. "I've seen a lot of people do this stuff, and I can't stand over somebody's shoulder and learn how to do it without a lot of verbal hints." Thus, White and others do not buy the argument that moderns were somehow cognitively superior, especially if Neandertals' inferiority meant that they lacked language. Instead it seems that moderns invented a culture that relied more heavily on material symbols.

Researchers have also looked to Neandertal brain morphology for clues to their cognitive ability. According to Ralph L. Holloway of Columbia University, all the brain asymmetries that characterize modern humans are found in Neandertals. "To be able to discriminate between the two," he remarks, "is, at the moment, impossible." As to whether Neandertal anatomy would have permitted speech, studies of the base of the skull conducted by Jeffrey T. Laitman of the Mount Sinai School of Medicine suggest that if they talked, Neandertals had a somewhat limited vocal repertoire. The significance of such physical constraints, however, is unclear.

FADING AWAY

If Neandertals possessed basically the same cognitive ability as moderns, it makes their disappearance additionally puzzling. But the recent redating of Neandertal remains from Vindija cave in Croatia emphasizes that this did not happen overnight. Smith and his colleagues have demonstrated that Neandertals still lived in central Europe 28,000 years ago, thousands of years after moderns had moved in [see "The Fate of the Neandertals," above]. Taking this into consideration, Stringer imagines that moderns, whom he views as a new species, replaced Neandertals in a long, slow process. "Gradually the Neandertals lost out because moderns were a bit more innovative, a bit better able to cope with rapid environmental change quickly, and they probably had bigger social networks," he supposes.

On the other hand, if Neandertals were an equally capable variant of our own species, as Smith and Wolpoff believe, long-term overlap of Neandertals and the new population moving into Europe would have left plenty of time for mingling, hence the mixed morphology that these scholars see in late Neandertals and early moderns in Europe. And if these groups were exchanging genes, they were probably exchanging cultural ideas, which might account for some of the similarity between, say, the Châtelperronian and the Aurignacian. Neandertals as entities disappeared, Wolpoff says, because they were outnumbered by the newcomers. Thousands of years of interbreeding between the small Neandertal population and the larger modern human population, he surmises, diluted the distinctive Neandertal features, which ultimately faded away.

"If we look at Australians a thousand years from now, we will see that the European features have predominated [over those of native Australians] by virtue of many more Europeans," Wolpoff asserts. "Not by virtue of better adaptation, not by virtue of different culture, not by virtue of anything except many more Europeans. And I really think that's what describes what we see in Europe—we see the predominance of more people."

From the morass of opinions in this notoriously contentious field, one consensus emerges: researchers have retired the old vision of the shuffling, cultureless Neandertal. Beyond that, whether these ancient hominids were among the ancestors of living people or a very closely related species that competed formidably with our own for the Eurasian territory and eventually lost remains to be seen. In either case, the details will most likely be extraordinarily complicated. "The more we learn, the more questions arise, the knottier it gets," muses archaeologist Lawrence G. Straus of the University of New Mexico. "That's why simple explanations just don't cut it."

Food for Thought

WILLIAM R. LEONARD

ORIGINALLY PUBLISHED IN DECEMBER 2002

We humans are strange primates. We walk on two legs, carry around enormous brains and have colonized every corner of the globe. Anthropologists and biologists have long sought to understand how our lineage came to differ so profoundly from the primate norm in these ways, and over the years all manner of hypotheses aimed at explaining each of these oddities have been put forth. But a growing body of evidence indicates that these miscellaneous quirks of humanity in fact have a common thread: they are largely the result of natural selection acting to maximize dietary quality and foraging efficiency. Changes in food availability over time, it seems, strongly influenced our hominid ancestors. Thus, in an evolutionary sense, we are very much what we ate.

Accordingly, what we eat is yet another way in which we differ from our primate kin. Contemporary human populations the world over have diets richer in calories and nutrients than those of our cousins, the great apes. So when and how did our ancestors' eating habits diverge from those of other primates? Further, to what extent have modern humans departed from the ancestral dietary pattern?

Scientific interest in the evolution of human nutritional requirements has a long history. But relevant investigations started gaining momentum after 1985, when S. Boyd Eaton and Melvin J. Konner of Emory University published a seminal paper in the *New England Journal of Medicine* entitled "Paleolithic Nutrition." They argued that the prevalence in modern societies of many chronic diseases—obesity, hypertension, coronary heart disease and diabetes, among them—is the consequence of a mismatch between modern dietary patterns and the type of diet that our species evolved to eat as prehistoric hunter-gatherers. Since then, however, understanding of the evolution of human nutritional needs has advanced considerably—thanks in large part to new comparative analyses of traditionally living human populations and other primates—and a more nuanced picture has emerged. We now know that humans have evolved not to subsist on a single, Paleolithic diet but to be flexible eaters, an insight that has important implications for the current debate over what people today should eat in order to be healthy.

To appreciate the role of diet in human evolution, we must remember that the search for food, its consumption and, ultimately, how it is used for biological processes are all critical aspects of an organism's ecology. The energy dynamic between organisms and their environments—that is, energy expended in relation to energy acquired—has important adaptive consequences for survival and reproduction. These two components of Darwinian fitness are reflected in the way we divide up an animal's energy budget. Maintenance energy is what keeps an animal alive on a day-to-day basis. Productive energy, on the other hand, is associated with producing and raising offspring for the next generation. For mammals like ourselves, this must cover the increased costs that mothers incur during pregnancy and lactation.

The type of environment a creature inhabits will influence the distribution of energy between these components, with harsher conditions creating higher maintenance demands. Nevertheless, the goal of all organisms is the same: to devote sufficient funds to reproduction to ensure the long-term success of the species. Thus, by looking at the way animals go about obtaining and then allocating food energy, we can better discern how natural selection produces evolutionary change.

BECOMING BIPEDS

Without exception, living nonhuman primates habitually move around on all fours, or quadrupedally, when they are on the ground. Scientists generally assume therefore that the last common ancestor of humans and chimpanzees (our closest living relative) was also a quadruped. Exactly when the last common ancestor lived is unknown, but clear indications of bipedalism—the trait that distinguished ancient humans from other apes—are evident in the oldest known species of *Australopithecus*, which lived in Africa roughly four million years ago. Ideas about why bipedalism evolved abound in the paleoanthropological literature. C. Owen Lovejoy of Kent State University proposed in 1981 that two-legged locomotion freed the arms to carry children and foraged goods. More recently, Kevin D. Hunt of Indiana University has posited that bipedalism emerged as a feeding posture that enabled access to foods that had previously been out of reach. Peter Wheeler of Liverpool John Moores University submits that moving upright allowed early humans to better regulate their

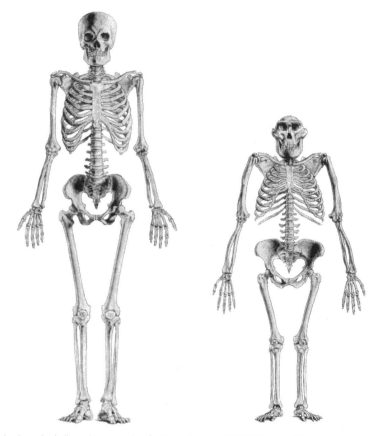

Skeletal remains indicate that our ancient forebears the australopithecines were bipedal by four million years ago. In the case of *A. afarensis* (*right*), one of the earliest hominids, telltale features include the arch in the foot, the nonopposable big toe, and certain characteristics of the knee and pelvis. But these hominids retained some apelike traits—short legs, long arms and curved toes, among others—suggesting both that they probably did not walk exactly like we do and that they spent some time in the trees. It wasn't until the emergence of our own genus, *Homo* (a contemporary representative of which appears on the left), that the fully modern limb and foot proportions and pelvis form required for upright walking as we know it evolved.

body temperature by exposing less surface area to the blazing African sun.

The list goes on. In reality, a number of factors probably selected for this type of locomotion. My own research, conducted in collaboration with my wife, Marcia L. Robertson, suggests that bipedalism evolved in our ancestors at least in part because it is less energetically expensive than quadrupedalism. Our analyses of the energy costs of movement in living animals of all sizes have shown that, in general, the strongest predictors of cost are the weight of the animal and the speed at which it travels. What is striking about human bipedal movement is that it is notably more economical than quadrupedal locomotion at walking rates.

Apes, in contrast, are not economical when moving on the ground. For instance, chimpanzees, which employ a peculiar form of quadrupedalism known as knuckle walking, spend some 35 percent more calories during locomotion than does a typical mammalian quadruped of the same size—a large dog, for example. Differences in the settings in which humans and apes evolved may help explain the variation in costs of movement. Chimps, gorillas and orangutans evolved in and continue to occupy dense forests where only a mile or so of trekking over the course of the day is all that is needed to find enough to eat. Much of early hominid evolution, on the other hand, took place in more open woodland and grassland, where sustenance is harder to come by. Indeed, modern human hunter-gatherers living in these environments, who provide us with the best available model of early human subsistence patterns, often travel six to eight miles daily in search of food.

These differences in day range have important locomotor implications. Because apes travel only short distances each day, the potential energetic benefits of moving more efficiently are very small. For far-ranging foragers, however, cost-effective walking saves many calories in maintenance energy needs—calories that can instead go toward reproduction. Selection for energetically efficient locomotion is therefore likely to be more intense among far-ranging animals because they have the most to gain.

For hominids living between five million and 1.8 million years ago, during the Pliocene epoch, climate change spurred this morphological revolution. As the African continent grew drier, forests gave way to grasslands, leaving food resources patchily distributed. In this context, bipedalism can be viewed as one of the first strategies in human nutritional evolution, a pattern of movement that would have substantially reduced the number of calories spent in collecting increasingly dispersed food resources.

BIG BRAINS AND HUNGRY HOMINIDS

No sooner had humans perfected their stride than the next pivotal event in human evolution—the dramatic enlargement of the brain—began. According to the fossil record, the australopithecines never became much brainier than living apes, showing only a modest increase in brain size, from around 400 cubic centimeters four million years ago to 500 cubic centimeters two million years later. *Homo* brain sizes, in contrast, ballooned

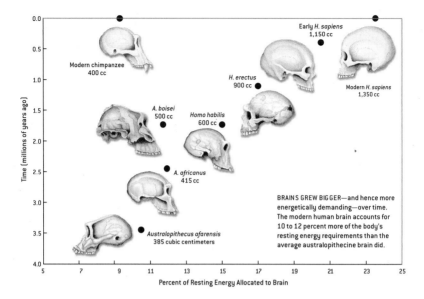

BRAINS GREW BIGGER—and hence more energetically demanding—over time. The modern human brain accounts for 10 to 12 percent more of the body's resting energy requirements than the average australopithecine brain did.

from 600 cubic centimeters in *H. habilis* some two million years ago up to 900 cubic centimeters in early *H. erectus* just 300,000 years later. The *H. erectus* brain did not attain modern human proportions (1,350 cubic centimeters on average), but it exceeded that of living nonhuman primates.

From a nutritional perspective, what is extraordinary about our large brain is how much energy it consumes—roughly 16 times as much as muscle tissue per unit weight. Yet although humans have much bigger brains relative to body weight than do other primates (three times larger than expected), the total resting energy requirements of the human body are no greater than those of any other mammal of the same size. We therefore use a much greater share of our daily energy budget to feed our voracious brains. In fact, at rest brain metabolism accounts for a whopping 20 to 25 percent of an adult human's energy needs—far more than the 8 to 10 percent observed in nonhuman primates, and more still than the 3 to 5 percent allotted to the brain by other mammals.

By using estimates of hominid body size compiled by Henry M. McHenry of the University of California at Davis, Robertson and I have reconstructed the proportion of resting energy needs that would have been required to support the brains of our ancient ancestors. Our calculations suggest that a typical, 80- to 85-pound australopithecine with a brain size of 450 cubic centimeters would have devoted about 11 percent of its resting energy to the brain. For its part, *H. erectus,* which weighed in at 125 to

130 pounds and had a brain size of some 900 cubic centimeters, would have earmarked about 17 percent of its resting energy—that is, about 260 out of 1,500 kilocalories a day—for the organ.

How did such an energetically costly brain evolve? One theory, developed by Dean Falk of Florida State University, holds that bipedalism enabled hominids to cool their cranial blood, thereby freeing the heat-sensitive brain of the temperature constraints that had kept its size in check. I suspect that, as with bipedalism, a number of selective factors were probably at work. But brain expansion almost certainly could not have occurred until hominids adopted a diet sufficiently rich in calories and nutrients to meet the associated costs.

Comparative studies of living animals support that assertion. Across all primates, species with bigger brains dine on richer foods, and humans are the extreme example of this correlation, boasting the largest relative brain size and the choicest diet [see "Diet and Primate Evolution," by Katharine Milton; *Scientific American,* August 1993]. According to recent analyses by Loren Cordain of Colorado State University, contemporary hunter-gatherers derive, on average, 40 to 60 percent of their dietary energy from animal foods (meat, milk and other products). Modern chimps, in comparison, obtain only 5 to 7 percent of their calories from these comestibles. Animal foods are far denser in calories and nutrients than most plant foods. For example, 3.5 ounces of meat provides upward of 200 kilocalories. But the same amount of fruit provides only 50 to 100 kilocalories. And a comparable serving of foliage yields just 10 to 20 kilocalories. It stands to reason, then, that for early *Homo,* acquiring more gray matter meant seeking out more of the energy-dense fare.

Fossils, too, indicate that improvements to dietary quality accompanied evolutionary brain growth. All australopithecines had skeletal and dental features built for processing tough, low-quality plant foods. The later, robust australopithecines—a dead-end branch of the human family tree that lived alongside members of our own genus—had especially pronounced adaptations for grinding up fibrous plant foods, including massive, dish-shaped faces; heavily built mandibles; ridges, or sagittal crests, atop the skull for the attachment of powerful chewing muscles; and huge, thickly enameled molar teeth. (This is not to say that australopithecines never ate meat. They almost certainly did on occasion, just as chimps do today.) In contrast, early members of the genus *Homo,* which descended from the gracile australopithecines, had much smaller faces, more delicate jaws, smaller molars and no sagittal crests—despite being far larger in terms of overall body size than their predecessors. Together these features suggest that early *Homo* was consuming less plant material and more animal foods.

As to what prompted *Homo*'s initial shift toward the higher-quality diet necessary for brain growth, environmental change appears to have once more set the stage for evolutionary change. The continued desiccation of the African landscape limited the amount and variety of edible plant foods available to hominids. Those on the line leading to the robust australopithecines coped with this problem morphologically, evolving anatomical specializations that enabled them to subsist on more widely available, difficult-to-chew foods. *Homo* took a different path. As it turns out, the spread of grasslands also led to an increase in the relative abundance of grazing mammals such as antelope and gazelle, creating opportunities for hominids capable of exploiting them. *H. erectus* did just that, developing the first hunting-and-gathering economy in which game animals became a significant part of the diet and resources were shared among members of the foraging groups. Signs of this behavioral revolution are visible in the archaeological record, which shows an increase in animal bones at hominid sites during this period, along with evidence that the beasts were butchered using stone tools.

These changes in diet and foraging behavior did not turn our ancestors into strict carnivores; however, the addition of modest amounts of animal foods to the menu, combined with the sharing of resources that is typical of hunter-gatherer groups, would have significantly increased the quality and stability of hominid diets. Improved dietary quality alone cannot explain *why* hominid brains grew, but it appears to have played a critical role in enabling that change. After the initial spurt in brain growth, diet and brain expansion probably interacted synergistically: bigger brains produced more complex social behavior, which led to further shifts in foraging tactics and improved diet, which in turn fostered additional brain evolution.

A MOVABLE FEAST

The evolution of *H. erectus* in Africa 1.8 million years ago also marked a third turning point in human evolution: the initial movement of hominids out of Africa. Until recently, the locations and ages of known fossil sites suggested that early *Homo* stayed put for a few hundred thousand years before venturing out of the motherland and slowly fanning out into the rest of the Old World. Earlier work hinted that improvements in tool technology around 1.4 million years ago—namely, the advent of the

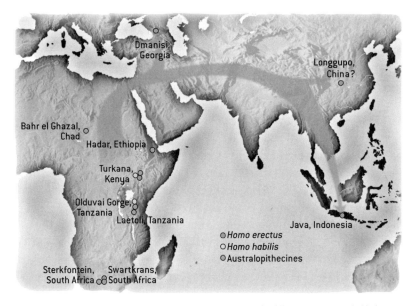

Bahr el Ghazal, Chad

Hadar, Ethiopia

Dmanisi, Georgia

Longgupo, China?

Turkana, Kenya

Olduvai Gorge, Tanzania

Laetoli, Tanzania

Java, Indonesia

Sterkfontein, South Africa

Swartkrans, South Africa

○ *Homo erectus*
○ *Homo habilis*
○ Australopithecines

African exodus began as soon as *H. erectus* evolved, around 1.8 million years ago, probably in part because it needed a larger home range than that of its smaller-bodied predecessors.

Acheulean hand ax—allowed hominids to leave Africa. But new discoveries indicate that *H. erectus* hit the ground running, so to speak. Rutgers University geochronologist Carl Swisher III and his colleagues have shown that the earliest *H. erectus* sites outside of Africa, which are in Indonesia and the Republic of Georgia, date to between 1.8 million and 1.7 million years ago. It seems that the first appearance of *H. erectus* and its initial spread from Africa were almost simultaneous.

The impetus behind this newfound wanderlust again appears to be food. What an animal eats dictates to a large extent how much territory it needs to survive. Carnivorous animals generally require far bigger home ranges than do herbivores of comparable size because they have fewer total calories available to them per unit area.

Large-bodied and increasingly dependent on animal foods, *H. erectus* most likely needed much more turf than the smaller, more vegetarian australopithecines did. Using data on contemporary primates and human hunter-gatherers as a guide, Robertson, Susan C. Antón of Rutgers University and I have estimated that the larger body size of *H. erectus,* combined with a moderate increase in meat consumption, would have necessitated an eightfold to 10-fold increase in home range size compared with that of the late australopithecines—enough, in fact, to account for the abrupt expansion of the species out of Africa. Exactly how far beyond the continent that shift would have taken *H. erectus* remains unclear, but migrating animal herds may have helped lead it to these distant lands.

As humans moved into more northern latitudes, they encountered new dietary challenges. The Neandertals, who lived during the last ice ages of Europe, were among the first humans to inhabit arctic environments, and they almost certainly would have needed ample calories to endure under those circumstances. Hints at what their energy requirements might have been come from data on traditional human populations that live in northern settings today. The Siberian reindeer-herding populations known as the Evenki, which I have studied with Peter Katzmarzyk of Queen's University in Ontario and Victoria A. Galloway of the University of Toronto, and the Inuit (Eskimo) populations of the Canadian Arctic have resting metabolic rates that are about 15 percent higher than those of people of similar size living in temperate environments. The energetically expensive activities associated with living in a northern climate ratchet their caloric cost of living up further still. Indeed, whereas a 160-pound American male with a typical urban way of life requires about 2,600 kilocalories a day, a diminutive, 125-pound Evenki man needs more than 3,000 kilocalories a day to sustain himself. Using these modern northern populations as benchmarks, Mark Sorensen of Northwestern University and I have estimated that Neandertals most likely would have required as many as 4,000 kilocalories a day to survive. That they were able to meet these demands for as long as they did speaks to their skills as foragers.

MODERN QUANDARIES

Just as pressures to improve dietary quality influenced early human evolution, so, too, have these factors played a crucial role in the more recent increases in population size. Innovations such as cooking, agriculture and even aspects of modern food technology can all be considered tactics for boosting the quality of the human diet. Cooking, for one, augmented the energy available in wild plant foods. With the advent of agriculture, humans began to manipulate marginal plant species to increase their productivity, digestibility and nutritional content—essentially making plants more like animal foods. This kind of tinkering continues today, with genetic modification of crop species to make "better" fruits, vegetables and grains. Similarly, the development of liquid nutritional supplements and meal replacement bars is a continuation of the trend that our ancient ancestors started: gaining as much nutritional return from our food in as little volume and with as little physical effort as possible.

Overall, that strategy has evidently worked: humans are here today and in record numbers to boot. But perhaps the strongest testament to the importance of energy- and nutrient-rich foods in human evolution lies in the observation that so many health concerns facing societies around the globe stem from deviations from the energy dynamic that our ancestors established. For children in rural populations of the developing world, low-quality diets lead to poor physical growth and high rates of mortality during early life. In these cases, the foods fed to youngsters during and after weaning are often not sufficiently dense in energy and nutrients to meet the high nutritional needs associated with this period of rapid growth and development. Although these children are typically similar in length and weight to their U.S. counterparts at birth, they are much shorter and lighter by the age of three, often resembling the smallest 2 to 3 percent of American children of the same age and sex.

In the industrial world, we are facing the opposite problem: rates of childhood and adult obesity are rising because the energy-rich foods we crave—notably those packed with fat and sugar—have become widely available and relatively inexpensive. According to recent estimates, more than half of adult Americans are overweight or obese. Obesity has also appeared in parts of the developing world where it was virtually unknown less than a generation ago. This seeming paradox has emerged as people who grew up malnourished move from rural areas to urban settings where food is more readily available. In some sense, obesity and other common diseases of the modern world are continuations of a tenor that started millions of years ago. We are victims of our own evolutionary success, having developed a calorie-packed diet while minimizing the amount of maintenance energy expended on physical activity.

The magnitude of this imbalance becomes clear when we look at traditionally living human populations. Studies of the Evenki reindeer herders that I have conducted in collaboration with Michael Crawford of the University of Kansas and Ludmila Osipova of the Russian Academy of Sciences in Novosibirsk indicate that the Evenki derive almost half their daily calories from meat, more than 2.5 times the amount consumed by the average American. Yet when we compare Evenki men with their U.S. peers, they are 20 percent leaner and have cholesterol levels that are 30 percent lower.

These differences partly reflect the compositions of the diets. Although the Evenki diet is high in meat, it is relatively low in fat (about 20 percent

Population	Energy Intake (kilocalories/day)	Energy from Animal Foods (percent)	Energy from Plant Foods (percent)	Total Blood Cholesterol (milligrams/deciliter)	Body Mass Index (weight/height squared)
HUNTER-GATHERERS					
!Kung (Botswana)	2,100	33	67	121	19
Inuit (North America)	2,350	96	4	141	24
PASTORALISTS					
Turkana (Kenya)	1,411	80	20	186	18
Evenki (Russia)	2,820	41	59	142	22
AGRICULTURALISTS					
Quechua (Highland Peru)	2,002	5	95	150	21
INDUSTRIAL SOCIETIES					
U.S.	2,250	23	77	204	26

Note: Energy intake figures reflect the adult average (males and females); blood cholesterol and body mass index (BMI) figures are given for males. Healthy BMI = 18.5–24.9; overweight = 25.0–29.9; obese = 30 and higher. BMI is weight (kilograms)/height (meters) squared.

Various diets can satisfy human nutritional requirements. Some populations subsist almost entirely on plant foods; others eat mostly animal foods. Although Americans consume less meat than do a number of the traditionally living people described here, they have on average higher cholesterol levels and higher levels of obesity (as indicated by body mass index) because they consume more energy than they expend and eat meat that is higher in fat.

of their dietary energy comes from fat, compared with 35 percent in the average U.S. diet), because free-ranging animals such as reindeer have less body fat than cattle and other feedlot animals do. The composition of the fat is also different in free-ranging animals, tending to be lower in saturated fats and higher in the polyunsaturated fatty acids that protect against heart disease. More important, however, the Evenki way of life necessitates a much higher level of energy expenditure.

Thus, it is not just changes in diet that have created many of our pervasive health problems but the interaction of shifting diets and changing lifestyles. Too often modern health problems are portrayed as the result of eating "bad" foods that are departures from *the* natural human diet—an oversimplification embodied by the current debate over the relative merits of a high-protein, high-fat Atkins-type diet or a low-fat one that emphasizes complex carbohydrates. This is a fundamentally flawed approach to assessing human nutritional needs. Our species was not designed to subsist on a single, optimal diet. What is remarkable about human beings is the extraordinary variety of what we eat. We have been able to thrive in almost every ecosystem on the earth, consuming diets ranging from almost all animal foods among populations of the Arctic to primarily tubers and cereal grains among populations in the high Andes. Indeed, the hallmarks of human evolution have been the diversity of strategies that we have developed to create diets that meet our distinctive metabolic requirements and the ever increasing efficiency with which we extract energy and nutrients from the environment. The challenge our modern societies now face is balancing the calories we consume with the calories we burn.

FURTHER READING

William R. Leonard and Marcia L. Robertson. "Evolutionary Perspectives on Human Nutrition: The Influence of Brain and Body Size on Diet and Metabolism" in *American Journal of Human Biology,* Vol. 6, No. 1, pages 77–88; January 1994.

William R. Leonard and Marcia L. Robertson. "Rethinking the Energetics of Bipedality" in *Current Anthropology,* Vol. 38, No.2, pages 304–309; April 1997.

Sara Stinson, Barry Bogin, Rebecca Huss-Ashmore and Dennis O'Rourke, eds. *Human Biology: An Evolutionary and Biocultural Approach.* Wiley-Liss, 2000.

William R. Leonard, Victoria A. Galloway, Evgueni Ivakine, Ludmila Osipova and Marina Kazakovtseva. "Ecology, Health and Lifestyle Change among the Evenki Herders of Siberia" in *Human Biology of Pastoral Populations.* Edited by William R. Leonard and Michael H. Crawford. Cambridge University Press, 2002.

Susan C. Antón, William R. Leonard and Marcia L. Robertson. "An Ecomorphological Model of the Initial Hominid Dispersal from Africa" in *Journal of Human Evolution,* vol. 43, pages 773–785; 2002.

Skin Deep

NINA G. JABLONSKI AND GEORGE CHAPLIN

ORIGINALLY PUBLISHED IN DECEMBER 2002

Among primates, only humans have a mostly naked skin that comes in different colors. Geographers and anthropologists have long recognized that the distribution of skin colors among indigenous populations is not random: darker peoples tend to be found nearer the equator, lighter ones closer to the poles. For years, the prevailing theory has been that darker skins evolved to protect against skin cancer. But a series of discoveries has led us to construct a new framework for understanding the evolutionary basis of variations in human skin color. Recent epidemiological and physiological evidence suggests to us that the worldwide pattern of human skin color is the product of natural selection acting to regulate the effects of the sun's ultraviolet (UV) radiation on key nutrients crucial to reproductive success.

FROM HIRSUTE TO HAIRLESS

The evolution of skin pigmentation is linked with that of hairlessness, and to comprehend both these stories, we need to page back in human history. Human beings have been evolving as an independent lineage of apes since at least seven million years ago, when our immediate ancestors diverged from those of our closest relatives, chimpanzees. Because chimpanzees have changed less over time than humans have, they can provide an idea of what human anatomy and physiology must have been like. Chimpanzees' skin is light in color and is covered by hair over most of their bodies. Young animals have pink faces, hands, and feet and become freckled or dark in these areas only as they are exposed to sun with age. The earliest humans almost certainly had a light skin covered with hair. Presumably hair loss occurred first, then skin color changed. But that leads to the question, When did we lose our hair?

The skeletons of ancient humans—such as the well-known skeleton of Lucy, which dates to about 3.2 million years ago—give us a good idea of the build and the way of life of our ancestors. The daily activities of Lucy and other hominids that lived before about three million years ago appear to have been similar to those of primates living on the open savannas of Africa today. They probably spent much of their day foraging for food over three to four miles before retiring to the safety of trees to sleep.

By 1.6 million years ago, however, we see evidence that this pattern had begun to change dramatically. The famous skeleton of Turkana Boy—which belonged to the species *Homo ergaster*—is that of a long-legged, striding biped that probably walked long distances. These more active early humans faced the problem of staying cool and protecting their brains from overheating. Peter Wheeler of John Moores University in Liverpool, England, has shown that this was accomplished through an increase in the number of sweat glands on the surface of the body and a reduction in the covering of body hair. Once rid of most of their hair, early members of the genus *Homo* then encountered the challenge of protecting their skin from the damaging effects of sunlight, especially UV rays.

BUILT-IN SUNSCREEN

In chimpanzees, the skin on the hairless parts of the body contains cells called melanocytes that are capable of synthesizing the dark-brown pigment melanin in response to exposure to UV radiation. When humans became mostly hairless, the ability of the skin to produce melanin assumed new importance. Melanin is nature's sunscreen: it is a large organic molecule that serves the dual purpose of physically and chemically filtering the harmful effects of UV radiation; it absorbs UV rays, causing them to lose energy, and it neutralizes harmful chemicals called free radicals that form in the skin after damage by UV radiation.

Anthropologists and biologists have generally reasoned that high concentrations of melanin arose in the skin of peoples in tropical areas because it protected them against skin cancer. James E. Cleaver of the University of California at San Francisco, for instance, has shown that people with the disease xeroderma pigmentosum, in which melanocytes are destroyed by exposure to the sun, suffer from significantly higher than normal rates of squamous and basal cell carcinomas, which are usually easily treated. Malignant melanomas are more frequently fatal, but they are rare (representing 4 percent of skin cancer diagnoses) and tend to strike only light-skinned people. But all skin cancers typically arise later in life, in most cases after the first reproductive years, so they could not have exerted enough evolutionary pressure for skin protection alone to account for darker skin colors. Accordingly, we began to ask what role melanin might play in human evolution.

THE FOLATE CONNECTION

In 1991 one of us (Jablonski) ran across what turned out to be a critical paper published in 1978 by Richard F. Branda and John W. Eaton, now at the University of Vermont and the University of Louisville, respectively. These investigators showed that light-skinned people who had been exposed to simulated strong sunlight had abnormally low levels of the essential B vitamin folate in their blood. The scientists also observed that subjecting human blood serum to the same conditions resulted in a 50-percent loss of folate content within one hour.

The significance of these findings to reproduction—and hence evolution—became clear when we learned of research being conducted on a major class of birth defects by our colleagues at the University of Western Australia. There Fiona J. Stanley and Carol Bower had established by the late 1980s that folate deficiency in pregnant women is related to an increased risk of neural tube defects such as spina bifida, in which the arches of the spinal vertebrae fail to close around the spinal cord. Many research groups throughout the world have since confirmed this correlation, and efforts to supplement foods with folate and to educate women about the importance of the nutrient have become widespread.

We discovered soon afterward that folate is important not only in preventing neural tube defects but also in a host of other processes. Because folate is essential for the synthesis of DNA in dividing cells, anything that involves rapid cell proliferation, such as spermatogenesis (the production of sperm cells), requires folate. Male rats and mice with chemically induced folate deficiency have impaired spermatogenesis and are infertile. Although no comparable studies of humans have been conducted, Wai Yee Wong and his colleagues at the University Medical Center of Nijmegen in the Netherlands have recently reported that folic acid treatment can boost the sperm counts of men with fertility problems.

Such observations led us to hypothesize that dark skin evolved to protect the body's folate stores from destruction. Our idea was supported by a report published in 1996 by Argentine pediatrician Pablo Lapunzina, who found that three young and otherwise healthy women whom he had attended gave birth to infants with neural tube defects after using sun beds to tan themselves in the early weeks of pregnancy. Our evidence about the breakdown of folate by UV radiation thus supplements what is already known about the harmful (skin-cancer-causing) effects of UV radiation on DNA.

HUMAN SKIN ON THE MOVE

The earliest members of *Homo sapiens,* or modern humans, evolved in Africa between 120,000 and 100,000 years ago and had darkly pigmented skin adapted to the conditions of UV radiation and heat that existed near the equator. As modern humans began to venture out of the tropics, however, they encountered environments in which they received significantly less UV radiation during the year. Under these conditions their high concentrations of natural sunscreen probably proved detrimental. Dark skin contains so much melanin that very little UV radiation, and specifically very little of the shorter-wavelength UVB radiation, can penetrate the skin. Although most of the effects of UVB are harmful, the rays perform one indispensable function: initiating the formation of vitamin D in the skin. Dark-skinned people living in the tropics generally receive sufficient UV radiation during the year for UVB to penetrate the skin and allow them to make vitamin D. Outside the tropics this is not the case. The solution, across evolutionary time, has been for migrants to northern latitudes to lose skin pigmentation.

The connection between the evolution of lightly pigmented skin and vitamin D synthesis was elaborated by W. Farnsworth Loomis of Brandeis University in 1967. He established the importance of vitamin D to reproductive success because of its role in enabling calcium absorption by the intestines, which in turn makes possible the normal development of the skeleton and the maintenance of a healthy immune system. Research led by Michael Holick of the Boston University School of Medicine has, over the past 20 years, further cemented the significance of vitamin D in development and immunity. His team also showed that not all sunlight contains enough UVB to stimulate vitamin D production. In Boston, for instance, which is located at about 42 degrees north latitude, human skin cells begin to produce vitamin D only after mid-March. In the wintertime there isn't enough UVB to do the job. We realized that this was another piece of evidence essential to the skin color story.

During the course of our research in the early 1990s, we sought in vain to find sources of data on actual UV radiation levels at the earth's surface. We were rewarded in 1996, when we contacted Elizabeth Weatherhead of the Cooperative Institute for Research in Environmental Sciences at the University of Colorado at Boulder. She shared with us a database of measurements of UV radiation at the earth's surface taken by NASA's Total Ozone Mapping Spectrophotometer satellite between 1978 and 1993. We were then able to model the distribution of UV radiation on the earth and relate the satellite data to the amount of UVB necessary to produce vitamin D.

We found that the earth's surface could be divided into three vitamin D zones: one comprising the tropics, one the subtropics and temperate regions, and the last the circumpolar regions north and south of about 45 degrees latitude. In the first, the dosage of UVB throughout the year is high enough that humans have ample opportunity to synthesize vitamin D all year. In the second, at least one month during the year has insufficient UVB radiation, and in the third area not enough UVB arrives on average during the entire year to prompt vitamin D synthesis. This distribution could explain why indigenous peoples in the tropics generally have dark skin, whereas people in the subtropics and temperate regions are lighter-skinned but have the ability to tan, and those who live in regions near the poles tend to be very light skinned and burn easily.

One of the most interesting aspects of this investigation was the examination of groups that did not precisely fit the predicted skin-color pattern. An example is the Inuit people of Alaska and northern Canada. The Inuit exhibit skin color that is somewhat darker than would be predicted given the UV levels at their latitude. This is probably caused by two factors. The first is that they are relatively recent inhabitants of these climes, having migrated to North America only roughly 5,000 years ago. The second is that the traditional diet of the Inuit is extremely high in foods containing vitamin D, especially fish and marine mammals. This vitamin D–rich diet offsets the problem that they would otherwise have with vitamin D synthesis in their skin at northern latitudes and permits them to remain more darkly pigmented.

Our analysis of the potential to synthesize vitamin D allowed us to understand another trait related to human skin color: women in all populations are generally lighter-skinned than men. (Our data show that women tend to be between 3 and 4 percent lighter than men.) Scientists have often speculated on the reasons, and most have argued that the phenomenon stems from sexual selection—the preference of men for women of lighter color. We contend that although this is probably part of the story, it is not the original reason for the sexual difference. Females have significantly greater needs for calcium throughout their reproductive lives, especially during pregnancy and lactation, and must be able to make the most of the calcium contained in food. We propose, therefore, that women tend to be lighter-skinned than men to allow slightly more UVB rays to penetrate their skin and thereby increase their ability to produce vitamin D. In areas of the world that receive a large amount of UV radiation, women are indeed at the knife's edge of natural selection, needing to maximize the photoprotective function of their skin on the one hand and the ability to synthesize vitamin D on the other.

WHERE CULTURE AND BIOLOGY MEET

As modern humans moved throughout the Old World about 100,000 years ago, their skin adapted to the environmental conditions that prevailed in different regions. The skin color of the indigenous people of Africa has had the longest time to adapt because anatomically modern humans first evolved there. The skin-color changes that modern humans underwent as they moved from one continent to another—first Asia, then Austro-Melanesia, then Europe and, finally, the Americas—can be reconstructed to some extent. It is important to remember, however, that those humans had clothing and shelter to help protect them from the elements. In some places, they also had the ability to harvest foods that were extraordinarily rich in vitamin D, as in the case of the Inuit. These two factors had profound effects on the tempo and degree of skin-color evolution in human populations.

Africa is an environmentally heterogeneous continent. A number of the earliest movements of contemporary humans outside equatorial Africa were into southern Africa. The descendants of some of these early colonizers, the Khoisan (previously known as Hottentots), are still found in southern Africa and have significantly lighter skin than indigenous equatorial Africans do—a clear adaptation to the lower levels of UV radiation that prevail at the southern extremity of the continent.

Interestingly, however, human skin color in southern Africa is not uniform. Populations of Bantu-language speakers who live in southern Africa today are far darker than the Khoisan. We know from the history of this region that Bantu speakers migrated into this region recently—probably within the past 1,000 years—from parts of West Africa near the equator. The skin-color difference between the Khoisan and Bantu speakers such as the Zulu indicates that the length of time that a group has inhabited a particular region is important in understanding why they have the color they do.

Cultural behaviors have probably also strongly influenced the evolution of skin color in recent human history. This effect can be seen in the indigenous peoples who live on the eastern and western banks of the Red Sea. The tribes on the western side, which speak so-called Nilo-Hamitic languages, are thought to have inhabited this region for as long as 6,000 years. These individuals are distinguished by very darkly pigmented skin and long, thin bodies with long limbs, which are excellent biological adaptations for dissipating heat and intense UV radiation. In contrast, modern agricultural and pastoral groups on the eastern bank of the Red Sea, on the Arabian Peninsula, have lived there for only about 2,000 years. These earliest Arab people, of European origin, have adapted to very similar environmental conditions by almost exclusively cultural means—wearing heavy protective clothing and devising portable shade in the form of tents. (Without such clothing, one would have expected their skin to have begun to darken.) Generally speaking, the more recently a group has migrated into an area, the more extensive its cultural, as opposed to biological, adaptations to the area will be.

PERILS OF RECENT MIGRATIONS

Despite great improvements in overall human health in the past century, some diseases have appeared or reemerged in populations that had previously been little affected by them. One of these is skin cancer, especially basal and squamous cell carcinomas, among light-skinned peoples. Another is rickets, brought about by severe vitamin D deficiency, in dark-skinned peoples. Why are we seeing these conditions?

As people move from an area with one pattern of UV radiation to another region, biological and cultural adaptations have not been able to keep pace. The light-skinned people of northern European origin who bask in the sun of Florida or northern Australia increasingly pay the price in the form of premature aging of the skin and skin cancers, not to mention the unknown cost in human life of folate depletion. Conversely, a number of dark-skinned people of southern Asian and African origin now living in the northern U.K., northern Europe or the northeastern U.S. suffer from a lack of UV radiation and vitamin D, an insidious problem that manifests itself in high rates of rickets and other diseases related to vitamin D deficiency.

The ability of skin color to adapt over long periods to the various environments to which humans have moved reflects the importance of skin color to our survival. But its unstable nature also makes it one of the least useful characteristics in determining the evolutionary relations between human groups. Early Western scientists used skin color improperly to delineate human races, but the beauty of science is that it can and does correct itself. Our current knowledge of the evolution of human skin indicates that variations in skin color, like most of our physical attributes, can be explained by adaptation to the environment through natural selection. We look ahead to the day when the vestiges of old scientific mistakes will be erased and replaced by a better understanding of human origins and diversity. Our variation in skin color should be celebrated as one of the most visible manifestations of our evolution as a species.

FURTHER READING

Nina G. Jablonski and George Chaplin. "The Evolution of Human Skin Coloration" in *Journal of Human Evolution*, Vol. 39, No. 1, pages 57–106; July 1, 2000. An abstract of the article is available online at www.idealibrary.com/links/doi/10.1006/jhev.2000.0403

Blake Edgar. "Why Skin Comes in Colors" in *California Wild*, Vol. 53, No. 1, pages 6–7; Winter 2000. The article is also available at www.calacademy.org/calwild/winter2000/html/horizons.html

Gina Kirchweger. "The Biology of Skin Color: Black and White" in *Discover*, Vol. 22, No. 2, pages 32–33; February 2001. The article is also available at www.discover.com/feb_01/featbiology.html

The Evolution of Human Birth

KAREN R. ROSENBERG AND WENDA R. TREVATHAN

ORIGINALLY PUBLISHED IN NOVEMBER 2001

Giving birth in the treetops is not the normal human way of doing things, but that is exactly what Sophia Pedro was forced to do during the height of the floods that ravaged southern Mozambique in March 2000. Pedro had survived for four days perched high above the raging floodwaters that killed more than 700 people in the region. The day after her delivery, television broadcasts and newspapers all over the world featured images of Pedro and her newborn child being plucked from the tree during a dramatic helicopter rescue.

Treetop delivery rooms are unusual for humans but not for other primate species. For millions of years, primates have secluded themselves in treetops or bushes to give birth. Human beings are the only primate species that regularly seeks assistance during labor and delivery. So when and why did our female ancestors abandon their unassisted and solitary habit? The answers lie in the difficult and risky nature of human birth.

Many women know from experience that pushing a baby through the birth canal is no easy task. It's the price we pay for our large brains and intelligence: humans have exceptionally big heads relative to the size of their bodies. Those who have delved deeper into the subject know that the opening in the human pelvis through which the baby must pass is limited in size by our upright posture. But only recently have anthropologists begun to realize that the complex twists and turns that human babies make as they travel through the birth canal have troubled humans and their ancestors for at least 100,000 years. Fossil clues also indicate that anatomy, not just our social nature, has led human mothers—in contrast to our closest primate relatives and almost all other mammals—to ask for help during childbirth. Indeed, this practice of seeking assistance may have been in place when the earliest members of our genus, *Homo*, emerged and may possibly date back to five million years ago, when our ancestors first began to walk upright on a regular basis.

TIGHT SQUEEZE

To test our theory that the practice of assisted birth may have been around for millennia, we considered first what scientists know about the way a primate baby fits through the mother's birth canal. Viewed from above, the infant's head is basically an oval, longest from the forehead to the back of the head and narrowest from ear to ear. Conveniently, the birth canal—the bony opening in the pelvis through which the baby must travel to get from the uterus to the outside world—is also an oval shape. The challenge of birth for many primates is that the size of the infant's head is close to the size of that opening.

For humans, this tight squeeze is complicated by the birth canal's not being a constant shape in cross section. The entrance of the birth canal, where the baby begins its journey, is widest from side to side relative to the mother's body. Midway through, however, this orientation shifts 90 degrees, and the long axis of the oval extends from the front of the mother's body to her back. This means that the human infant must negotiate a series of turns as it works its way through the birth canal so that the two parts of its body with the largest dimensions—the head and the shoulders—are always aligned with the largest dimension of the birth canal.

To understand the birth process from the mother's point of view, imagine you are about to give birth. The baby is most likely upside down, facing your side, when its head enters the birth canal. Midway through the canal, however, it must turn to face your back, and the back of its head is pressed against your pubic bones. At that time, its shoulders are oriented side to side. When the baby exits your body it is still facing backward, but it will turn its head slightly to the side. This rotation helps to turn the baby's shoulders so that they can also fit between your pubic bones and tailbone. To appreciate the close correspondence of the maternal and fetal dimensions, consider that the average pelvic opening in human females is 13 centimeters at its largest diameter and 10 centimeters at its smallest. The average infant head is 10 centimeters from front to back, and the shoulders are 12 centimeters across. This journey through a passageway of changing cross-sectional shape makes human birth difficult and risky for the vast majority of mothers and babies.

If we retreat far enough back along the family tree of human ancestors, we would eventually reach a point where birth was not so difficult. Although humans are more closely related to apes genetically, monkeys may present a better model for birth in prehuman primates. One line of reasoning to support this assertion is as follows: Of the primate fossils

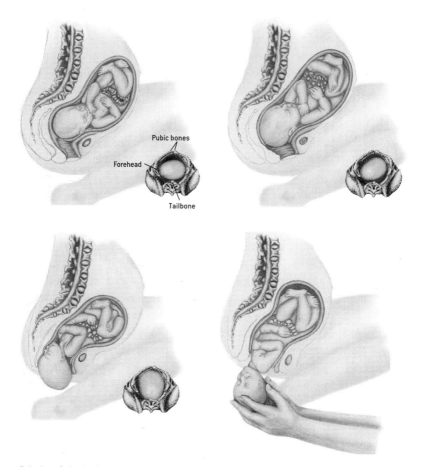

Baby born facing backward, with the back of its head against the mother's pubic bones, makes it difficult for a human female to guide the infant from the birth canal—the opening in the mother's pelvis *(insets)* — without assistance.

discovered from the time before the first known hominid, *Australopithecus,* one possible remote ancestor is *Proconsul,* a primate fossil dated to about 25 million years ago. This tailless creature probably looked like an ape, but its skeleton suggests that it moved more like a monkey. Its pelvis, too, was more monkeylike. The heads of modern monkey infants are typically about 98 percent the diameter of the mother's birth canal—a situation more comparable with that of humans than that of chimps, whose birth canals are relatively spacious.

Despite the monkey infant's tight squeeze, its entrance into the world is less challenging than that of a human baby. In contrast to the twisted birth canal of modern humans, monkeys' birth canals maintain the same cross-sectional shape from entrance to exit. The longest diameter of this oval shape is oriented front to back, and the broadest part of the oval is against the mother's back. A monkey infant enters the birth canal headfirst, with the broad back of its skull against the roomy back of the mother's pelvis and tailbone. That means the baby monkey emerges from the birth canal face forward—in other words, facing the same direction as the mother.

Firsthand observations of monkey deliveries have revealed a great advantage in babies' being born facing forward. Monkeys give birth squatting on their hind legs or crouching on all fours. As the infant is born, the mother reaches down to guide it out of the birth canal and toward her nipples. In many cases, she also wipes mucus from the baby's mouth and nose to aid its breathing. Infants are strong enough at birth to take part in their own deliveries. Once their hands are free, they can grab their mother's body and pull themselves out.

If human babies were also born face forward, their mothers would have a much easier time. Instead the evolutionary modifications of the human pelvis that enabled hominids to walk upright necessitate that most infants exit the birth canal with the back of their heads against the pubic bones, facing in the opposite direction as the mother (in a position obstetricians call "occiput anterior"). For this reason, it is difficult for the laboring human mother—whether squatting, sitting, or lying on her back—to reach down and guide the baby as it emerges. This configuration also greatly inhibits the mother's ability to clear a breathing passage for the infant, to remove the umbilical cord from around its neck or even to lift the baby up to her breast. If she tries to accelerate the delivery by grabbing the baby and guiding it from the birth canal, she risks bending its back awkwardly against the natural curve of its spine. Pulling on a newborn at this angle risks injury to its spinal cord, nerves and muscles.

For contemporary humans, the response to these challenges is to seek assistance during labor and delivery. Whether a technology-oriented professional, a lay midwife or a family member who is familiar with the birth process, the assistant can help the human mother do all the things

Baby born facing forward makes it possible for a monkey mother to reach down and guide the infant out of the birth canal. She can also wipe mucus from the baby's face to assist its breathing.

the monkey mother does by herself. The assistant can also compensate for the limited motor abilities of the relatively helpless human infant. The advantages of even simple forms of assistance have reduced maternal and infant mortality throughout history.

ASSISTED BIRTH

Of course, our ancestors and even women today can and do give birth alone successfully. Many fictional accounts portray stalwart peasant women giving birth alone in the fields, perhaps most famously in the novel *The Good Earth*, by Pearl S. Buck. Such images give the impression that delivering babies is easy. But anthropologists who have studied childbirth in cultures around the world report that these perceptions are highly romanticized and that human birth is seldom easy and rarely unattended. Today virtually all women in all societies seek assistance at delivery. Even among the !Kung of southern Africa's Kalahari Desert—who are well known for viewing solitary birth as a cultural ideal—women do not usually manage to give birth alone until they have delivered several babies at which mothers, sisters or other women are present. So, though rare exceptions do exist, assisted birth comes close to being a universal custom in human cultures.

Knowing this—and believing that this practice is driven by the difficulty and risk that accompany human birth—we began to think that midwifery is not unique to contemporary humans but instead has its roots deep in our ancestry. Our analysis of the birth process throughout human evolution has led us to suggest that the practice of midwifery might have appeared as early as five million years ago, when the advent of bipedalism first constricted the size and shape of the pelvis and birth canal.

A behavior pattern as complex as midwifery obviously does not fossilize, but pelvic bones do. The tight fit between the infant's head and the mother's birth canal in humans means that the mechanism of birth can be reconstructed if we know the relative sizes of each. Pelvic anatomy is now fairly well known from most time periods in the human fossil record, and we can estimate infant brain and skull size based on our extensive knowledge of adult skull sizes. (The delicate skulls of infants are not commonly found preserved until the point when humans began to bury their dead about 100,000 years ago.) Knowing the size and shape of the skulls and pelvises has also helped us and other researchers to understand whether infants were born facing forward or backward relative to their mothers—in turn revealing how challenging the birth might have been.

WALKING ON TWO LEGS

In modern humans, both bipedalism and enlarged brains constrain birth in important ways, but the first fundamental shift away from a nonhuman primate way of birth came about because of bipedalism alone. This unique way of walking appeared in early human ancestors of the genus *Australopithecus* at least four million years ago [see "Evolution of Human Walking," by C. Owen Lovejoy; *Scientific American,* November 1988]. Despite their upright posture, australopithecines typically stood no more than four feet tall, and their brains were not much bigger than those of living chimpanzees. Recent evidence has called into question which of the several australopithecine species were part of the lineage that led to *Homo*. Understanding the way any of them gave birth is still important, however, because walking on two legs would have constricted the maximum size of the pelvis and birth canal in similar ways among related species.

The anatomy of the female pelvis from this time period is well known from two complete fossils. Anthropologists unearthed the first (known as Sts 14 and presumed to be 2.5 million years old) in Sterkfontein, a site in the Transvaal region of South Africa. The second is best known as Lucy, a fossil discovered in the Hadar region of Ethiopia and dated at just over three million years old. Based on these specimens and on estimates of newborns' head size, C. Owen Lovejoy of Kent State University and Robert G. Tague of Louisiana State University concluded in the mid-1980s that birth in early hominids was unlike that known for any living species of primate.

The shape of the australopithecine birth canal is a flattened oval with the greatest dimension from side to side at both the entrance and exit. This shape appears to require a birth pattern different from that of monkeys, apes or modern humans. The head would not have rotated within the birth canal, but we think that in order for the shoulders to fit through, the baby might have had to turn its head once it emerged. In other words, if the baby's head entered the birth canal facing the side of the mother's body, its shoulders would have been oriented in a line from the mother's belly to her back. This starting position would have meant that the shoulders probably also had to turn sideways to squeeze through the birth canal.

This simple rotation could have introduced a kind of difficulty in australopithecine deliveries that no other known primate species had ever experienced. Depending on which way the baby's shoulders turned, its head could have exited the birth canal facing either forward or backward relative to the mother. Because the australopithecine birth canal is a symmetrical opening of unchanging shape, the baby could have just as easily turned its shoulders toward the front or back of its body, giving it about a 50–50 chance of emerging in the easier, face-forward position. If the infant were born facing backward, the australopithecine mother—like modern human mothers—may well have benefited from some kind of assistance.

GROWING BIGGER BRAINS

If bipedalism alone did not introduce into the process of childbirth enough difficulty for mothers to benefit from assistance, then the expanding size of the hominid brain certainly did. The most significant expansion in adult and infant brain size evolved subsequent to the australopithecines, particularly in the genus *Homo*. Fossil remains of the pelvis of early *Homo* are quite rare, and the best-preserved specimen, the 1.5-million-year-old Nariokotome fossil from Kenya, is an adolescent often referred to as Turkana Boy. Researchers have estimated that the boy's adult relatives probably had brains about twice as large as those of australopithecines but still only two thirds the size of modern human brains.

By reconstructing the shape of the boy's pelvis from fragments, Christopher B. Ruff of Johns Hopkins University and Alan Walker of Pennsylvania State University have estimated what he would have looked like had he reached adulthood. Using predictable differences between male and female pelvises in more recent hominid species, they could also infer what a female of that species would have looked like and could estimate the shape of the birth canal. That shape turns out to be a flattened oval similar to that of the australopithecines. Based on these reconstructions, the researchers determined that Turkana Boy's kin probably had a birth mechanism like that seen in australopithecines.

In recent years, scientists have been testing an important hypothesis that follows from Ruff and Walker's assertion: the pelvic anatomy of early *Homo* may have limited the growth of the human brain until the evolutionary point at which the birth canal expanded enough to allow a larger infant head to pass. This assertion implies that bigger brains and roomier pelvises were linked from an evolutionary perspective. Individuals who displayed both characteristics were more successful at giving birth to offspring who survived to pass on the traits. These changes in pelvic anatomy, accompanied by assisted birth, may have allowed the dramatic increase in human brain size that took place from two million to 100,000 years ago.

Fossils that span the past 300,000 years of human evolution support the connection between the expansion of brain size and changes in pelvic anatomy. In the past 20 years, scientists have uncovered three pelvic fossils of archaic *Homo sapiens*: a male from Sima de los Huesos in Sierra

Atapuerca, Spain (more than 200,000 years old); a female from Jinniushan, China (280,000 years old); and the male Kebara Neandertal—which is also an archaic *H. sapiens*—from Israel (about 60,000 years old). These specimens all have the twisted pelvic openings characteristic of modern humans, which suggests that their large-brained babies would most likely have had to rotate the head and shoulders within the birth canal and would thus have emerged facing away from the mother—a major challenge that human mothers face in delivering their babies safely.

The triple challenge of big-brained infants, a pelvis designed for walking upright, and a rotational delivery in which the baby emerges facing backward is not merely a contemporary circumstance. For this reason, we suggest that natural selection long ago favored the behavior of seeking assistance during birth because such help compensated for these difficulties. Mothers probably did not seek assistance solely because they predicted the risk that childbirth poses, however. Pain, fear and anxiety more likely drove their desire for companionship and security.

Psychiatrists have argued that natural selection might have favored such emotions—also common during illness and injury—because they led individuals who experienced them to seek the protection of companions, which would have given them a better chance of surviving [see "Evolution and the Origins of Disease," by Randolph M. Nesse and George C. Williams; *Scientific American,* November 1998]. The offspring of the survivors would then also have an enhanced tendency to experience such emotions during times of pain or disease. Taking into consideration the evolutionary advantage that fear and anxiety impart, it is no surprise that women commonly experience these emotions during labor and delivery.

Modern women giving birth have a dual evolutionary legacy: the need for physical as well as emotional support. When Sophia Pedro gave birth in a tree surrounded by raging floodwaters, she may have had both kinds of assistance. In an interview several months after her helicopter rescue, she told reporters that her mother-in-law, who was also in the tree, helped her during delivery. Desire for this kind of support, it appears, may well be as ancient as humanity itself.

FURTHER READING

Wenda R. Trevathan. *Human Birth: An Evolutionary Perspective.* Aldine de Gruyter, 1987.
Robbie Davis-Floyd. *Birth as an American Rite of Passage.* University of California Press, 1993.
Karen R. Rosenberg and Wenda R. Trevathan. "Bipedalism and Human Birth: The Obstetrical Dilemma Revisited" in *Evolutionary Anthropology,* Vol. 4, No. 5, pages 161–168; 1996.
Peter T. Ellison. *On Fertile Ground: A Natural History of Human Reproduction.* Harvard University Press, 2001.

Once Were Cannibals

TIM D. WHITE

ORIGINALLY PUBLISHED IN AUGUST 2001

I t can shock, disgust and fascinate in equal measure, whether through tales of starved pioneers and airplane crash survivors eating the deceased among them or accounts of rituals in Papua New Guinea. It is the stuff of headlines and horror films, drawing people in and mesmerizing them despite their aversion. Cannibalism represents the ultimate taboo for many in Western societies—something to relegate to other cultures, other times, other places. Yet the understanding of cannibalism derived from the past few centuries of anthropological investigation has been too unclear and incomplete to allow either a categorical rejection of the practice or a fuller appreciation of when, where and why it might have taken place.

New scientific evidence is now bringing to light the truth about cannibalism. It has become obvious that long before the invention of metals, before Egypt's pyramids were built, before the origins of agriculture, before the explosion of Upper Paleolithic cave art, cannibalism could be found among many different peoples—as well as among many of our ancestors. Broken and scattered human bones, in some cases thousands of them, have been discovered from the prehistoric pueblos of the American Southwest to the islands of the Pacific. The osteologists and archaeologists studying these ancient occurrences are using increasingly sophisticated analytical tools and methods. In the past several years, the results of their studies have finally provided convincing evidence of prehistoric cannibalism.

Human cannibalism has long intrigued anthropologists, and they have worked for decades to classify the phenomenon. Some divide the behavior according to the affiliation of the consumed. Thus, endocannibalism refers to the consumption of individuals within a group, exocannibalism indicates the consumption of outsiders, and autocannibalism covers everything from nail biting to torture-induced self-consumption. In addition, anthropologists have come up with classifications to describe perceived or known motivations. Survival cannibalism is driven by starvation. Historically documented cases include the Donner Party—whose members were trapped during the harsh winter of 1846–47 in the Sierra Nevada—and people marooned in the Andes or the Arctic with no other food. In contrast, ritual cannibalism occurs when members of a family or community consume their dead during funerary rites in order to inherit their qualities or honor their memory. And pathological cannibalism is generally reserved for criminals who consume their victims or, more often, for fictional characters such as Hannibal Lecter in *The Silence of the Lambs.*

Despite these distinctions, however, most anthropologists simply equate the term "cannibalism" with the regular, culturally encouraged consumption of human flesh. This dietary, customary, gourmet, gustatory or gastronomic cannibalism, as it is variously called, is the phenomenon on which ethnographers have focused much of their attention. In the age of ethnographic exploration—which lasted from the time of Greek historian Herodotus in about 400 B.C. to the early 20th century—the non-Western world and its inhabitants were scrutinized by travelers, missionaries, military personnel and anthropologists. These observers told tales of gastronomic human cannibalism in different places, from Mesoamerica to the Pacific islands to central Africa.

Controversy has often accompanied these claims. Professional anthropologists participated in only the last few waves of these cultural contacts—those that began in the late 1800s. As a result, many of the historical accounts of cannibalism have come to be viewed skeptically. In 1937 anthropologist Ashley Montagu stated that cannibalism was "pure traveler's myth."

In 1979 anthropologist William Arens of the State University of New York at Stony Brook extended this argument by reviewing the ethnographic record of cannibalism in his book *The Man-Eating Myth.* Arens concluded that accounts of cannibalism among people from the Aztec to the Maori to the Zulu were either false or inadequately documented. His skeptical assertion has subsequently been seriously questioned, yet he nonetheless succeeded in identifying a significant gulf between these stories and evidence of cannibalism: "Anthropology has not maintained the usual standards of documentation and intellectual rigor expected when other topics are being considered. Instead, it has chosen uncritically to lend its support to the collective representations and thinly disguised prejudices of western culture about others."

The anthropologists whom Arens and Montagu were criticizing had not limited themselves to commenting solely on contemporary peoples. Some had projected their prejudices even more deeply—into the archaeological record. Interpretations of cannibalism inevitably followed many discoveries of prehistoric remains. Archaeological findings in Europe and elsewhere led to rampant speculation about cannibalism. And by 1871 American author Mark Twain had weighed in on the subject in an essay later published in *Life as I Find It:* "Here is a pile of bones of primeval man and beast all mixed together, with no more damning evidence that the man ate the bears than that the bears ate the man—yet paleontology holds a coroner's inquest in the fifth geologic period on an 'unpleasantness' which transpired in the quaternary, and calmly lays it on the MAN, and then adds to it what purports to be evidence of CANNIBALISM. I ask the candid reader, Does not this look like taking advantage of a gentleman who has been dead two million years. . . ."

In the century after Twain's remarks, archaeologists and physical anthropologists described the hominids *Australopithecus africanus, Homo erectus* and *H. neanderthalensis* as cannibalistic. According to some views, human prehistory from about three million years ago until very recently was rife with cannibalism.

In the early 1980s, however, an important critical assessment of these conclusions appeared. Archaeologist Lewis Binford's book *Bones: Ancient Men and Modern Myths* argued that claims for early hominid cannibalism were unsound. He built on the work of other prehistorians concerned with the composition, context and modifications of Paleolithic bone assemblages. Binford emphasized the need to draw accurate inferences about past behaviors by grounding knowledge of the past on experiment and observation in the present. His influential work coupled skepticism with a plea for methodological rigor in studies of prehistoric cannibalism.

HIGHER STANDARDS OF EVIDENCE

It would be helpful if we could turn to modern-day cannibals with our questions, but such opportunities have largely disappeared. So today's study of this intriguing behavior must be accomplished through a historical science. Archaeology has therefore become the primary means of investigating the existence and extent of human cannibalism.

One of the challenges facing archaeologists, however, is the amazing variety of ways in which people dispose of their dead. Bodies may be buried, burned, placed on scaffolding, set adrift, put in tree trunks or fed to scavengers. Bones may be disinterred, washed, painted, buried in bundles or scattered on stones. In parts of Tibet, future archaeologists will have difficulty recognizing any mortuary practice at all. There most corpses are dismembered and fed to vultures and other carnivores. The bones are then collected, ground into powder, mixed with barley and flour and again fed to vultures. Given the various fates of bones and bodies, distinguishing cannibalism from other mortuary practices can be quite tricky.

Consequently, scientists have set the standard for recognizing ancient cannibalism very high. They confirm the activity when the processing patterns seen on human remains match those seen on the bones of other animals consumed for food. Archaeologists have long argued for such a comparison between human and faunal remains at a site. They reason that damage to *animal* bones and their arrangement can clearly show that the animals had been slaughtered and eaten for food. And when *human* remains are unearthed in similar cultural contexts, with similar patterns of damage, discard and preservation, they may reasonably be interpreted as evidence of cannibalism.

When one mammal eats another, it usually leaves a record of its activities in the form of modifications to the consumed animal's skeleton. During life, varying amounts of soft tissue, much of it with nutritive value, cover mammalian bones. When the tissue is removed and prepared, the bones often retain a record of this processing in the form of gnawing marks and fractures. When humans eat other animals, however, they mark bones with more than just their teeth. They process carcasses with tools of stone or metal. In so doing, they leave imprints of their presence and actions in the form of scars on the bones. These same imprints can be seen on butchered human skeletal remains.

The key to recognizing human cannibalism is to identify the patterns of processing—that is, the cut marks, hammering damage, fractures or burns seen on the remains—as well as the survival of different bones and parts of bones. Nutritionally valuable tissues, such as brains and marrow, reside within the bones and can be removed only with forceful hammering—and such forced entry leaves revealing patterns of bone damage. When human bones from archaeological sites show patterns of damage uniquely linked to butchery by other humans, the inference of cannibalism is strengthened. Judging which patterns are consistent with dietary butchery can be based on the associated archaeological record—particularly the nonhuman food-animal remains discovered in sites formed by the same culture—and checked against predictions embedded in ethnohistorical accounts.

This comparative system of determining cannibalism emphasizes multiple lines of osteological damage and contextual evidence. And, as noted earlier, it sets the standard for recognizing cannibalism very high. With this approach, for instance, the presence of cut marks on bones would not by themselves be considered evidence of cannibalism. For example, an American Civil War cemetery would contain skeletal remains with cut marks made by bayonets and swords, but this would not constitute evidence of cannibalism. Medical school cadavers are dissected, their bones cut-marked, but cannibalism is not part of this ritual.

With the threshold set so conservatively, most instances of past cannibalism will necessarily go unrecognized. A practice from Papua New Guinea, where cannibalism was recorded ethnographically, illustrates this point. There skulls of the deceased were carefully cleaned and the brains removed. The dry, mostly intact skulls were then handled extensively, often creating a polish on their projecting parts. They were sometimes painted and even mounted on poles for display and worship. Soft tissue, including brain matter, was eaten at the beginning of this process; thus, the practice would be identified as ritual cannibalism. If such skulls were encountered in an archaeological context without modern informants describing the cannibalism, they would not constitute direct evidence for cannibalism under the stringent criteria that my colleagues and I advocate.

Nevertheless, adoption of these standards of evidence has led us to some clear determinations in other, older situations. The best indication of prehistoric cannibalism now comes from the archaeological record of the American Southwest, where archaeologists have interpreted dozens of assemblages of human remains as providing evidence of cannibalism. Compelling evidence has also been found in Neolithic and Bronze Age Europe. Even Europe's earliest hominid site has yielded convincing evidence of cannibalism.

EARLY EUROPEAN CANNIBALS

The most important paleoanthropological site in Europe lies in northern Spain, in the foothills of the Sierra de Atapuerca. Prehistoric habitation of the caves in these hills created myriad sites, but the oldest known so far is the Gran Dolina, currently under excavation. The team working there has recovered evidence of occupation some 800,000 years ago by what may prove to be a new species of human ancestor, *H. antecessor.* The hominid bones were discovered in one horizon of the cave's sediment, intermingled with stone tools and the remains of prehistoric game animals such as deer, bison and rhinoceros. The hominid remains consist of 92 fragments from six individuals. They bear unmistakable traces of butchery with stone tools, including skinning and removal of flesh, as well as processing of the braincase and the long bones for marrow. This pattern of butchery matches that seen on the nearby animal bones. This is the earliest evidence of hominid cannibalism.

Cannibalism among Europe's much younger Neandertals—who lived between 35,000 and 150,000 years ago—has been debated since the late 1800s, when the great Croatian paleoanthropologist Dragutin Gorjanović-Kramberger found the broken, cut-marked and scattered remains of more than 20 Neandertals entombed in the sands of the Krapina rockshelter. Unfortunately, these soft fossil bones were roughly extracted by today's standards and then covered with thick layers of preservative, which obscured evidence of processing by stone tools and made interpretation of the remains exceedingly difficult. Some workers believe the Krapina Neandertal bones show clear signs of cannibalism; others have attributed the patterns of bone damage to falling rocks from the cave's ceiling, to carnivore chewing or to some form of Neandertal burial. But recent analysis of the Krapina bones as well as those from another Croatian cave, Vindija—which has younger Neandertal and animal remains—indicates that cannibalism was practiced at both sites.

In the past few years, yet another Neandertal site has offered support for the idea that some of these hominids practiced cannibalism. On the banks of the Rhône River in southeastern France, Alban Defleur of the University of the Mediterranean at Marseilles has been excavating the cave of Moula-Guercy for the past nine years. Neandertals occupied this small cave approximately 100,000 years ago. In one layer the team unearthed the remains of at least six Neandertals, ranging in age from six years to adult. Defleur's meticulous excavation and recovery standards have yielded data every bit the equivalent of a modern forensic crime scene investigation. Each fragment of fauna and Neandertal bone, each macrobotanical clue, each stone tool has been precisely plotted three-dimensionally. This care has allowed an understanding of how the bones were spread around a hearth that has been cold for 1,000 centuries.

Microscopic analysis of the Neandertal bone fragments and the faunal remains has led to the same conclusion that Spanish workers at the older Gran Dolina site have drawn: cannibalism was practiced by some Paleolithic Europeans. But determining how often it was practiced and under what conditions represents a far more difficult challenge. Nevertheless, the frequency of cannibalism is striking. We know of just one very early European site with hominid remains, and those were cannibalized. The two Croatian Neandertal sites are separated by hundreds of generations, yet analyses suggest that cannibalism was practiced at both. And now a Neandertal site in France has supported the same interpretation. These findings are built on exacting standards of evidence. Because of this, most paleoanthropologists these days are asking "Why cannibalism?" rather than "Was this cannibalism?"

Similarly, recent discoveries at much younger sites in the American Southwest have altered the way anthropologists think of Anasazi culture in this area. Corn agriculturists have inhabited the Four Corners region of the American Southwest for centuries, building their pueblos and spectacular cliff dwellings and leaving one of the richest and most fine-grained archaeological records anywhere on earth. Christy G. Turner II of Arizona State University conducted pioneering work on unusual sets of broken and burned human skeletal remains from Anasazi sites in Arizona, New Mexico and Colorado in the 1960s and 1970s. He saw a pattern suggestive of cannibalism: site after site containing human remains with the telltale signs. Yet little in the history of the area's more recent Puebloan peoples suggested that cannibalism was a widespread practice, and some modern tribes who claim descent from the Anasazi have found claims of cannibalism among their ancestors disturbing.

The vast majority of Anasazi burials involve whole, articulated skeletons frequently accompanied by decorated ceramic vessels that have become a favorite target of pot hunters in this area. But, as Turner recorded, several dozen sites had fragmented, often burned human remains, and a larger pattern began to emerge. Over the past three decades the total number of human bone specimens from these sites has grown to tens of thousands, representing dozens of individuals spread across 800 years of prehistory and tens of thousands of square kilometers of the American Southwest. The assemblage that I analyzed 10 years ago from an Anasazi site in the Mancos Canyon of southwestern Colorado, for instance, contained 2,106 pieces of bone from at least 29 Native American men, women and children.

These assemblages have been found in settlements ranging from small pueblos to large towns and were often contemporaneous with the abandonment of the dwellings. The bones frequently show evidence of roasting before the flesh was removed. They invariably indicate that people extracted the brain and cracked the limb bones for marrow after removing the muscle tissue. And some of the long bone splinters even show end-polishing, a phenomenon associated with cooking in ceramic vessels. The bone fragments from Mancos revealed modifications that matched the marks left by Anasazi processing of game animals such as deer and bighorn sheep. The osteological evidence clearly demonstrated that humans were skinned and roasted, their muscles cut away, their joints severed, their long bones broken on anvils with hammerstones, their spongy bones crushed and the fragments circulated in ceramic vessels. But articles outlining the results have proved controversial. Opposition to interpretations of cannibalism has sometimes seemed motivated more by politics than by science. Many practicing anthropologists believe that scientific findings should defer to social sensitivities. For such anthropologists, cannibalism is so culturally delicate, so politically incorrect, that they find any evidence for it impossible to swallow.

The most compelling evidence in support of human cannibalism at Anasazi sites in the American Southwest was published last fall by Richard A. Marlar of the University of Colorado School of Medicine and his colleagues. The workers excavated three Anasazi pit dwellings dating to approximately A.D. 1150 at a site called Cowboy Wash near Mesa Verde in southwestern Colorado. The same pattern that had been documented at other sites such as Mancos was present: disarticulated, broken, scattered human bones in nonburial contexts. Excellent preservation, careful excavation and thoughtful sampling provided a chemical dimension to the analysis and, finally, direct evidence of human cannibalism.

Marlar and his colleagues discovered residues of human myoglobin—a protein present in heart and skeletal muscle—on a ceramic vessel, suggesting that human flesh had been cooked in the pot. An unburned human coprolite, or ancient feces, found in the fireplace of one of the abandoned dwellings also tested positive for human myoglobin. Thus, osteological, archaeological and biochemical data indicate that prehistoric cannibalism occurred at Cowboy Wash. The biochemical data for processing and consumption of human tissue offer strong additional support for numerous osteological and archaeological findings across the Southwest.

UNDERSTANDING CANNIBALISM

It remains much more challenging to establish why cannibalism took place than to establish that it did. People usually eat because they are hungry, and most prehistoric cannibals were therefore probably hungry. But discerning more than that—such as whether the taste of human flesh was pleasing or whether cannibalism presented a way to get through the lean times or a satisfying way to get rid of outsiders—requires knowledge not yet available to archaeologists. Even in the case of the Anasazi, who have been well studied, it is impossible to determine whether cannibalism resulted from starvation, religious beliefs or some combination of these and other things. What is becoming clear through the refinement of the science of archaeology, however, is that cannibalism is part of our collective past.

FURTHER READING

T. D. White. *Prehistoric Cannibalism at Mancos 5MTUMR-2346*. Princeton University Press, 1992.
L. Osborne. "Does Man Eat Man? Inside the Great Cannibalism Controversy" in *Lingua Franca*, Vol. 7, No. 4, pages 28–38; April/May 1997.
D. DeGusta. "Fijian Cannibalism: Osteological Evidence from Navatu" in *American Journal of Physical Anthropology*, Vol. 110, pages 215–241; October 1999.
A. Defleur, T. D. White, P. Valensi, L. Slimak and E. Crégut-Bonnoure. "Neanderthal Cannibalism at Moula-Guercy, Ardèche, France" in *Science*, Vol. 286, pages 128–131; October 1, 1999.
R. A. Marlar, B. L. Leonard, B. R. Billman, P. M. Lambert and J. E. Marler. "Biochemical Evidence of Cannibalism at a Prehistoric Puebloan Site in Southwestern Colorado" in *Nature*, Vol. 407, pages 74–78; September 7, 2000.

If Humans Were Built to Last

S. JAY OLSHANSKY, BRUCE A. CARNES AND ROBERT N. BUTLER

ORIGINALLY PUBLISHED IN MARCH 2001

Bulging disks, fragile bones, fractured hips, torn ligaments, varicose veins, cataracts, hearing loss, hernias and hemorrhoids: the list of bodily malfunctions that plague us as we age is long and all too familiar. Why do we fall apart just as we reach what should be the prime of life?

The living machines we call our bodies deteriorate because they were not designed for extended operation and because we now push them to function long past their warranty period. The human body is artistically beautiful and worthy of all the wonder and amazement it invokes. But from an engineer's perspective, it is a complex network of bones, muscles, tendons, valves and joints that are directly analogous to the fallible pulleys, pumps, levers and hinges in machines. As we plunge further into our postreproductive years, our joints and other anatomical features that serve us well or cause no problems at younger ages reveal their imperfections. They wear out or otherwise contribute to the health problems that become common in the later years.

In evolutionary terms, we harbor flaws because natural selection, the force that molds our genetically controlled traits, does not aim for perfection or endless good health. If a body plan allows individuals to survive long enough to reproduce (and, in humans and various other organisms, to raise their young), then that plan will be selected. That is, individuals robust enough to reproduce will pass their genes—and therefore their body design—to the next generation. Designs that seriously hamper survival in youth will be weeded out (selected against) because most affected individuals will die before having a chance to produce offspring. More important, anatomical and physiological quirks that become disabling only after someone has reproduced will spread. For example, if a body plan leads to total collapse at age 50 but does not interfere with earlier reproduction, the arrangement will get passed along despite the harmful consequences late in life.

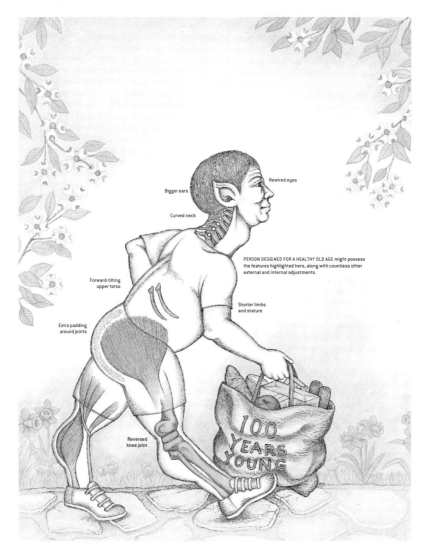

Bigger ears

Rewired eyes

Curved neck

Forward-tilting
upper torso

PERSON DESIGNED FOR A HEALTHY OLD AGE might possess
the features highlighted here, along with countless other
external and internal adjustments.

Shorter limbs
and stature

Extra padding
around joints

Reversed
knee joint

100
YEARS
YOUNG

Had we been crafted for extended operation, we would have fewer flaws
capable of making us miserable in our later days. Evolution does not work
that way, however. Instead it cobbles together new features by tinkering
with existing ones in a way that would have made Rube Goldberg proud.

The upright posture of humans is a case in point. It was adapted from
a body plan that had mammals walking on all fours. This tinkering undoubtedly aided our early hominid ancestors: standing on our own two
feet is thought to have promoted tool use and enhanced intelligence. Our

backbone has since adapted somewhat to the awkward change: the lower vertebrae have grown bigger to cope with the increased vertical pressure, and our spine has curved a bit to keep us from toppling over. Yet these fixes do not ward off an array of problems that arise from our bipedal stance.

WHAT IF?

Recently the three of us began pondering what the human body would look like had it been constructed specifically for a healthy long life. The anatomical revisions depicted on the pages that follow are fanciful and incomplete. Nevertheless, we present them to draw attention to a serious point. Aging is frequently described as a disease that can be reversed or eliminated. Indeed, many purveyors of youth-in-a-bottle would have us believe that medical problems associated with aging are our own fault, arising primarily from our decadent lifestyles. Certainly any fool can shorten his or her life. But it is grossly unfair to blame people for the health consequences of inheriting a body that lacks perfect maintenance and repair systems and was not built for extended use or perpetual health. We would still wear out over time even if some mythical, ideal lifestyle could be identified and adopted.

This reality means that aging and many of its accompanying disorders are neither unnatural nor avoidable. No simple interventions can make up for the countless imperfections that permeate our anatomy and are revealed by the passage of time. We are confident, however, that biomedical science will be able to ease certain of the maladies that result. Investigators are rapidly identifying (and discerning the function of) our myriad genes, developing pharmaceuticals to control them, and learning how to harness and enhance the extraordinary repair capabilities that already exist inside our bodies. These profound advances will eventually help compensate for many of the design flaws contained within us all.

A number of the debilitating and even some of the fatal disorders of aging stem in part from bipedal locomotion and an upright posture—ironically, the same features that have enabled the human species to flourish. Every step we take places extraordinary pressure on our feet, ankles, knees and back—structures that support the weight of the whole body above them. Over the course of just a single day, disks in the lower back are subjected to pressures equivalent to several tons per square inch. Over a lifetime, all this pressure takes its toll, as does repetitive use of our joints and the constant tugging of gravity on our tissues.

FLAWS

BONES THAT LOSE MINERALS AFTER AGE 30
Demineralization makes bones susceptible to
fractures and, in extreme cases, can cause
osteoporosis (severe bone degeneration),
curvature of the spine and "dowager's hump"

FALLIBLE SPINAL DISKS
Years of pressure on the spongy disks that
separate the vertebrae can cause them to slip,
rupture or bulge; then they, or the
vertebrae themselves, can press
painfully on nerves

MUSCLES THAT LOSE MASS AND TONE
Such atrophy can impede all activities,
including walking. In the abdomen, hernias
can arise as the intestines (always pulled by
gravity) protrude through weak spots
in the abdominal wall. Flaccid abdominal
muscles also contribute to lower-back pain

LEG VEINS PRONE TO VARICOSITY
Veins in the legs become enlarged
and twisted when small valves that should
snap shut between heartbeats (to keep blood
moving up toward the heart)
malfunction, causing blood to pool.
Severe varicosities can lead
to swelling and pain
and, on rare occasions,
to life-threatening blood clots

Normal direction
of blood flow

Malfunctioning
check valve

Pooled blood

**RELATIVELY SHORT
RIB CAGE**
Current cage
does not fully
enclose and protect
most internal organs

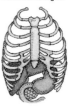

JOINTS THAT WEAR
As joints are used repetitively
through the years, their lubricants
can grow thin, causing the bones
to grind against each other.
The resulting pain may be
exacerbated by osteoarthritis
and other inflammatory disorders

FIXES

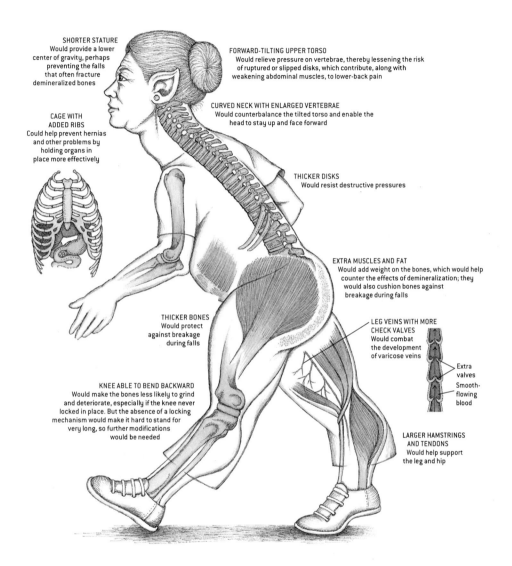

SHORTER STATURE
Would provide a lower center of gravity, perhaps preventing the falls that often fracture demineralized bones

FORWARD-TILTING UPPER TORSO
Would relieve pressure on vertebrae, thereby lessening the risk of ruptured or slipped disks, which contribute, along with weakening abdominal muscles, to lower-back pain

CURVED NECK WITH ENLARGED VERTEBRAE
Would counterbalance the tilted torso and enable the head to stay up and face forward

CAGE WITH ADDED RIBS
Could help prevent hernias and other problems by holding organs in place more effectively

THICKER DISKS
Would resist destructive pressures

EXTRA MUSCLES AND FAT
Would add weight on the bones, which would help counter the effects of demineralization; they would also cushion bones against breakage during falls

THICKER BONES
Would protect against breakage during falls

LEG VEINS WITH MORE CHECK VALVES
Would combat the development of varicose veins

Extra valves

Smooth-flowing blood

KNEE ABLE TO BEND BACKWARD
Would make the bones less likely to grind and deteriorate, especially if the knee never locked in place. But the absence of a locking mechanism would make it hard to stand for very long, so further modifications would be needed

LARGER HAMSTRINGS AND TENDONS
Would help support the leg and hip

Although gravity tends to bring us down in the end, we do possess some features that combat its ever present pull. For instance, an intricate network of tendons helps to tether our organs to the spine, keeping them from slumping down and crushing one another.

But these anatomical fixes—like the body in general—were never meant to work forever. Had longevity and persistent good health been the overarching aim of evolution, arrangements such as those depicted here might have become commonplace.

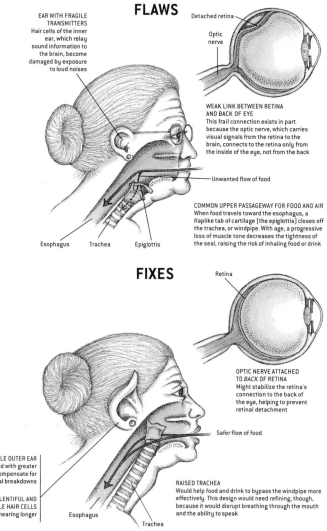

FLAWS

EAR WITH FRAGILE TRANSMITTERS
Hair cells of the inner ear, which relay sound information to the brain, become damaged by exposure to loud noises

Detached retina

Optic nerve

WEAK LINK BETWEEN RETINA AND BACK OF EYE
This frail connection exists in part because the optic nerve, which carries visual signals from the retina to the brain, connects to the retina only from the inside of the eye, not from the back

Unwanted flow of food

Esophagus   Trachea   Epiglottis

COMMON UPPER PASSAGEWAY FOR FOOD AND AIR
When food travels toward the esophagus, a flaplike tab of cartilage (the epiglottis) closes off the trachea, or windpipe. With age, a progressive loss of muscle tone decreases the tightness of the seal, raising the risk of inhaling food or drink

FIXES

Retina

OPTIC NERVE ATTACHED TO *BACK* OF RETINA
Might stabilize the retina's connection to the back of the eye, helping to prevent retinal detachment

Safer flow of food

ENLARGED, MOBILE OUTER EAR
Would collect sound with greater efficiency, to compensate for internal breakdowns

MORE PLENTIFUL AND DURABLE HAIR CELLS
Would preserve hearing longer

Esophagus

Trachea

RAISED TRACHEA
Would help food and drink to bypass the windpipe more effectively. This design would need refining, though, because it would disrupt breathing through the mouth and the ability to speak

Various parts of the head and neck become problematic with disturbing regularity as people age. Consider the eye. The human version is an evolutionary marvel, but its complexity provides many opportunities for things to go wrong over a long lifetime.

Our vision diminishes as the protective fluid of the cornea becomes less transparent over time. The muscles that control the opening of the iris and the focusing of the lens atrophy and lose responsiveness, and the lens thickens and yellows, impairing visual acuity and color perception. Further, the retina—responsible for transmitting images to the brain—can detach fairly easily from the back of the eye, leading to blindness.

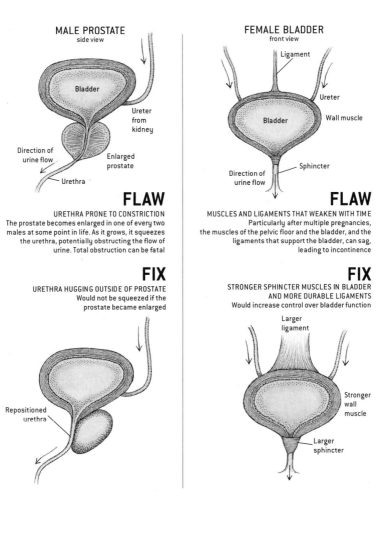

MALE PROSTATE
side view

Bladder

Ureter
from
kidney

Direction of
urine flow

Enlarged
prostate

Urethra

FEMALE BLADDER
front view

Ligament

Ureter

Bladder

Wall muscle

Direction of
urine flow

Sphincter

FLAW

URETHRA PRONE TO CONSTRICTION
The prostate becomes enlarged in one of every two males at some point in life. As it grows, it squeezes the urethra, potentially obstructing the flow of urine. Total obstruction can be fatal

FLAW

MUSCLES AND LIGAMENTS THAT WEAKEN WITH TIME
Particularly after multiple pregnancies, the muscles of the pelvic floor and the bladder, and the ligaments that support the bladder, can sag, leading to incontinence

FIX

URETHRA HUGGING OUTSIDE OF PROSTATE
Would not be squeezed if the prostate became enlarged

FIX

STRONGER SPHINCTER MUSCLES IN BLADDER AND MORE DURABLE LIGAMENTS
Would increase control over bladder function

Repositioned
urethra

Larger
ligament

Stronger
wall
muscle

Larger
sphincter

Many of those problems would be difficult to design away, but the squid eye suggests an arrangement that could have reduced the likelihood of retinal detachment. A few anatomical tweaks could also have preserved hearing in the elderly.

Suboptimal design of the upper respiratory and digestive systems makes choking another risk for older people. A simple rearrangement would have fixed that problem, albeit at the cost of severe trade-offs.

An experienced plumber looking at the anatomy of a man's prostate might suspect the work of a young apprentice, because the urethra, the tube leading from the bladder, passes straight through the inside of the gland. This configuration may have as yet unknown benefits, but it eventually causes urinary problems in many men, including weak flow and a frequent need to void.

Women also cope with plumbing problems as they age, particularly incontinence. Both sexes could have been spared much discomfort if evolution had made some simple modifications in anatomical design.

FURTHER READING

D'Arcy Wentworth Thompson. *On Growth and Form.* Expanded edition, 1942. (Reprinted by Dover Publications, 1992.)

Stephen Jay Gould. *The Panda's Thumb: More Reflections in Natural History.* W. W. Norton, 1980.

Richard Dawkins. *The Blind Watchmaker: Why the Evidence of Evolution Reveals a Universe without Design.* W. W. Norton, 1986.

Elaine Morgan. *The Scars of Evolution: What Our Bodies Tell Us about Human Origins.* Souvenir Press, 1990. (Reprinted by Oxford University Press, 1994.)

Randolph M. Nesse and George C. Williams. *Why We Get Sick: The New Science of Darwinian Medicine.* Random House, 1994.

The Olshansky and Carnes Web site is www.thequestforimmortality.com

The International Longevity Center Web site is www.ilcusa.org

ILLUSTRATION CREDITS

Pages 16, 17, and 19: Don Dixon. Pages 26–27: Alfred T. Kamajian. Pages 41 and 45: Don Dixon and George Musser. Pages 43 and 48: Laurie Grace and George Musser. Pages 64, 66, 68, and 69: Alfred Kamajian. Pages 76–77: Roberto Osti; graphs by Laurie Grace. Pages 78 and 81: Ian Worpole. Pages 88 and 97: Jana Brenning. Pages 91 and 95: Christoph Blumrich. Page 93: Christoph Blumrich; source: David F. Spencer, Dalhousie University. Pages 100–101, 102–103, and 106–107: Roberto Osti. Page 110: Dimitry Schidlovsky. Page 112: S. E. Frederick, University of Wisconsin–Madison. Pages 129, 132, and 134: Dimitry Schidlovsky. Pages 142 and 146: Bryan Christie. Page 144: Julie Forman-Kay, University of Toronto, and Tony Pawson. Pages 153, 156, and 159; Dimitry Schidlovsky. Page 172: Tomo Narashima and Cleo Vilett. Page 175: Ed Heck. Page 177: Karen Carr; Adrienne Smucker (vertebrae). Page 179; Tomo Narashima (animals); Edward Bell (sclerotic ring). Pages 186–187: Portia Sloan. Page 194: Patricia J. Wynne. Page 205: Sara Chen. Pages 215, 216–217, 218–219, 226, 232, 239, and 242: Patricia J. Wynne. Page 236: David Starwood. Page 262: Slim Films. Page 264: Slim Films; source: Frank Brown and Craig Feibel, 1991. Page 271; Laurie Grace; map source: F. Rögl, *The Miocene Mammals of Europe*. Page 275: Portia Sloan. Pages 282 and 283: Jay H. Matternes. Page 285: Patricia J. Wynne. Page 299: Laurie Grace; source: Ian Tattersall, *The Last Neanderthal*. Page 306: Michael Rothman. Page 312: John Gurche. Page 314: Cornelia Blik. Page 317: Laurie Grace. Pages 332 and 333: Nina Finkel. Pages 348, 350, 351, 352, and 353: Patricia J. Wynne.